Advances in
THE STUDY OF BEHAVIOR
VOLUME 10

Contributors to This Volume

RICHARD E. BROWN
ANNE D. MAYER
MILDRED MOELK
F. J. ODLING-SMEE
H. C. PLOTKIN
JAY S. ROSENBLATT
DOUGLAS Y. SHAPIRO
HAROLD I. SIEGEL

Advances in
THE STUDY OF
BEHAVIOR

Edited by

JAY S. ROSENBLATT
Institute of Animal Behavior
Rutgers University
Newark, New Jersey

ROBERT A. HINDE
Medical Research Council
Unit on the Development and Integration of Behaviour
University Sub-Department of Animal Behaviour
Madingley, Cambridge, England

COLIN BEER
Institute of Animal Behavior
Rutgers University
Newark, New Jersey

MARIE-CLAIRE BUSNEL
Laboratoire de Physiologie Acoustique
Institut National de la Recherche Agronomique
Ministère de l'Agriculture
Jouy en Josas (S. et O.), France

———————————— VOLUME 10 ————————————

ACADEMIC PRESS   New York   London   Toronto   Sydney   San Francisco   1979
A Subsidiary of Harcourt Brace Jovanovich, Publishers

591. 51
A 244
v. 10

178031

COPYRIGHT © 1979, BY ACADEMIC PRESS, INC.
ALL RIGHTS RESERVED.
NO PART OF THIS PUBLICATION MAY BE REPRODUCED OR
TRANSMITTED IN ANY FORM OR BY ANY MEANS, ELECTRONIC
OR MECHANICAL, INCLUDING PHOTOCOPY, RECORDING, OR ANY
INFORMATION STORAGE AND RETRIEVAL SYSTEM, WITHOUT
PERMISSION IN WRITING FROM THE PUBLISHER.

ACADEMIC PRESS, INC.
111 Fifth Avenue, New York, New York 10003

United Kingdom Edition published by
ACADEMIC PRESS, INC. (LONDON) LTD.
24/28 Oval Road, London NW1 7DX

LIBRARY OF CONGRESS CATALOG CARD NUMBER: 64–8031

ISBN 0–12–004510–9

PRINTED IN THE UNITED STATES OF AMERICA

79 80 81 82   9 8 7 6 5 4 3 2 1

Contents

Learning, Change, and Evolution: An Enquiry into the Teleonomy of Learning

H. C. PLOTKIN AND F. J. ODLING-SMEE

Social Behavior, Group Structure, and the Control of Sex Reversal in Hermaphroditic Fish

DOUGLAS Y. SHAPIRO

Mammalian Social Odors: A Critical Review

RICHARD E. BROWN

The Development of Friendly Approach Behavior in the Cat: A Study of Kitten–Mother Relations and the Cognitive Development of the Kitten from Birth to Eight Weeks

MILDRED MOELK

Progress in the Study of Maternal Behavior in the Rat: Hormonal, Nonhormonal, Sensory, and Developmental Aspects

JAY S. ROSENBLATT, HAROLD I. SIEGEL, AND ANNE D. MAYER

List of Contributors

Numbers in parentheses indicate the pages on which the authors' contributions begin.

RICHARD E. BROWN, *Department of Psychology, Dalhousie University, Halifax, Nova Scotia, Canada B3H 4J1 (103)*

ANNE D. MAYER, *Institute of Animal Behavior, Rutgers—The State University of New Jersey, Newark, New Jersey 07102 (225)*

MILDRED MOELK, *11 Winston Woods, Brockport, New York 14420 (163)*

F. J. ODLING-SMEE, *Department of Psychology, Brunel University, Middlesex, Uxbridge, England (1)*

H. C. PLOTKIN, *Department of Psychology, University College London, London WC1E 6BT, England (1)*

JAY S. ROSENBLATT, *Institute of Animal Behavior, Rutgers—The State University of New Jersey, Newark, New Jersey 07102 (225)*

DOUGLAS Y. SHAPIRO, *Department of Marine Sciences, University of Puerto Rico, Mayaguez, Puerto Rico 00708 (43)*

HAROLD I. SIEGEL, *Institute of Animal Behavior, Rutgers—The State University of New Jersey, Newark, New Jersey 07102 (225)*

Preface

With the publication of the tenth volume of *Advances in the Study of Behavior*, we wish to restate in more contemporary terms the aims stated in the original preface, namely, to serve ". . . as a contribution to the development of cooperation and communication among scientists in our field." Since that preface was written in 1965, an increasing number of scientists from disciplines as widely separated as behavioral ecology and the biochemistry of behavior have become engaged in the study of animal behavior, employing the specialized techniques and concepts of their disciplines. Even then, the boundaries of ethology and comparative psychology were no longer distinct: now they have been merged with broader syntheses of social and individual functioning and have together provided the bases for studies of the neural and biochemical mechanisms of behavior. New vigor has been given to traditional fields of animal behavior by their coalescence with closely related fields and by the closer relationship that now exists between those studying animal and human subjects. Scientists engaged in studying animal behavior now range from ecologists through evolutionary biologists, geneticists, endocrinologists, ethologists, and comparative and developmental psychologists, to neurophysiologists and neuropharmacologists. The task of developing cooperation and communication among scientists whose skills and concepts necessarily differ in accordance with the diversity of the phenomena they study has become more difficult than it was at the inception of this publication. Yet the need to do so has become even greater as it has become more difficult. The Editors and publisher of *Advances in the Study of Behavior* will continue to provide the means by publishing critical reviews of research in our field, by inviting extended presentations of significant research programs, by encouraging the writing of theoretical syntheses and reformulations of persistent problems, and by highlighting especially penetrating research that introduces important new concepts.

Contents of Previous Volumes

Advances in
THE STUDY OF BEHAVIOR
VOLUME 10

ADVANCES IN THE STUDY OF BEHAVIOR, VOL. 10

Learning, Change, and Evolution: An Enquiry into the Teleonomy of Learning

H. C. PLOTKIN AND F. J. ODLING-SMEE*

DEPARTMENT OF PSYCHOLOGY
UNIVERSITY COLLEGE LONDON
LONDON, ENGLAND
AND
DEPARTMENT OF PSYCHOLOGY
BRUNEL UNIVERSITY
MIDDLESEX, UXBRIDGE, ENGLAND

I. INTRODUCTION

Behavior is being increasingly incorporated into the synthesis of modern evolutionary theory. The recognition that natural selection acts on phenotypes and not on genotypes (Mayr, 1963; Waddington, 1969) and the acknowledgement that behavior constitutes a significant part of phenotypic expression (Mayr, 1974) has made such incorporation inevitable, but by no means easy. Waddington (1975) has pointed out that recognizing the relevance of behavior to evolution and vice versa is insufficient, and a key to the successful incorporation of be-

*There is no senior author. The order was determined by the toss of a coin.

Copyright © 1979 by Academic Press, Inc.
All rights of reproduction in any form reserved.
ISBN 0-12-004510-9

havior into modern biology lies in working out the "logic" of the relationship between the behavior of individual phenotypes and phylogenesis. This task is still to be achieved.

In psychology a prominent casualty of the exclusion of behavior from the modern synthesis for about last half a century has been learning theory. A considerable amount of behavior in some animals is influenced and shaped by learning. Thus, if behavior in general must be incorporated into evolutionary theory, then so too must learning. However, until recently most learning theorists have not found it necessary either to look to evolutionary theory for guidance or to reconcile their theories with evolution. With the advantage of hindsight, that attitude seems to have contributed to a conceptual crisis that has overtaken the study of learning in the last decade, and centers around the contemporary inability of theorists to sort out what is species specific and what is universal about learning (Bitterman, 1975; Hinde, 1973; Mason and Lott, 1976; Razran, 1971; Revusky, 1974; Schwartz, 1974).

By now the origins of this crisis are clear. Currently much more attention is being paid than formerly to performance differences among different species of learners. Thus far, interest has focused mainly upon the outcomes of learning (Shettleworth, 1972), but to some extent it has also been concerned with the possibility of differences in learning processes (Bitterman, 1975). This cognizance of difference contrasts sharply with a basic assumption about learning that most psychologists formerly made (Lockard, 1971; Hinde, 1973), namely, that the processes of learning are the same in all species that can learn, and the course of learning in different species is relatively free of any interactions arising out of the nature of the stimuli, responses, and reinforcers involved in learning (Jenkins, 1975). [We shall follow Seligman (1970) by referring to all approaches based on this assumption as general process theory, and we shall call the assumption itself one of unity.]

One result of this history is a continuing paradox of assumed unity versus observed natural diversity. A simple way of resolving this paradox would be to abandon the unity assumption, and with it general process theory, and to proceed as if each species needed its own species-specific learning model. Shettleworth's (1972) advocacy of a "multiplicity of principles" appears to be a step in this direction. This solution, however, does not have parsimony on its side. It also carries with it the penalty of leaving unexplained all those learning phenomena that general process theory did handle rather well, for example, the generality across several phyla of a sensitivity to habituation or Pavlovian conditioning. The premature rejection of the unity assumption therefore could turn out to be at least as damaging as was its original premature acceptance. To make sure that the unity assumption is not rejected precipitously, it is our contention that psychologists should at least try to develop a theory that can handle both the unity and diversity of learning simultaneously. This is unlikely to be possible unless learning can find its rightful place within the evolutionary synthesis. If it can, however, then, given the proven capacity of evolutionary theory to reconcile the unity and diversity of life in general, it may become possible to devise such a theory.

What follows then is an attempt to take the first step toward establishing a truly general theory of learning, one based on evolution, and able to resolve the central paradox of unity versus diversity. This step means acceptance of Waddington's challenge by trying to work out the logic of the relationship between learning and phylogenesis. It leads to an exploration of both the evolutionary origins and functions of learning and to a consideration of what we, like others (e.g., Slobodkin and Rapoport, 1974), believe to be central to both evolution and learning, namely, environmental change. As one by-product of this analysis we shall also be touching on that perennial and troubled issue of nature, nurture, and the provenance of behavior (Bateson, 1976a). Finally, we shall indicate certain lines of study for the future. Throughout we shall borrow heavily from the ideas of others, but our hope is that the particular synthesis of ideas advocated here is sufficiently original to move the debate on.

Clarification of Terms

Provisionally *learning* will be defined as one out of several classes of change in the internal organization of an animal that are consequent upon changes in external environmental order. The particular class that learning represents is characterized by the fact that it excludes all changes in internal organization that result from such processes as fatigue and sensory adaptation (Hinde, 1970). Further, the internal changes resulting from learning typically form the basis for future adaptive behavior. This definition introduces two terms that will frequently be used in the following pages. Both *organization* and *order* refer to a complex of relations among component parts. In this context the distinction between the two will be based simply and arbitrarily on whether these component parts are within the animal or are in the external environment. If the former is the case, we shall refer to organization; if the latter is the case, then we shall refer to order. The processes by which within-animal organizational changes occur relative to changes in the external environment we shall collectively refer to as *information gain,* without committing ourselves to any convention as to what is meant by information beyond that it must in general imply an increase in biological fitness.

A second definition is also required. Insofar as learning represents an adaptive change in the internal organization of a learner in response to changing environmental events, then learning by an animal bears a close conceptual relationship to the provisioning of an animal with adaptations by phylogenesis (Pittendrigh, 1958). Both learning and phylogenesis are forms of information gain that affect the fitness of animals. However, there is a crucial difference between these two processes. The biological "referent" is not the same in the two cases.

By referent here is meant an organization-gaining unit defined by specific information-gaining and -storing mechanisms, supporting a specific process or set of processes. Thus, the referent for phylogenesis consists of a set of individual organisms that collectively constitute a species or intrabreeding population (for convenience these terms will be used synonymously although erroneously, the error not being one that bears upon the present argument). In fact, a population is so defined by its possession of phylogenetic information gaining and storing mechanisms and by the fact that natural selection cannot operate except upon a set of organisms. In practice this limitation of natural selection makes it impossible for phylogenesis to supply any individual organism directly with adaptations. Phylogenesis can only feed adaptations to an individual organism as a function of that organism's membership of some evolving population.

There is, however, nothing to stop every individual member of a population from being equipped by phylogenesis with adaptations that themselves take the form of subsidiary, and autonomous, mechanisms of information gain and storage. This actually occurs in every animal species that can learn. In all such cases the existence of these subsidiary within-organism mechanisms, and of the additional process of learning that they support, automatically defines a second distinct referent. Thus, the referent for learning comprises the specific organ subsystems that underlie learning in individual animals. Moreover, since these subsystems do lie within the scope of individual animals, then learning, unlike phylogenesis, can supply individuals with behavioral adaptations directly.

In fact, as many referents are assumed to exist as there are logically distinct classes of information-gaining and -storing mechanisms and processes. This discussion will, nevertheless, be confined to consideration of just three referents. The first of these is a population or gene pool, defined by phylogenesis. The second is an individual member of a population regarded as a developing genotype. Here the referent consists of an individual genotype that is actively translating into a phenotype, and gaining additional information about the environment while doing so. It is defined by epigenesis. The third referent consists of various specific organ subsystems within each individual phenotype, which, in their turn, can gain still further information about the environment. This referent is defined by various processes including learning. In the future these three referents will be termed the primary, secondary, and tertiary referents, respectively (Table I).

The distinction between referents is basic to the following analysis. Evolution is not regarded here as a process that is subserved by only a single referent. Rather it is considered to be the outcome of several processes and to be subserved by several interacting referents. Hence, evolution as a whole is seen as both a process of dynamic interactions between distinct referents, with information about the world being transmitted across these referents, and as the interaction

TABLE I

THE THREE INFORMATION-GAINING REFERENTS, AND THEIR ASSOCIATED PROCESSES

Referent	Fitness-Gaining Unit	Process
Primary referent	Gene pool of a breeding population	Phylogenesis
Secondary referent	Genotype–phenotype developmental system	Epigenesis
Tertiary referent	Specialist physiological and behavioral sub-systems within a phenotype	A variety of information-gaining processes, including learning

between each particular referent and its environment. It may also be noted that there is no simple identity between an individual organism and any single referent. An individual organism is always part of the primary referent. It may or may not include the secondary and/or tertiary referents, depending on whether it possesses secondary and/or tertiary information-gaining mechanisms.

One further point also requires some comment. Our concept of a plurality of interacting referents appears to differ markedly from the notion that the basic unit of evolution is the gene. Both Williams (1966) and Dawkins (1976), for example, following in the tradition set by Fisher (1930), Haldane (1931), and Wright (1931), advocate this view. In doing so they draw attention to the fact that the ultimate unit of life selected by natural selection is the gene, even though its selection is mediated by the selection of phenotypes. Thus, William and Dawkins are drawing attention to the one unit of life that strongly tends to persist unchanged across time. Our concern, however, is to lay the emphasis elsewhere. Since we are discussing information-gaining processes, we are naturally most concerned with those units of life that are actually doing the information gaining, and are, in addition, changing across time in accordance with the information they gain. In fact, the present concept of an information-gaining referent is not in conflict with the position held by either Williams or Dawkins. Instead, it is a necessary adjunct to it, the notion of referents deriving from the reverse rather than the obverse side of the evolutionary story.

One caution is necessary, however. In treating a population as the primary referent of evolution, we are referring exclusively to a population of adapted individuals. We never mean to imply an adapted population. The distinction between these two descriptions is vital. Williams (1966) points out that, whereas a population of adapted individuals demands no more than classical natural selection, the notion of a population that is itself an adapted entity implies group selection. We have no wish to enter arguments as to whether either biotic adaptation or group selection processes exist. Thus, whenever an evolving population is referred to here, or else whenever adaptation is discussed relative to some population, the word "population" must be regarded as simply a shortened form of "the set of individuals which are members of an evolving population." For our purposes the latter is synonomous with the gene pool of that population.

Finally a brief explanation is needed as to what we mean by "teleonomy." This word was introduced by Pittendrigh (1958) in order to distinguish from teleology the apparent end directedness of adaptation. Adaptation is something that, in principle, is fully understandable in terms of the history of selection pressures acting on a range of adaptive forms. Thus, teleonomy refers to the past and present of adaptive structure and function, and is not to be confused with the finalism of teleology, whose explanatory recourse is to agents projecting into the future.

II.   THE TELEONOMY OF LEARNING: WHY SOME ANIMALS LEARN

If the derivation of the origins and functions of learning is to proceed other than on an arbitrary basis, then we must start by assuming a general principle of parsimony. Accordingly, it is assumed that phylogenesis is the primary process of information gain, and that the evolution of the secondary and tertiary processes only ever occur if and when a primary referent (a breeding population) reaches an upper limit with respect to the ability of its members to adapt exclusively by phylogenesis. If this principle is adopted, then the search for the origins of learning becomes nonarbitrary. It also becomes largely synonymous with a search for the limitations of phylogenesis. In practice the discovery of these limits will not be possible without some reexamination of evolutionary theory. Since the modern theory of evolution is well known, however, we shall try as far as possible not to reiterate familiar material. Instead, we shall concentrate only on issues that are central to the present argument, namely, the problem of adaptation, the nature of the primary referent, and the logic of phylogenesis. We then shall attempt to bring all three of these concepts together in an effort to see how far a primary referent can be expected to solve its adaptation problem exclusively by means of phylogenesis. Then we shall derive the origins of the secondary and tertiary referents from whatever limits of phylogenesis are discovered.

A.   ADAPTATION

1.   The Components of Fitness and Change

The fitness of the members of a population refers to their capacity to persist across space and time and to their capacity to reproduce. Hence, traditional definitions of fitness invoke the probability of survival and differential rates of reproduction of individual organisms (e.g., Maynard Smith, 1966). Since all organisms require energy to live, every individual organism must acquire sufficient organization to be able to exploit energy sources in its external environment. The biological organization that an organism may acquire is its adaptations, and it is possible for these adaptations to occur at all levels within the organism, for example, genetic, biochemical, embryological, anatomical, physiological, and behavioral (Conrad, 1972; Slobodkin and Rapoport, 1974; Thoday, 1964). Strictly speaking, the concepts of fitness and adaptation are distinct; for instance, in population genetics (e.g., Dobzhansky, 1968) fitness is always taken to refer to gene frequencies. However, here there is no need for any such distinction, and we shall treat fitness and adaptation as functionally equivalent.

Thoday (1953) has argued that there are two components of fitness and not just one. The first component is immediate fitness and is defined as adaptation to the contemporary environment. The second component is potential fitness and consists of adaptation to future environments. Both of these components are necessary because it is never sufficient for an organism to be adapted to its contemporary environment only. If a population of organisms is actually to persist across space and time, then its members must acquire and carry some additional potential adaptations that are likely to render them fit relative to their future as well.

Thoday's distinction between these components of fitness is basic. Its importance lies in its introducing into evolutionary theory a second all-pervasive notion, that of environmental change. If a population's environment were perfectly homogeneous throughout space and time, then there would be no need for any demarcation between the two separate components of fitness. In these circumstances, adaptation to the future environment would be identical to adaptation to the present environment. In reality, however, a population's future environment is always likely to differ in at least some respects from its contemporary environment, and this future environment will therefore always make some different adaptive demands upon the members of that population. It may also be noted that this distinction between the two components of fitness is not fundamentally affected by whether the change that a particular population encounters results from temporal or spatial environmental diversity, or both. If an environment is temporally diverse, then any population that continues to contact the same spatial region of that environment will encounter change in its future as a function of the changes that occur in that particular spatial region over time. Conversely, if an environment is spatially diverse, then any population that encounters different spatial regions at different times will encounter change in its future as a function of its own changing dispersal patterns, even though in this case the environment itself may not change at all.

There is one general point about environmental change that, although implicit, must be made explicit. As far as any living system is concerned, environmental change is not simply "objective." It is in part "subjective." By this is meant that the distinction between the two components of fitness does not refer to the total change that may have occurred "objectively" in an environment over an interval of time. Rather it refers only to that subset of current environmental events that have changed relative to the particular record of past environments carried by that particular living system. With respect to an evolving population, the record carried from the past to the present always consists of its previously acquired adaptations. These adaptations indirectly record the past selection pressures that arose from former environments. Thus, change for a population actually consists of a discrepancy between the selection pressures that currently impinge upon it, and the selection pressures that formerly impinged upon it, and formerly induced the currently expressed adaptations of its members. Any other environmental changes that may also have occurred, but do not fall into this restricted category, are irrelevant. They do not make any contribution to the distinction between the two components of fitness.

It is helpful to classify the principal kinds of environmental change to which a population is liable to be subjected. Table II is a preliminary attempt, which combines Thoday's (1953, 1964) three forms of change with two main sources, the latter division being in terms of whether the population members themselves are the agent of change or whether the causal agent is independent of the population in question. There is a further division that can be made for every cell of Table II. This division is into the categories of reversible or irreversible change. By reversible change is meant a succession of environmental states in which at least one state that previously has been encountered and registered by a population recurs, and so matches some adaptive feature of the members of that popula-

TABLE II

MAJOR SOURCES AND TYPES OF ENVIRONMENTAL CHANGE[a]

	Population independent		
	Inorganic	Organic	Population dependent
Singular	Volcanic upheaval	Viral or bacterial infection	Sudden dispersion to new environment
Cyclic	Seasonal effects	Migratory interactions	Migration
Directional	Continental drift	Evolution of other populations	Evolution of own population

[a]The examples shown in each cell are arbitrary illustrations.

tion. A simple example of such reversible change that is periodic is the diurnal cycle and its effects upon a population whose individual members have an average life span of more than 24 hr. By irreversible change is meant a succession of environmental states, none of which is held within the record of the population. For example, a succession of ice ages is "objectively" a periodic change, but since no living population (other than man, whose cultural–scientific information store does not belong to any of the three referents being discussed here) is likely to have the capacity to maintain a record in the gene pool of sufficient coherence over the required period of tens of thousands of years, such change becomes "subjectively" directional and irreversible. Thus, an important determiner of the distinction between reversible and irreversible change is the interaction between the intervals over which the changes occur and the period of time over which the environment is monitored and recorded by successive generations of a population, together with the coherence of the maintained record.

The examples given in Table II are illustrative and arbitrary. They easily could be replaced by countless alternatives. Thus, under population independent directional change, long-term climatic effects could serve as well. Similarly, in the cell corresponding to inorganic cyclic change, diurnal, or monthly tidal effects would be equally apt.

Apart from source and type, there are two other major parameters of change that are not shown in Table II, but also affect the discrepancy between the two components of fitness. They are amplitude and rate. *Amplitude* of change will be defined here as the difference between the adaptations that an animal already has and those that the environmental change require. The greater this difference the greater will be the amplitude of the change, and the greater will be the discrepancy between the two components of fitness. *Rate of change* refers to the frequency with which changes occur relative to any specified referent-dependent temporal unit. Rate of change is a parameter that will be referred to repeatedly in the following pages in a variety of connections, but here there is one general point that needs commenting on from the start. Waddington (1969) has drawn attention to the observation that there is a general tendency in nature for rates of environmental change to increase over time because all living systems act upon their own environments, and in doing so change them. Furthermore, they change them not only with respect to themselves, but potentially with respect to all other living systems as well. Hence, any environmental change wrought by the activities of one population is liable to alter the selection pressures that act, not only upon itself, but also upon every other population that contacts the same environment. In their turn any such induced environmental changes are likely to induce further adaptive changes in these populations. The new changes may then elicit still further environmental changes and so on. In this way positive feedback is built into organic evolution with the consequence that change itself generates change and tends to do so at ever accelerating rates.

2. Constraints on Adaptation in a Changing Environment

The principal effect that these two components of fitness have on the members of a population is to subject them to two constraints that act so as to limit severely the adaptive possibilities available to them. The first of these constraints was originally pointed out by Darlington (1939). It concerns the fact that the two components of fitness make antagonistic demands on evolving populations. The first component of fitness requires specialization and stability for the attainment of optimal adaptation to the contemporary environment. The second component of fitness demands a capacity for variability, or a so-called generalist capacity, for optimal adaptation to future and changing environments. In this context specialization takes the form of specific adaptations in response to "known" specific environmental demands, whereas generalization takes the form of multipurpose adaptations. As long as the environment remains stable, a population of specialists is likely to be the better adapted, relative to both components of fitness, but in a changing environment its former adaptive specializations may become a disadvantage to it and may impede it. In contrast, a population of generalists is always likely to have some capacity to adapt to changing environments as a function of its multipurpose adaptations, so, if faced with a changing environment, it should actually do better than a population of specialists.

The antagonism between these demands arises because there is always a degree of opposition between biological investment in specialization and that in generalization. There are limits to the extent to which any given population can be both specialist and generalist at the same time. The limits have been described in abstract terms by Conrad (1972), who points out that specialization always implies a system that requires the operation of many different parts before the system as a whole can operate, whereas generalization implies a system that is redundant, and therefore requires only the operation of some of its parts before the system as a whole can work. Consequently, the only way in which it is possible for a system to be both specialist and generalist simultaneously is for it to carry its redundancy in the form of multiple specializations. However, to do this on a scale that could have a significant chance of success would be prohibitively expensive biologically. Conrad refers to the two different kinds of biological solution that specialization and generalization eventually map into, as "states of adaptedness" on the one hand, versus "states of adaptability" on the other. We shall use these terms with Conrad's meanings in this contribution. Like Thoday and Darlington, Conrad also stresses that there is inevitably some "interference" between adaptedness and adaptability.

The second constraint on the adaptation of the members of a population is conceptually much more straightforward. The capacity of the members of a population to cope with the second component of fitness is contingent upon their capacity to cope with the first component (Thoday, 1953). This means that a population must always be fit relative to its contemporary environment, regardless of what may be in store for it in the future. If it is not fit in the present, it will have no future anyway. Therefore, no population can afford to invest too heavily in general-purpose capacity, however adaptive such capacity might prove to be later on, if by so doing it loses its adaptive hold on its present environment relative to some more specialized competitor.

3. The Primary Adaptive Strategies

The preceding points may now be summarized as follows:
Let

$t_1 - t_0 =$ some period of time,
$C =$ change occurring during $t_1 - t_0$,
$S =$ the minimal adaptedness that must be maintained by a population throughout $t_1 - t_0$,
$V =$ the adaptability imposed by C.

The problem of fitness is that (1) S is primary, (2) S and V are inversely related, and (3) the environment imposes certain levels of S and of V that must both be met if the hypothetical population is to be successful.

The principal inference that can be drawn from this brief statement is that there can be only two logically distinct classes of adaptive strategy available to any evolving population that is confronted by a changing environment. The members of that population can either avoid change or they can track change by generating adaptive changes in the gene pool. Moreover, the latter changes must always be synchronized to, and must necessarily track behind, the environmental changes. In terms of the preceding statement, these two primary strategies can be summarized as follows. If change can be avoided (or more correctly minimized), then V is reduced to a negligible level, which means that S can be raised to a high level. In effect, this allows the members of a population to achieve a high state of specialized adaptedness. Conversely, if change cannot be avoided, then V must rise to nonnegligible levels, even though this will always be at some cost to S. It follows that a population which tracks change will only be successful if it can evolve mechanisms that do so at only minimal cost to S.

In practice there are probably very many different ways of pursuing either of these two strategies. Moreover, the two strategies are by no means mutually exclusive. Any population can, and the vast majority certainly do, follow both strategies simultaneously.

B. THE PRIMARY REFERENT: A POPULATION

The second notion that also needs some brief elaboration is that of the primary referent. Up to this point the primary referent has been identified simply as the members of any intrabreeding population. However, it is necessary to describe more fully one of the fundamental characteristics of this unit.

At any given moment a population consists of a set of currently living organisms. Each of these organisms, however, is itself organized around a genotype–phenotype dichotomy. The chief significance of this dichotomy is that it allows adaptations to occur at either the genotypic or the phenotypic level. Since natural selection acts on phenotypes and not on genotypes, the adaptations that may occur at either of these two levels are potentially equally effective determinants of the fitness of the organism, and therefore of the population. Thoday (1953) has shown that the existence of this genotype–phenotype dichotomy means that members of a population have available to them a component of within-population variability in the form of phenotypic flexibility, which is independent of genetic variability, at least to the extent to which it can occur independently of immediate genetic changes. He stresses that considerable biological advantages can accrue to a population because of this arrangement. Potentially it allows the members of a population to respond to environmental change at the phenotypic level, without necessarily having to make any new adjustments at the genetic level. The genotype–phenotype dichotomy thus becomes a means by which members of a population can increase their adaptability (V) at the phenotypic level at either very low or zero cost to their adaptedness (S) at the genetic level. It thereby goes a long way toward allowing a population to resolve the antagonism between the two components of fitness. The same point has been made by Waddinton (1969) although in slightly different terms.

Thoday (1953) distinguishes two distinct forms of phenotypic variability. The first he refers to as developmental flexibility. The second he refers to as behavioral flexibility. An individual is said to possess *developmental flexibility* if either its genotype is capable of developing into different phenotypes in different environments, or if its genotype is so balanced that it is capable of developing into apparently the same phenotype in spite of different environments. In contrast, an individual is said to possess *behavioral flexibility* if it can react temporarily to reversible environmental change. (While this terminology is adequate for our purposes, it should be noted that "behavioral" is unduly restrictive. Many forms of phenotypic flexibility are neither developmental nor behavioral. For example, tbe response of water conservation at high temperatures in mammals is physiological.)

The main difference between these two forms of phenotypic flexibility is that developmental flexibility is necessarily linked to one-way ontogenetic trajectories. Organisms do not grow younger; they only grow older. Developmental flexibility is therefore governed by controls that are largely irreversible. This does not, of course, mean that it will always be impossible for a phenotype to cope with some reversed change in its environment by means of developmental flexibility once it has passed the appropriate stage of development. What it does mean is that the phenotype will then only be able to cope with such a change by secondary processes of compensation during the course of its still future development. The phenotype will not be able to go back and undo the development that has already occurred. Behavioral flexibility, in contrast, is subserved by within-animal subsystems that, while themselves a product of ontogenesis, can become largely independent of the one-way ontogenetic trajectories once they have reached maturity. They therefore can act reversibly as Thoday (1953) states.

Thoday's distinction between forms of phenotypic flexibility is maintained here, but with one significant modification. A distinction that Thoday does not make is between those forms of phenotypic flexibility that depend exclusively upon phylogenetic information gain (primary referent) and those other forms of phenotypic flexibility that depend upon supplementary information-gaining processes (the second and third referents), as well as upon phylogenesis. This distinction is central to this argument and will be adhered to strictly here. Thus, to begin with we shall be referring only to forms of phenotypic flexibility that are entirely the result of phylogenesis, and can, in consequence, be regarded as entirely predetermined or innate.

C. THE LOGIC OF PHYLOGENESIS

Phylogenesis is the primary information-gaining process in a biological system. Lorenz (1969) refers to it as a process of "trans-information" to emphasize that it represents the primary channel of communication between living systems and their environments, and that it is the ultimate means by which all living systems gain their fitness. Campbell (1974) has suggested that the processes of phylogenesis can be seen in terms of three distinct subprocesses in an overall information-gaining procedure. These are, first, a group of processes that generate variety within a population, with the variety taking the form of variation among the the individual organisms that make up the population. Second, there are relatively consistent selection processes that relate to the order which obtains in the population's environment and which comprise natural selection pressures. Last, there are processes that store and propagate selected variants within the population, the propagating processes including both retrieval and transmission systems. For convenience we shall refer to all three subprocesses as collectively constituting a "generate–test–regenerate" (g–t–r) heuristic, but in doing so it should be clearly noted that the regenerate phase in this heuristic also represents the next generate phase of a continuously ongoing process. The regenerate phase must, in fact, be taken to refer to both the retrieval and propagation of previously selected variants, and to the generation of some new variants that as yet still have to be acted on by any subsequent test phase selection pressures.

Popper (1961, 1966) has remarked that one of the principal characteristics of this g-t-r heuristic is that it acts so as to simulate induction. It proceeds from particulars to particulars, but it always does so via a generalizing step. The particulars are the currently living batch of phenotypes that the population is expressing. The generalizing step consists of the regeneration and dispersal of new phenotypes by the population across time and space.

A second characteristic is equally fundamental. The g-t-r heuristic is based on a mix of conservative pragmatism and radical chance. The conservative pragmatic component lies in the fact that the regenerate phase is partly constrained by the test phase. The pragmatism takes the form of a "blind" assumption (Campbell, 1974), which holds that the future of the population will be the same as its past. Hence, the regenerate phase effectively carries forward into a population's future genetically coded information that at some previous time has been tested by its past environments. The radical chance component lies in the gene and chromosomal mutations, recombinations, and interactions that occur first at the generate phase, and then recur at each subsequent regenerate phase. Collectively these genetic variations act so as to secure the partial release of the regenerate phase from the constraints of the test phase. Waddington (1969) points out that this mix of chance and constraint consists of a game-theory strategy. In effect the g-t-r heuristic "gambles" that the future will be the same as the past. At the same time it hedges its bets with aleatoric jumps, just in case it is not.

A third characteristic of phylogenesis is that it acts stochastically. Every population is necessarily localized in space and time. It therefore can encounter only samples of the universal environment. This means that the conservatism of the g-t-r heuristic will always be based only upon those selection pressures that have actually arisen from those particular samples of the environment that the members of the population have already encountered. In a very real sense, therefore, the heuristic is prone to make "sampling errors." In fact the adaptive value of the conserved information transmitted from generation to generation will vary according to two standard sampling variables. These are, first, the degree of dispersal of the intrabreeding population across time and space, and, second, the degree of environmental heterogeneity. The greater the dispersal of the population, and the more homogeneous the environment, then the more likely it will be that the conservatism of the heuristic will be based on a "representative" sampling of the environment, in which case the greater will be the probability that the adaptations that are conserved will continue to prove adaptive as the population disperses into other temporal and spatial regions of the environment. Conversely, the more localized the population, and the more heterogeneous the environment, then the more likely it will be that the conservatism of the heuristic will be based on a "biased" sample of the environment, in which case the greater will be the probability that the previously acquired adaptations of the members of the population will no longer be adaptive as the population moves into future environments.

D. The Limits of Phylogenesis

We now can bring the components of fitness and the logic of phylogenesis to bear upon the key question of the extent to which it is possible for an evolving population to solve its adaptation problems exclusively by means of phylogenesis, and without its having to evolve additional information-gaining and -storing mechanisms. The limits of phylogenesis will be considered in terms of both the avoid-change and track-change strategies.

1. Phylogenesis and the Avoid-Change Strategy

It is probably never possible for members of a population to avoid environmental change completely. Change can be minimized, however, and, if it is, the antagonism between the two components of fitness is accordingly reduced. Because avoiding change means that the future of a population is going to be similar to the past, full advantage can be taken of the conservative pragmatic aspect of the g–t–r heuristic. In such circumstances the g and r phases of the heuristic should continue to supply the population with new variants, but the selection pressures at the t phase should remain relatively constant and so continually select for the same or similar variants. Thus, the total variance within the population should be progressively reduced and be accompanied by a parallel increase in the specialization of the members of the population and their fitness relative to that constant environment. Evolution, in short, should proceed to the point of an optimal set of genotypes transforming into an optimal set of phenotypes, and then stop. There will always be the possiblity of new variants being generated by chance being still better adapted to a relatively unchanging environment, but in time even this possibility will become remote. Thus, in a static world, single valued in Waddington's (1969) terms, evolution should eventually result in a maximum state of adaptedness (S) and a minimum state of adaptability (V).

Whether such extreme conditions ever occur in reality is hard to assess, although species are known for which evolution does appear to have stopped. The horseshoe crab, oyster, and sphenodon are examples cited by Simpson (1967). *Latimeria* (Smith, 1962), a contemporary representative of the coelocanths dredged from deep-sea trenches off the south-east coast of Africa, is another. So too is *Lingula,* a brachiopod that has survived unchanged since the Cambrian, a period of over 500 million years. If there are species that have survived largely by avoiding change, then they are likely to be found among such examples.

There are two limiting factors to the extent to which an avoid-change strategy can be followed successfully. The first is whether population members have available to them niches or habitats in which change is negligible. Whether an unchanging niche is available is not something over which population members have total control. Indeed, the number of such niches must be reducing constantly over time because of the positive acceleration of biological evolution itself. On these grounds it is probable that the avoid-strategy was more prevalent in the past than it is today. However, even if an unchanging niche is available, the second limit is the capacity of members of a population to make and maintain contact with it. A population's contact with its environment is determined by the interaction of a range of isolating and dispersal mechanisms, including those responsible for speciation itself (Mayr, 1963). At less differentiated levels of organization the members of a population can only react passively to conditions that are imposed on it from without, and so contact with any new region of the environment is fortuitous. Every population capable of expressing more complex phenotypic adaptations, however, may select to some extent the environmental niche that it will encounter and exploit. Thus, in part such a population is selecting the selection pressures that will act upon its members, a subtle feature of evolutionary theory to which Waddington (1975) has repeatedly drawn attention. He refers to it as the *exploitive system* and he includes not only habitat selection but also subordinate factors such as food and mate selection. Insofar as the mechanisms of habitat selection, themselves phenotypic adaptations, act conservatively, as most do (Mayr, 1963), then their effect is to ensure that population members will face relatively constant selection pressures. In this case, even though the world in general may be becoming progressively more heterogeneous, the capacity of a population to select only those niches that, relative to the requirements of its members, do not change may act as an effective counter against increasing change.

There are many examples of conservative habitat selection in nature that depend on phenotypic adaptations (Alcock, 1975; Mayr, 1963; Wilson, 1975). Some are inappropriate to a discussion oriented toward the limits of phylogenesis, since they involve supplementary information-gaining processes such as learning. However, others do appear to depend exclusively upon phylogenesis. The lizard *Aporosaura anchietas* extensively discussed by McFarland (1976) is one such example. Certain forms of Curculionid beetles, which can only live in the bark of a single species of tree (Evans, 1975), are another. On the macro scale, the migratory flights of insects (Wilson, 1971) provide a third example. Another less obvious form of habitat selection results from parental care. Parents may select favorable initial habitats for their offspring, say in egg laying, or they may themselves act as vital and predictable resource-supplying components in the environments of their offspring (Milne and Milne, 1976; Wilson, 1971).

In spite of such diverse instances, the extent to which change can be avoided by conservative habitat selection is limited because, first, as long as the mechanisms of habitat selection depend exclusively upon phylogenesis, then their sensitivity and discriminative capacities will be limited to what can be preprogrammed by phylogenesis. Second, the conservatism of habitat selection will only remain successful as long as there are relatively static regions of the environment within reach of the adaptive capacities of that population. Finally, the buffering function of parental care in certain species, notably birds and mammals, may be offset by the parents themselves constituting powerful, if controlled, agents of change, a fact that becomes of great importance in the biology of learning.

2. Phylogenesis and the Track Change Strategy

If a population of organisms cannot avoid change, then it must track change. Tbe track strategy introduces several extra complications, however, some of which have already been introduced. In a changing environment the conservative pragmatic component of the phylogenetic heuristic ceases to be sufficient. Both the population's present and future became different from its past, so that which used to be fit in the past may no longer be fit in the present. There is therefore some irreducible conflict between two components of fitness. In addition, no members of a population can ever know what their future will bring, so that in the last analysis the population can only proceed blindly (Campbell, 1974). Finally there is a time lag implicit in the tracking process itself. To track change the gene pool must be able to generate changes in the members of the population, but this is something that takes time, and it is time during which further environmental change may be occurring. In the light of so many difficulties how can a population of animals track change successfully? Undoubtedly populations do succeed, but the question is far from trivial and raises the issue of whether success is achieved solely by the mechanisms of phylogenesis:

> The systematic exploration of the evolutionary strategies in facing an unknown, but usually not wholly unforecasteable, future would take us into a realm of thought which is the most challenging and very characteristic of the basic problem of biology. The main issue in evolution is how populations deal with unknown futures (Waddington, 1969, p. 122).

We shall follow Waddington by referring to this issue as the uncertain futures problem.

With respect to the g–t–r heuristic, the main effects of introducing a changing environment are essentially those described by Darwinian theory. A changing environment means changing natural selection pressures at the "t" phase. Some variants within the population that used to be fit become unfit. Eventually these are eliminated and cease to form a part of the genetic record that underlies the conservative pragmatic component of the g–t–r heuristic. Conversely, some other variants that hitherto were either neutral, unfit, or not previously in existence may now become fit. In any case, the loss of unfit variants from the population has to be compensated for by the continuous introduction of new variants into the population. Moreover, in this respect the population can only depend for its supply of new variants upon the "g" and "r" phases and therefore upon its radical chance component. Whether the supply of new variants proves adequate in both number and variety is subsequently determined by whether or not the population always has available to it sufficient variants that are fit to ensure its continuous survival, regardless of how great the changes at the "t" phase may be.

The ability of a population gene pool to track change depends primarily upon two closely related capacities, neither of which are demanded of a population that is merely avoiding change. First, the members of a population must have between them a capacity to maintain and continuously update an adequate reserve of variants. These variants may take the form of either actual or potential alternative forms of phenotypic expression. Second, the present and future members of the population must also have between them a capacity to switch between alternative forms either of phenotypic expression or of flexible phenotypic response, as is made necessary by the changing environmental order itself. Together these two capacities will constitute the population's adaptability (V). The greater this adaptability, the greater will be the chance that both the current and future members of that population will have of surviving their unknown future. However, the greater the value of V, then the greater also is likely to be the cost to the population members in terms of their adaptedness (S).

We now can examine each of these two capacities separately in order to see how a population's ability to respond to change is limited as long as the evolution of that population is wholly dependent upon phylogenesis. First, there is the generation of variance by devices that are well known. In every population there are the fundamental mechanisms of genetic mutation for replenishing its reserve of variants. There are then a number of supplementary processes that allow a population both to increase realized variance as well as to carry cryptic variance within the gene pool (that is, a reserve of variants that are either not expressed, or are only partly expressed in the current generation). These are the processes of sexual reproduction itself, gene interactions, and polymorphism including those due to neutral alleles, epistasis, and genetic homeostasis (Dobzhansky, 1970; Lerner, 1954; Thoday, 1975). However, the amount of variance that can be built in to genotypes by any of these means is limited by the point at which excessive variance begins to undermine the internal stability of the genotype (Whyte, 1965). Thus, a special case of the general limitation imposed by the primacy of S operates upon the variance that can be supplied by genetic mechanisms since the internal stability of the genotypes must be maintained at all times if the members of the population are to remain viable.

A secondary means of increasing variants is simply to increase the number of offspring in each generation and thus to increase both the genotypic and phenotypic range. But here too there are limits to such a solution. First, the environment must be able to support the increase in numbers. Second, any increase in competition for environmental resources will increase the likelihood of population fragmentation and thus increase the rates of environmental change.

The second form of tracking change, switching between alternatives, can occur either at the gene-pool or individual organism level. The former is embodied in such concepts as "fitness surfaces" (Wright, 1949; Waddington, 1959) and involves the use of redundancy accumulated within the reserves of variants of the population gene pool to switch from one form of phenotypic expression to another as a function of prevailing environmental conditions. The latter is equivalent to Thoday's (1953) notion of phenotypic flexibility, but with our additional requirement that the information upon which such flexibility is based is confined to the primary referent. Thus, the individual phenotype responds adaptively to changes in the environment but does so without altering the store of information upon which such flexibility is based. An example of such developmental flexibility is seen in the various larval instar stages, pupation, and final molt to adult form that occurs in many insects as a response to varying external conditions such as temperature and light. Examples of behavioral flexibility need not be confined to animals. Nyctinastic movement in plants (Jones and Mansfield, 1975) conform to such a definition of adaptive response to change. But so too do kineses, taxes, and action patterns in animals [in fact, all of those adaptive responses termed "cognitive processes not involving learning" by Lorenz (1969)]. Overall, such switching may take two forms. Either the process or response remains relatively constant, but the timing of its appearance will vary. Thus, metamorphosis, for example, is an invariant sequence of processes. The adaptability derives from the environmental changes that trigger the events and thus determine the timing of the processes. Or a quite distinct phenotypic form might be determined by prevailing environmental conditions, as in phase determination in the locust.

Again, however, all of these switching devices are limited. As before, many depend upon genetic adaptations, which are limited primarily by the constraints that arise from the relationship between the two components of fitness. Those devices that depend on phenotypic flexibility may escape from these basic constraints to a considerable degree since phenotypic flexibility costs so little in terms of genetic stability (Thoday, 1953). Nevertheless, phenotypic flexibility is in its turn limited by a further constraint, which in this case arises directly from the operating characteristics of the phylogenetic heuristic itself (that is, the uncertain futures problem). Moreover, the latter constraint is not confined to phenotypic flexibility alone but affects every form of adaptation, genetic and phenotypic, that is dependent upon phylogenesis.

This constraint arises from the time lag that is inherent in every tracking process, and in the specific case of phylogenesis, results from the time it takes for members of a population to regenerate. Every organism that reproduces, whether by way of gamete production and exchange, or by some more primitive means, produces offspring that themselves take time to develop to the point where they in their turn become reproductively competent. Therefore, a period of time must elapse between the moment of reproduction by a parent generation (designated time t_0 in line with our previous statement of the adaptation problem) and the moment of reproduction by the offspring (designated time t_1). For each individual organism it is always likely that the environmental conditions prevailing at time t_1 will be different from those prevailing at time t_0. In Waddington's (1969) words, "... it will not in general be true that the environmental influences which contribute to the formation of the phenotype are identical with those which exert the most important selection upon it" (p. 120). Similarly, Lorenz (1969) observes that genetic mechanisms of information gain are subject to "generational dead time" and cannot know about or cope with "quick" changes in the environment.

Applied to genetic adaptation the main effect that generational dead time has is to make it impossible for members of a population to respond to environmental changes within generations. A population can only respond to the environment on a between-generation basis. The most obvious way of dealing with this limit is for the members to speed up their rate of generational turnover. If they do so, it effectively reduces the interval $(t_1 - t_0)$ and thereby reduces the amount of change that is likely to occur within any one generation. However, the extent to which a population of animals may follow this option will be limited in practice by its prior history. Thus, a population of microorganisms might well be free to reproduce every 20 min, but a population of mammals may be forced by the kind of biological system that it has already become to take weeks, months, or even years to regenerate.

The effect that generational dead time has on phenotypic flexibility is less direct and more subtle. As stated, phenotypic flexibility deals specifically with changes and events that occur during the lifetimes of individual organisms. In fact, it permits organisms to switch between alternative forms of expression in response to within-lifetime events. On the face of it, therefore, phenotypic flexibility can deal with within-generation change. Nevertheless, the dead time limit still applies. The reason why this is so is that insofar as an organism's phenotypic flexibility depends exclusively on phylogenesis the organism will only be able to respond adaptively to those within-generation changes for which it has previously been prepared or "preprogrammed" by phylogenesis. Thus, the organism must be equipped by phylogenesis with appropriate innate mechanisms that will, in effect, constitute specialized states of adaptedness. However, if an organism is to be preprogrammed by phylogenesis in this way, then the population to which it belongs must previously have been exposed to consistent selection pressures for many generations by the environment. So to the extent to which an organism relies on forms of phenotypic flexibility that are exclusively dependent for their fitness upon phylogenesis, it should only be possible for that organism to respond adaptively to a restricted subset of within-lifetime events. Specifically, the organism should only be able to cope with those changes and events that, although possibly only occurring as little as once per generation, have, nevertheless, occurred sufficiently regularly across multiple generations of the population to have constituted coherent selection pressures. The organism should not be able to cope with any within-lifetime changes and events that have not been able to act as coherent selection pressures in the past history of its population. This position can be simply summarized by stating that for an organism that is exclusively dependent upon phylogenesis for its fitness, phenotypic flexibility should only be able to cope with those within-lifetime changes that have been "predicted" or "forecast" by phylogenesis as likely to occur. It should not be able to cope with any phylogenetically "unpredicted" events. It has to be stressed, however, that in the present context the words "predicted" and "forecast" are being used deliberately and only as convenient teleonomic shorthand. Here, and on all subsequent occasions the words merely signify the presence of innate mechanisms in organisms that have previously been selected for by consistent selection pressures acting on evolving populations.

We now have shown that the capacity of members of a population both to build and maintain a reserve of variants, and to switch between alternative forms of expression in response to changes in its environment, is ultimately limited in every direction. As long as a population is wholly dependent for its fitness upon phylogenesis, there therefore will always be some upper limit to the amount or rate of environmental change that can be tracked.

3. The Limits of Phylogenesis versus Unlimited Environmental Change

The central point of the preceding argument is that members of a population are limited in the extent to which changing environmental order can be either avoided or tracked. There is no equivalent limit, however, to the amount of change that may arise in a population's environment. Because positive feedback is built into biological evolution with change inevitably generating further change, the potential lability of the environment is unlimited. This means that the demands for adaptations that an environment may make upon a population are also unlimited. In this case, what happens to a population that has already reached the upper limits to the amount or rate of change that it can cope with via phylogenesis, yet which is faced with a further increment in environmental change? One clear answer is the extinction of the population. Another lies in the evolution by its members of various forms of supplementary information-gaining process. These supplementary processes, by acting in support of the primary phylogenetic process, may then permit the continuing survival of the population in a much more labile environment than it could possibly cope with by phylogenesis alone. It is to these supplementary processes that we turn next.

E.   THE EVOLUTION OF THE SECONDARY AND TERTIARY
     REFERENTS

No supplementary information-gaining process can escape from the constraints that arise from the antagonistic relationship between the two components of fitness. This antagonism is fundamental and affects all kinds of adaptation and all information-gaining processes. But supplementary forms of information gain can overcome the constraints arising from the generation dead-time limit imposed by phylogenesis. In principle there is nothing to stop the members of a population from evolving new ways of gaining information that act more rapidly than phylogenesis by working on a within-generation and within-lifetime basis, instead of on a between-generation basis. Such processes should be effective in coping with at least some of those "unpredictable" within-generation changes, which phylogenesis, with its relatively long time lags, cannot cope with.

Whether or not supplementary information gaining processes do evolve in any given population depends on the course already taken by phylogenesis because phylogenetic information gaining is primary and because there is no other available evolutionary source. Therefore, supplementary information-gaining processes can only evolve in response to appropriate selection pressures acting at the "t" phase of the phylogenetic g–t–r heuristic. This raises the question as to what kind of environmental events could possibly constitute appropriate selection pressures. The answer, clearly, is within-generation changes, which, though phylogenetically "unpredictable," still constitute coherent selection pressures acting across generations, and are therefore capable of affecting the evolution of a population. However, to be effective the "unpredictability" of these within-generation events must be maintained across multiple generations at approximately the same level of uncertainty, and with respect to approximately the same components of the environment. To use Alcock's (1975) phrase, "environmental unpredictability" must be "reliably present in certain situations." When this condition is satisfied, then the "unpredictability" of the within-generation events itself becomes "predictable" by phylogenesis.

Selection pressures of the latter type must, as always, select for adaptations that take the form of innate mechanisms within each individual member of a population. However, unlike selection pressures that result from phylogenetically "predictable" events, sustained environmental "unpredictability" cannot select for specialist mechanisms that are fully preprogrammed or "closed" by phylogenesis in order to deal with specific predetermined events. It cannot do this because, by definition, the environmental events that these mechanisms will have to cope with cannot be specified in advance by phylogenesis. Hence, sustained "unpredictability" can, at best, only select for adaptability mechanisms that are left "open" by phylogenesis. Only if they are open, might they cope with the phylogenetically unspecifiable events that will, nevertheless, "predictably" occur during the lifetimes of individual organisms. Put in other terms, uncertainty in a population's environment cannot select for specialised states of its member's adaptedness. As usual, it can only select for multipurpose adaptability. Any mechanisms of adaptability that are selected for in this way may then be expected to demonstrate two basic characteristics. First, it is parsimonious to assume that even though innate adaptability mechanisms are left open by phylogenesis, they will still make maximum use of any useful information that can be specified in advance by phylogenesis. They therefore should demonstrate phylogenetically predetermined operating characteristics and tolerance limits with respect to the degree of their openess, up to the limits to which the strategy of using phylogentically acquired information is adaptive. Second, and still more important, to the extent to which any adaptability mechanism is left open by phylogenesis, then it must also be equipped by phylogenesis with a capacity for autonomous information gaining. Adaptability mechanisms cannot simply be left open with no capacity for self-closure. If they were, then they would be useless organs, and with respect to evolution, sterile. They therefore must have access to enough additional information to convert those events left unspecified and "unpredicted" by phylogenesis into newly specified and newly predicted events. Only then might they be able to steer the individual organisms within which they are active toward the necessary additional adaptations. But additional information relative to within-generation and within-lifetime events can only come from the adaptability mechanisms themselves.

At this stage the concept of further evolutionary units, or additional referents as they are called here, becomes unavoidable. The existence within individual organisms of mechanisms that are themselves capable of autonomous information gaining actually defines these new referents. The reason for this is that the unit which is in direct receipt of the information provided by these mechanisms is no longer a primary referent. It cannot be, because there is no channel for feeding back the information gained by these adaptability mechanisms into the population's gene pool. This is simply a restatement of the Weismann maxim concerning the noninheritance of acquired characters. The adaptability mechanisms therefore may only feed their information into a subsidiary unit or units, which are necessarily restricted to the scope of an individual organism, and its individual lifetime.

One interesting consequence of the inability of innate mechanisms of adaptability to communicate directly with a population's gene pool is that they are cut off from any information that might be directly obtained by the adaptability mechanisms of other members of the same population. This means that the information gained by the adaptability mechanisms of a single individual must necessarily be derived exclusively from information available in those samples of the environment that happen to be encountered by that organism. They cannot gain information from other samples of the environment encountered by other organisms.* In practice this restriction need not be serious and could even hold advantages. The samples of the environment that are encountered by an individual organism are likely to be especially relevant to its particular adaptation. Those encountered by other individuals in other environmental regions are likely to be less relevant, and could even introduce some interference. An organism's innate adaptability mechanisms therefore can be legitimately regarded as fine tuning devices, which are responsible for the unique adaptation of that individual.

1. Epigenesis: The Secondary Referent

By epigenesis is meant the provision of a relatively plastic developmental program responsive, within limits, to a range of environmental circumstances (Waddington, 1957). In this case the unit that receives the additional information is a developing individual organism. Specifically it is the system that includes a genotype actively translating into a phenotype. This unit is shown as the secondary referent in Table I. Epigenesis actually leads to a gain in information by allowing interactions to occur between the environment and a range of alternative possible developmental trajectories. The alternative possibilities themselves are carried in the form of open gene programs by individual genotypes (Mayr, 1974) and are a form of genetic redundancy and adaptability. An open gene program permits the environment itself to select between alternative possible phenotypic traits whose maturation is modified by epigenesis and therefore represents the end product of two distinct information-gaining processes, phylogenesis and epigenesis.

One of the basic functions of epigenesis is to act as a source of additional developmental phenotypic flexibility. Epigenetic mechanisms can respond adaptively to at least some of the within-generation environmental events that cannot be "predicted" by phylogenesis. Specifically, they can cope with changing environmental order, which impinges upon the organism during its development, and which is sufficiently stable across the life span of that organism that it becomes predictable and permits useful within-life span adaptations. They thereby can increase the capacity of an individual organism to track change, and in some cases they may also increase its capacity to avoid change by sharpening

---

*This statement is true unless and until a population evolves some second nongenetic channel of communication between its individual members. If it does so, then some information gained by an individual's adaptability mechanisms may be spread throughout the population as a whole. This possibility introduces all the additional complexities associated with social and cultural evolution. It also introduces yet another referent. However, in the present context such complexities are ignored. This is not because we underestimate the widespread occurrence and importance of such an additional referent, but simply because we are limiting ourselves here to the evolution of those categories of learning that depend upon an animal's direct encounters with its environment rather than information passed to it by other animals.

up the sensitivity of phenotypic isolating mechanisms, and by increasing the capacity of the organism for active conservative habitat selection. However, the capacity of epigenesis to deal with change is still limited. Epigenesis cannot respond adaptively to any environmental changes that recur and reverse, possibly rapidly and possibly frequently, within the lifetime of an individual organism because epigenetic development, like every other form of development, is firmly linked to one-way developmental trajectories. Waddington (1957, 1969) refers to these trajectories as "chreods," and he points out that their underlying control processes are "homeorhetic" in character rather than homeostatic. That is, although they may be capable of compensatory activities, they are not capable of going into reverse in responses to environmental changes that themselves repeat or reverse within a single lifetime.

2. Specific Subsystems: The Tertiary Referent

The other class of innate adaptability mechanisms that concern us are the subsystems underlying a whole variety of additional information-gaining processes. In general these are specific organ systems that act as "defence mechanisms which do not have to know in advance what they will have to defend against" (Waddington, 1969, p. 122). One example is the vertebrate immune system. Another is the learning system. In each case the unit that is actually gaining information is an organic subsystem, even though the ultimate beneficiary in terms of adaptation is a phenotype. For this reason every subsystem actually capable of autonomous information gaining is regarded here as a separate referent although collectively all such subsystems are referred to as tertiary referents (see Table I).

One possible objection to this scheme concerns the occurrence of learning in Protozoa (Corning and Von Burg, 1973). Learning may not always be legitimately attributed to a specific organ system. However, whatever capacity for learning unicellular animals may possess, probably this capacity is still based upon specialized organelles or cellular regions within the animal (e.g., Applewhite and Gardner, 1971), and so with some small adjustment the same general scheme ought still to apply.

The basic function of tertiary referents is to act as a source of additional behavioral phenotypic flexibility. Thus, each tertiary referent should cope with at least some of the environmental changes that are both phylogenetically and epigenetically unpredictable, and therefore cannot be coped with by the members of a population in any other way. Because they consist of adaptability mechanisms that are left open by phylogenesis, tertiary referents can, in principle, cope with within-generation changes that cannot be "predicted" and cannot therefore be coped with by phylogenesis. Similarly, because they are organ subsystems that, subsequent to their own maturation, are largely independent of one-way developmental trajectories, then they also can cope with recurring and reversible events of a kind that cannot be coped with by epigenesis. To the extent that tertiary referents may themselves be limited, they should therefore only be limited by their own operating characteristics. The net adaptive output of any tertiary referent should be similar to that of the secondary referent. A tertiary referent should tend to act so as either to increase the capacity of an individual organism to track change, as a function of its increased behavior plasticity, or it should increase the organism's capacity to avoid change as a function of its more finely tuned, and behaviorally dependent, conservative habitat selection.

3.   The Relationship between the Three Referents

The relationship between the primary, the secondary, and the tertiary referents is illustrated in Fig. 1. In this figure it is shown that the secondary and tertiary referents are not mutually exclusive. Thus, it is quite possible for a specific subsystem of the tertiary referent to be modified by epigenetic processes that are result from the secondary referent (Fig. 1, Scheme 3). A distinguishing feature of

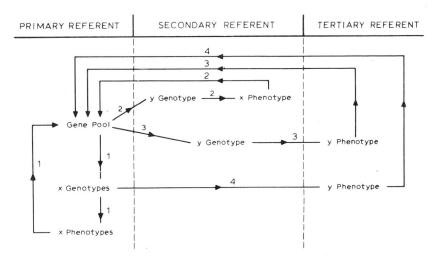

FIG. 1. The nested relationship between the primary, the secondary, and the tertiary referents. An x genotype is one with invariant translation into a phenotype. A y genotype has flexible development. An x phenotype is inflexible. A y phenotype has independent information-gaining processes. Scheme 1.   Genotype–phenotype translation is invariant and the phenotype inflexible. All information gain is confined to the primary referent. Scheme 2.   Genotype–phenotype translation is flexible, but postdevelopment the phenotype becomes inflexible. Information gain is confined to the primary and secondary referents. Scheme 3.   Genotype–phenotype translation is flexible, and postdevelopment the phenotype is also flexible. Information gain occurs at all three referent levels. Scheme 4.   Genotype–phenotype translation is invariant, but postdevelopment, the phenotype is flexible. Information gain is confined to the primary and tertiary referents.

epigenetic flexibility is that it is not confined to any one or few organs or organ systems. The study of epigenetics thus far has concentrated largely upon morphology and cell biology (e.g., Lovtrup, 1974), but there is little doubt that epigenetic adaptations also affect behavior (Bateson, 1976b; Gottlieb, 1976; Hamburger, 1963; Schneirla, 1965), including those behaviors and their physiological substrates that enter into learning. Schneirla (1956) recognized this complex interrelationship with his concept of ''experience'' as distinct from learning. It should also be possible for a behavioral subsystem to act as a tertiary referent without it being subject to any epigenetic modification (Fig. 1, Scheme 4). Conversely, it should be possible for an organism to be capable of epigenetic adaptations, but to be incapable of any adaptations resulting from any tertiary referent (Fig. 1, Scheme 2).

In Fig. 1 it also is shown that both the secondary and tertiary referents are themselves products of phylogenesis. As such, they fall within the genetic endowment of some population. This fact has been mentioned before, but it needs to be strongly stressed. It means that, although the secondary and tertiary referents will always depend on phylogenetically supplied innate generalist mechanisms, there is no logical reason why these mechanisms should turn out to be the same kind of mechanisms in every population in which they occur. On the contrary, all innate adaptability mechanisms must be population or species specific, which means that their operating characteristics and their tolerance limits must also be species specific. It is therefore to be expected that the innate adaptability mechanisms of all the members of a population are ''primed'' by phylogenesis to deal most efficiently with those particular features of the environment that are most relevant to their particular needs because it is probable that these features will, in general, still correspond to the principal sources of sustained ''unpredictability,'' and, therefore, to the principal selection pressures that have acted in favor of the evolution of adaptability mechanisms in the past history of that particular population. This is congruent with both Tinbergen's (1951) and Lorenz's (1965, 1969, 1977) repeated advocacy of the innate bases of learning, which are tailored by phylogenesis to the particular fitness needs of individual species.

A third point illustrated in Fig. 1 likewise has been introduced before, but it too needs stressing. The relationship between the primary referent, on the one hand, and the secondary and tertiary referents on the other is a two-way, not a one-way, relationship. Both the secondary and tertiary referents are products of phylogenesis, and both are species specific. At the same time the teleonomic justification for their existence is that they support phylogenesis, and stretch the capacity of the primary referent in terms of its reproductive success. In fact, these two subordinate referents support phylogenesis in two very closely related ways. First, they act as sources of extraphenotypic flexibility, the one of developmental flexibility, the other behavioral. Because any increase in phenotypic flexibility, regardless of its source can serve to increase a populations adaptability (*V*) at little or no extra cost to its adaptedness (*S*) (Thoday, 1953), then, even though it is not possible for the subordinate referents to escape from the constraints imposed by the two components of fitness, it is possible for them to increase the capacity of the primary referent to live within these constraints. Second, the repeated theme of this analysis is that these two subordinate referents actually achieve an increase in phenotypic flexibility by allowing the primary referent to escape from the generation dead time limit imposed by phylogenesis. Hence, they directly aid resolution of the uncertain futures problem. Both subordinate referents may therefore be regarded as having a status analagous to that of a subroutine within an algorithm. Each referent is a subroutine that has itself been produced by the primary referent or main evolutionary program, but each in turn acts so as to support and perpetuate the primary program. This relationship corresponds to that referred to as a nested hierarchy by Campbell (1974) in his discussion of the logic of evolutionary processes.

F. The Status and Function of Learning

The principal points which have emerged from the preceeding analysis can now be summarized.

1. Learning is a tertiary information gaining process. The capacity to learn is conferred upon individual organisms within some populations by innate adaptability mechanisms. These mechanisms constitute a separate referent from phylogenesis and epigenesis.

2. Innate adaptability mechanisms are a product of phylogenesis acting upon a primary referent (an evolving population). They are therefore population specific. They may also be modified by epigenesis acting upon a secondary referent (an individual developing organism). If they are, then to some extent they may be individual specific too.

3. The selection pressures responsible for the evolution of learning appear to have arisen from the limitations, primarily of phylogenesis, and secondarily of epigenesis, with respect to certain components of unavoidable environmental change. Specifically, it appears that selection pressures can arise from the inability of phylogenesis to cope with within-generation changes that are too irregular with respect to successive generations of a population to be "predicted" by phylogenesis, but will nevertheless "predictably" occur. They also can arise from the limited ability of epigenesis to cope with changes that reoccur and reverse within the lifetimes of individual organisms. In short, learning provides the fine behavioral tuning that cannot be supplied by either primary or secondary referents acting alone.

4. Learning is always in nested relationship with phylogenesis, and may also be in nested relationship with epigenesis. The basic function of all learning is to act, via this nested relationship, as an evolutionary subroutine in support of

phylogenesis. In practice learning fulfills this function by acting as a source of additional behavioral phenotypic flexibility. This phenotypic flexibility permits individual organisms an increased capacity either to avoid change or to track change, or both. It thereby increases the probability that an individual organism will contribute to the regenerate phase of the phylogenetic g–t–r heuristic, which, in turn, will increase the probability that the population to which the individual belongs will be successful.

III. IMPLICATIONS

What remains is to show that the present approach has a potential for hypothesis testing and prediction. This in itself is of some interest given that evolutionary theory is notoriously prone to the charge that, while it can explain biological phenomena, it cannot predict them (e.g., Mayr, 1961). Failure to make predictions is a classic cause of disenchantment in science, and has probably been one of the underlying reasons why so many psychologists have been content to see behavior left out of the evolutionary synthesis, and why in particular general process learning theory has been concerned solely with "process laws" (Brookshire, 1976). However, the preceding analysis does hold out some promise of prediction with respect to at least one major issue.

The problem of nature versus nurture is one of locating the sources of information gain within a biological system which has already evolved subsidiary referents. Cast in this form, Fig. 1 is a systems analytic approach to the problem. However, the preceding argument implies more than the simple replacement of classical terminology by our notion of referents. Consider again the parsimony assumption made at the beginning of this contribution. It states that phylogenesis is the primary process of information gain and that subsidiary referents with their own associated processes of information gain and storage only occur when the capacity of adapting to changing environmental order by the primary referent is outstripped by these environmental changes. This assumption alone carries a potential for hypothesis testing, but it does not distinguish between the secondary and tertiary referents. The analysis that followed from this assumption, however, showed that one of the principal constraints for any referent in its adapting to change is the time lag involved in the detection and response to change (that is, Waddington's uncertain futures problem). In the case of the primary referent, this time lag corresponds to Lorenz's generational dead time. The secondary referent is effective because it reduces this time lag by allowing phenotypic flexibility to occur up to certain points in individual development. Similary, the tertiary referent is able to reduce the time lag still further by allowing repeated and potentially reversible adaptations to occur in response to repeated and reversed changes in the environment, and throughout an even larger proportion of the animal's life span. Thus, with respect to the degree to which a referent can reduce the time lag inherent in the uncertain futures problem, there is a clear ordering of the referents. This order corresponds to the nomenclature of Table I. On the basis of this order, therefore, a further assumption can be made, which has the merit of being congruent with the expected degree of fine adaptive tuning supplied by each referent and also of allowing for further prediction and testing. This assumption is that over relatively long periods of evolutionary time (that is, a sufficient number of generations for selection pressures to give rise to new adaptations), and in the presence of relatively consistent selection pressures, an adaptation will occur at the most primary referent that is capable of forming it. The question then becomes, ''What are the factors that determine the incapacity of a more primary referent from forming an appropriate adaptation?'' A simplified answer is the frequency of environmental change relative to the rate of generational turnover. A more complex and certainly more correct answer would take into account rates of individual animal development as well.

A recent paper by Slobodkin and Rapoport (1974), which took the different route of game theory, arrived at a point very close to that of this paper: ''Note that the need for change in organisms is proportional to the stakes in an event, but the mechanism 'chosen' by the evolutionary process in order to make this change will relate to the frequency'' (Slobodkin and Rapoport, 1974, p. 195). The mechanisms that Slobodkin and Rapoport refer to are graded according to ''depth'' of response (for example, behavioral physiological, death-rate changes, selective mortality and fecundity, deep genetic changes in anatomy), where behavior is considered a shallow response and genetic changes are considered a deep response. It should be noted that this dimension of response depth does not map in any simple way onto our notion of referents. Further, Slobodkin and Rapoport give three possible schemes for the interaction of these different possible mechanisms: a completely hierarchical system, a completely nonhierarchical direct response system, and a mixed, multilinked response system that is a combination of hierarchical and direct response mechanisms. They opt for the mixed system, which of necessity reduces the stringency of the predictions that their analysis will lead to. We are taking a slightly different tack in adopting the strong assumption of the primacy of the referents. This assumption implies an exclusively hierarchical system, and it allows for unequivocal testing.

Consider a certain change C that occurs with a certain rate $= r$ (without regard to units of measure since the rate per se may be instantaneous and therefore it may be more meaningful to measure the interval between changes). If C impinges upon a species with a rate of generational turnover $= t_1 - t_0$ (where again rate refers to an interval), then

$$r/(t_1 - t_0) = \beta, \tag{1}$$

where β is the fundamental parameter determining which referent will operate the adaptation required by C.

Several points here need amplifying. First, it may be recalled that the change that is being referred to by the r term is not simply any alteration in environmental order. It is change that affects the animal concerned by introducing a mismatch between its previously acquired internal organization and the newly changed environmental order. In other words, the change that is occurring must refer to aspects of environmental order that (1) had previously been incorporated into the organization of members of a species, (2) in its altered form no longer matches the organization, and (3) in its altered form continues to affect the individual members of that species. This precludes from consideration any completely 'new' environmental events and confines consideration to relationships between order and organization that already have been established.

Second, the parameter β is concerned with which referent will operate the adaptation imposed by a specific change in the environment. As such, neither considerations of source nor of amplitude of change are directly relevant. Source and amplitude are relevant to the kind of mechanism within a referent that will be used to cope with the change, and much more nearly conform to the kind of analysis undertaken by Slobodkin and Rapoport (1974). But they fall outside the scope of the parameter β.

Third, fundamental to the present approach is the notion of "predictability." Purely random changes cannot be dealt with by any referent except by some chance combination of the nature and amplitude of the change being absorbed by whatever spare capacity for adaptation happens to be available at the time of the occurrence of the random change. If environmental order may change in the next moment, or the next day, or the next week, and if the nature of that change is also random and unpredictable, then no biological system and no referents can win advantage by adapting to these transient and unpredictable events. Slobodkin and Rapoport (1974) have made this point with special reference to learning, but the same point applies to any referent. Living systems cannot cope with chaos. They can, however, cope with a constrained form of chaos of the kind that we have already referred to as predictable unpredictableness. This phrase refers to changes that are pegged within certain limits, but within these limits can vary randomly. Depending on the value of β, this kind of change can be dealt with by a combination of one or other of the referents because what is being adapted to in the first instance are the limits. These limits may themselves be changing, but if they are, then the change must be regular and hence predictable. We shall refer to such random changes that occur within certain contrained and predictable limits as random changes around a regular feature. In such a condition Eq. (1) needs to be written twice. The expression $t_1 - t_0$ remains constant, but the term r will be different for the changes that are occurring to the regular feature (let us continue to call that r) and the mean rate of random change superimposed upon the regular feature, which we will refer to as r'. Random changes which occur around a regular feature will thus be describable by β and β'.

A. Change and the Three Referents

Consider now three species of animals, *A, B,* and *C.* Assume that all three species have already evolved both second and third referents, and assume further that the only differences between these species is in the rate of generational turnover $(t_1 - t_0)$. Figure 2, which we have modified from Thoday (1964), shows that *A* is a species of relatively short-lived individual organisms, *B* is intermediate in life span, and species *C* is made up of individuals of relatively long life spans. The question is asked as to whether it is possible to predict which referent will supply the adaptation for coping with different forms of change in these three populations. The forms of change shown in Fig. 2 are chosen so as to be congruent with those of Table II, with the addition of random change around a

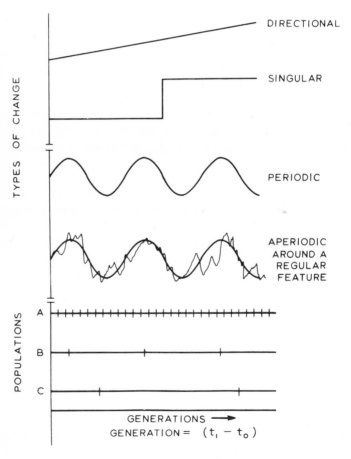

FIG. 2. Types and rates of environmental change and populations. For details see text.

regular feature. For our purposes it is sufficient that the third referent be considered only in terms of learning.

Consider each form of change in turn. Singular change, which is presented for completeness, really lies beyond the bounds of this analysis. It may take many forms, but the main characteristics of such change are that it is rapid, irreversible, and hence unpredictable. In essence it represents a change from one steady state to another. Provided that the magnitude of the change does not take it out of the adaptive reach of the secondary and teritary referents of the individual animals, according to our strong assumption, the new steady state should eventually be absorbed into the adaptations of the primary referent.

Cyclic change always extends with regularity beyond the generational time for each of the three species. As such, it is phylogenetically highly predictable, but what is predicted is very much a function of the rate of generational turnover. In Fig. 2, r is constant, being equal to a single cycle of change, while $t_1 - t_0$ varies across species. Thus, the parameter β is equal to 1 for Population B, more than 1 for Population A and less than 1 for Population C. If $\beta = 1$ (Population B), the reproductive cycle of the population is synchronized to the periodicity of r. Hence, for an individual member of such a population, the change it encounters will be predictably irreversible, and will only occur once during its lifetime. Therefore, change of this type should be dealt with by the primary referent in terms of phylogenetically determined developmental flexibility within each phenotype. If $\beta < 1$ (Population C), multiple cycles of environmental change will occur within the life span of the individual organism. In this case the change becomes predictably reversible. Such change should still be dealt with by the primary referent, but this time it must be dealt by the phylogenetically determined behavioral flexibility of each phenotype. In our third example (Population A) $\beta > 1$. Here the single life span of an individual organism covers only a portion of the periodic environmental cycle. Therefore, for each individual member of this population the form of change encountered becomes effectively directional. In this case the primary referent cannot predict at what point in the cycle an individual organism may encounter the environment, but it may still be possible to predict the limits and duration of each cycle. If the primary referent cannot predict the limits and duration of one cycle, then the changes encountered by the population will ultimately have to be dealt with by the primary referent in terms of the genetic flexibility that resides within the gene pool of that population. However, if the primary referent is able to predict the limits and duration of one environmental cycle, as a function of the regularity of that cycle across numerous generations, then it should become possible for a population to evolve an effective secondary referent. The population then should be able to cope with the change by means of open gene programs. Members of the population then could adapt in terms of their own individual development to the particular environmental conditions that they happen to encounter, regardless of the precise moment at which they enter the environmental cycle.

Directional change consists of environmental trends that are typically, although not necessarily, long term. In Fig. 2 the rate parameter r is indicated by the slope of the line. Several changes in slope might occur relative to the same trend, in which case r would be made up of both numbers of such changes, as well as the overall slope. Directional change differs from cyclic change in that it takes animals into conditions never before experienced by any member of the population to which they belong, whereas cyclic change, by definition, involves recurring environmental conditions. Directional change therefore cannot be predicted by phylogenesis in the same sense that cyclic change can be predicted. There is, however, a possibility that a consistent long-term trend in the environment may select some kind of genetic response within a population such that the population does become predisposed to deal with subsequent stages of the same directional change. In this restricted sense it may therefore be possible for directional change to be predicted by a primary referent and in line with our strong assumption it should be handled by that referent. Since the whole issue of the part played by long-term trends in evolution is particularly fraught (Simpson, 1967), we shall leave this possibility as an open question.

If directional change cannot be predicted by a population in any way, then in the case of every population that already has evolved subordinate referents, as is true of our three example populations, there should be a complicated response to the change. All the populations shown in Fig. 2 should first attempt to deal with the directional change they encounter in terms of either the secondary or tertiary referents, or both. If, however, the change still exceeds the adaptive reach of both of these subordinate referents, then all three populations should be forced to deal with the change at the level of the primary referent, and by means of phylogenesis. The response of phylogenesis might then be the evolution of additional adaptive capacity at the level of one or more of the subordinate referents or it might be some other response, depending on both the ecological circumstances and the prior evolutionary history of the population concerned, as described in the preceding pages.

In none of these instances discussed thus far have we considered the actual conditions leading to the evolution of the tertiary referent although it obviously has been assumed that if a tertiary referent has already evolved, then sometimes it will be used. But consider now the final form of change, namely, random changes around a regular feature. The regular feature may itself be changing, and may be either cyclic or directional. All of the previous considerations apply to this regular feature such that the parameter β will involve only the primary referent or both the primary and secondary referents. However, which referent will cope with the second element of change, that is, the random fluctuations around the regularity, is going to be a function or r' and hence of β'.

In general there are three possible outcomes. The first is shown by Population C. Here r' is so high relative to $t_1 - t_0$ that it reduces β' below some threshold γ, such that it can only be dealt with by the primary referent "assuming" minor irregular fluctuations around the mean and the evolution of various homeostatic devices that will maintain the animal close to the predicted regularity defined by β. Thus, the example of environmental temperature fluctuations given by Slobodkin and Rapoport (1974) should result in devices that are either physiological (for example, sweating) or behavioral (for example, selecting location within the environment as a function of temperature) in form but need draw only upon the resources of the primary referent. Such a solution conforms to Scheme 1 of Fig. 1.

The second possibility is illustrated by species A of Fig. 2. Here $\beta' \geqslant 1$, with the random fluctuations or changes occupying either very large parts of any individual animal's life span or occurring over an entire life span. In this case the change has to be coped with by both primary and secondary referents. The primary referent, acting on the basis of β, will act to establish possible developmental trajectories within the limits of the regular feature, while the secondary referent, as a result of the actual environmental conditions experienced by the developing genotype, will push the epigenesis in the direction of the information gained by the secondary referent. Such a solution conforms to Scheme 2 of Fig. 1.

The final possibility is illustrated by both populations B and C in Fig. 2. Again the nature of β will determine whether the regular feature is coped with by only the primary referent, or whether a secondary referent is also required. However, $\beta' > \gamma < 1$ means that the random changes are occurring, on average, several or many times during the life of each individual animal and thus cannot be adequately predicted by the secondary referent, but the rate of change is sufficiently low to allow the existence of relatively stable environmental conditions to occur several or many times within each animal's life span. In this case, β' can only be coped with by the tertiary referent. Depending upon how β is being dealt with we therefore have solutions that conform to Schemes 3 and 4 of Fig. 1.

Central to this whole approach is the parameter r. This raises very formidable problems of definition and measurement, both as to what the rate limits are relative to $t_1 - t_0$ and hence the triggering of subsidiary referents, as well as comparisons of rates of different forms of change. While acknowledging these difficulties, it nonetheless is suggested that, if the ecology of a species is known, then in theory r can be established for different forms of change, in which case the potential of the approach for real hypothesis testing should be realized. It should also be expanded as comparative information from different species allows for more information to be gained regarding the limits of these parameters and the interaction of the referents with different forms of change.

B.   The Functional Specificity of the Subsidiary Referents

The strong assumption that an adaptation will occur at the most primary referent capable of forming it does not imply, and should not be taken to imply, that, once a subsidiary referent has evolved, it then cannot be employed functionally for coping with changes that are different from those causing it to evolve. Although a separate matter from the central concern of this article, it raises the very basic issue of whether mechanisms of adaptability can ever be truly 'generalist' in their capacity for coping with change, or whether, having evolved for the specific purpose of dealing with some particular form of change, they can only ever be used for coping with that particular kind of change. Our uncertainty on this question was the reason for qualifying our comments about amplitude and novelty of change. In part it is a definitional problem and in part an empirical matter. One thing follows clearly from our analysis, however: even if these subsidiary referents are capable of truly generalist capacity, they should still always be bound by the parameter β, the value of which will always determine which referent will supply the appropriate adaptation. Thus, while functional specificity is recognized as an interesting and important matter in its own right, it does not escape from the limitations imposed on each referent by the rates of change and generational turnover.

IV. CONCLUDING COMMENTS

We began this article with Waddington's comments about the need to establish the logic of the relationships between evolutionary processes and behavior. We have attempted to do this with special reference to learning. One of the results, we believe, is an analytic framework with a real potential for prediction and hypothesis testing of the sort just outlined in the preceding section. However, the approach has a number of other interesting implications as well. First, it goes some way toward solving the problem that we suggested in the introduction is now one of the central issue in theories of learning, namely, the paradox of an assumed unity and manifest diversity. It does this by not simply assuming unity but by actually teasing out learning's teleonomic universals, and in so doing the source of diversity becomes obvious. The teleonomic universals are that (1) the evolutionary origins of learning are the same in all animals that can learn (that is, the limitations of the primary and secondary referents); (2) learning shares the same logical relationship to phylogenesis in all learners; and thus (3) learning always serves the same fundamental function (that is, supplying the finest possible temporal resolution to the uncertain futures problem). In so doing it always tends to act in support of phylogenesis. At the same time, the nested relationship that learning has to phylogenesis demands diversity of learning. All learning is a function of innate mechanisms of adaptability which are both provided and primed by phylogenesis with respect to some specific primary referent. As a result, all learning must be species specific, and must reflect the particular fitness needs of a particular population. Hence, the extent to which learning in different species is similar or dissimilar will depend not upon whether theorists choose to assume it. Rather it will depend upon the two principles discussed by Lockard (1971), viz. phylogenetic relatedness and ecological convergence.

There is another feature of the preceding argument that needs to be underlined. This is the considerable emphasis that has been placed upon naturally occurring environmental change. Such change takes in a very wide range of events including change confined to the inanimate world, change that is a consequence of the animal's own activity, and change that is concerned with kin recognition, relationships, and alliances. This emphasis upon change is a new point of departure for the study of learning in psychology. Typically, psychologists have sought to unravel the issues of learning armed only with laboratory-derived paradigms and experimenter-induced change. The thrust of our analysis is that the emphasis and reliance upon the paradigms has become excessive. There are many instances of learning, some of them such as bird song (Marler, 1970, 1976; Nottebohm, 1972) forming some of the most elegant studies in the learning literature, which do not find a place in most standard texts on learning presumably because they do not conform well to the paradigm approach. However, in the same way as Table II classifies change for the primary referent, so too can change be classified for the tertiary referent. Bird song, depending upon species, becomes an instance of learner-dependent change when there is need for the young bird to hear its own vocal output, or learner-independent organic change if it needs to hear its species song from other adults. Learning that occurs through play activities, place learning, the learning of motor skills, all equally ignored in texts of "animal" learning, can be similarly classified and analyzed. These examples can be multiplied many times. The important point, however, is that the analysis of learning through the analysis of change rather than paradigm brings with it the possibility of widely diverse forms of learning being studied within the same analytic framework, and this has never been achieved before.

The suggestion is therefore that there is now more than ever an urgent need to study learning from the basis of an analysis of the changes that occur in a species' natural habitat. This, of course, is precisely what ethologists have been asking for, for over 30 years. This does not make either the traditional paradigms of learning or laboratory experiments irrelevant, but it does mean that the paradigms need to be recast in terms of the fundamental parameters of environmental change as well as in terms of certain population parameters, most notably generational turnover.

Accepting that all that we may have achieved in this article is a framework within which future analyses, theoretic and experimental, can proceed, it should be pointed out that the further validation of a teleonomic approach to learning must come from its allowing a fruitful basis for investigating process laws as well. This has not yet been achieved. The most likely direction from which it will come is the argument that the processes of learning constitute a repetition of the phylogenetic g–t–r heuristic, but that this repetition is disguised by two complicating factors that greatly change the appearance of all three subprocesses within this heuristic. The factors are, first, the nested relationship between learning and phylogenesis, and, second, the change of referent that learning necessarily incurs. The nested relationship implies that learning processes, unlike phylogenetic processes, never start "de novo," but always start on the basis of phylogenetic, and possibly epigenetic, "priming." Likewise, the change of referent implies that all learning processes invariably act on a within-individual instead of a between-individual basis. In our view the only reason why the many former attempts to draw parallels between phylogenetic processes and learning processes (e.g., Breland and Breland, 1966; Campbell, 1974; Popper, 1972; Pringle, 1951; Russell, 1962; Skinner, 1966; Slobodkin and Rapoport, 1974; Staddon, 1975; Staddon and Simmelhag, 1971) have had so little impact is that in each case either one or other or both of these complicating factors has been overlooked or underestimated, and thus the analyses have not succeeded.

If, however, it does prove possible to revive this very old idea, and to demonstrate, in the light of the material presented here, that learning processes are simply phylogenetic processes writ small, then it may also be possible to establish a complete evolution-based theory of learning, one that should be more than capable of finally resolving the unity versus diversity paradox of learning.

Acknowledgments

We are grateful for constructive comments and criticisms from Drs. P. P. G. Bateson (Cambridge), R. Dawkins (Oxford), P. Hammond (British Museum of Natural History), D. A. Oakley (London), S. Shettleworth (Toronto), J. M. Thoday (Cambridge), and W. R. Thompson (Georgia Mental Health Institute).

References

Alcock, J. 1975. "Animal Behavior." Sinauer Assoc., Sunderland, Massachusetts.
Applewhite, P., and Gardner, E. T. 1971. A theory of protozoan habituation. *Nature (London)* **230,** 285–287.
Bateson, P. P. G. 1976a. Specificity and the origins of behavior. *Adv. Study Behav.* **6,** 1–20.
Bateson, P. P. G. 1976b. Rules and reciprocity in behavioral development. *In* "Growing Points in Ethology" (P. P. G. Bateson and R. A. Hinde, eds.), pp. 401–421. Cambridge Univ. Press, London and New York.
Bitterman, M. E. 1975. The comparative analysis of learning. *Science* **188,** 699–709.
Bitterman, M. E., and Woodard, W. T. 1976. Vertebrate learning: Common processes. *In* "Evolution of Brain and Behavior in Vertebrates" (R. B. Masterson, C. B. G. Campbell, M. E. Bitterman, and N. Hotton, eds.), pp. 169–189. Lawrence Erlbaum Assoc., Hillsdale, New Jersey.
Breland, K. and Breland, M. 1966. "Animal Behavior." Macmillan, New York.

Brookshire, K. H. 1976. Vertebrate learning: Evolutionary divergences. *In "Evolution of Brain and Behavior in Vertebrates"* (R. B. Masterson, C. B. G. Campbell, M. E. Bitterman, and N. Hotton, eds.), pp. 191–216. Lawrence Erlbaum Assoc., Hillsdale, New Jersey.

Campbell, D. T. 1974. Evolutionary epistemology. *In* "The Philosophy of Karl Popper" (P. A. Schilpp, ed.), pp. 413–463. Open Court Publ., Chicago, Illinois.

Conrad, M. 1972. Statistical and hierarchical aspects of biological organization. *In* "Towards a Theoretical Biology" (C. H. Waddington, ed.), Vol. 4, pp. 189–221. Edinburgh Univ. Press, Edinburgh.

Corning, W. C., and Von Burg, R. 1973. Protozoa. *In* "Invertebrate Learning" (W. C. Corning, J. A. Dy;al, and A. O. D. Willows, eds.), Vol. 1, pp. 49–122. Plenum, New York.

Darlington, C. D. 1939. "The Evolution of Genetic Systems." Cambridge Univ. Press, London.

Dawkins, R. 1976. "The Selfish Gene." Oxford Univ. Press, London and New York.

Dobzhansky, T. 1968. Adaptedness and fitness. *In* "Population Biology and Evolution" (R. C. Lewontin, ed.), pp. 109–121. Syracuse Univ. Press, Syracuse, New York.

Dobzhansky, T. 1970. "Genetics of the Evolutionary Process." Columbia Univ. Press, New York.

Evans, G. 1975. "The Life of Beetles." Allen & Unwin, London.

Fisher, R. A. 1930. The Genetical Theory of Natural Selection." Clarendon Press, Oxford.

Gottlieb, G. 1976. Conceptions of prenatal development: Behavioral embryology. *Psychol. Rev.* **83,** 215–234.

Haldane, J. B. S. 1931. A mathematical theory of natural selection, Part vii. Selection intensity as a function of mortality rate. *Proc. Cambridge Philos. Soc.* **27,** 131–142.

Hamburger, V. 1963. Some aspects of the embryology of behavior. *Q. Rev. Biol.* **38,** 342–365.

Hinde, R. A. 1970. "Animal Behavior." McGraw-Hill, New York.

Hinde, R. A. 1973. Constraints on learning. *In* "Constraints on Learning" (R. A. Hinde and J. Stevenson-Hinde, eds.), pp. 1–19. Academic Press, New York.

Jenkins, H. M. 1975. Behavior theory today: A return to fundamentals. *Mex. J. Behav. Anal.* **1,** 39–54.

Jones, M. B., and Mansfield, T. A. 1975. Circadian rhythms in plants. *Sci. Prog. (London)* **62,** 103–125.

Lerner, I. M. 1954. "Genetic Homeostasis." Oliver & Boyd, Edinburgh.

Lockard, R. B. 1971. Reflections on the fall of comparative psychology. *Am. Psychol.* **26,** 168–179.

Lorenz, K. 1965. "Evolution and Modification of Behavior." Univ. of Chicago Press, Chicago, Illinois.

Lorenz, K. 1969. Innate bases of learning. *In* "On the Biology of Learning" (K. Pribram, ed.), pp. 13–91. Harcourt, New York.

Lorenz, K. 1977. "Behind the Mirror." Methuen, London.

Lovtrup, S. 1974. "Epigenetics." Wiley, New York.

McFarland, D. J. 1976. Form and function in the temporal organization of behavior. *In* "Growing Points in Ethology" (P. P. T. Bateson and R. A. Hinde, eds.), pp. 55–93. Cambridge Univ. Press, London and New York.

Marler, P. 1970. A comparative approach to vocal learning. *Comp. Physiol. Psychol. Monog.* **71,** 1–25.

Marler, P. 1976. Sensory templates in species-specific behavior. *In* "Simpler Networks and Behavior" (J. C. Fentress, ed.), pp. 314–329. Sinauer Assoc., Sunderland, Massachusetts.

Mason, W. A., and Lott, D. F. 1976. Ethology and comparative psychology. *Annu. Rev. Psychol.* **27,** 129–154.

Maynard Smith, J. 1966. "The Theory of Evolution." Penguin Books, London.

Maynard Smith, J. 1976. Evolution and the theory of games. *Am. Sci.* **64,** 41–45.

Mayr, E. 1961. Cause and effect in biology. *Science* **134,** 1501–1505.

Mayr, E. 1963. "Animal Species and Evolution." Belknap Press, Cambridge, Massachusetts.

Mayr, E. 1974. Behavior programs and evolutionary strategies. *Am. Sci.* **62**, 650–659.

Milne, L. J., and Milne, M. 1976. The social behavior of burying beetles. *Sci. Am.* **235**, 84–89.

Nottebohm, F. 1972. The origins of vocal learning. *Am. Nat.* **106**, 116–140.

Pittendrigh, C. S. 1958. Adaptation, natural selection, and behavior. *In* "Behavior and Evolution" (A. Roe and G. G. Simpson, eds.), pp. 390–416. Yale Univ. Press, New Haven, Connecticut.

Popper, K. R. 1961. Evolution and the tree of knowledge. "Herbert Spencer Lecture." (Reprinted in Popper, 1972.)

Popper, K. R. 1966. Of clouds and clocks. "Second Archer Holly Compton Lecture." (Reprinted in Popper, 1972.)

Popper, K. R. 1972. "Objective Knowledge: An Evolutionary Approach." Oxford Univ. Press, London and New York.

Pringle, J. W. S. 1951. On the parallel between learning and evolution. *Behaviour* **3**, 174–215.

Razran, E. 1971. "Mind in Evolution." Houghton, Boston, Massachusetts.

Revusky, S. 1974. Alas, poor Learning Theory, I knew it well. *Contemp. Psychol.* **19**, 692–694.

Russell, W. M. S. 1962. Evolutionary concepts in behavioural science: IV. The analogy between organic and individual behavioral evolution, and the evolution of intelligence. *Gen. Syst.* **7**, 157–193.

Schneirla, T. C. 1956. Interelationships of the "innate" and "acquired" in instinctive behaviour. *In* "L'Instinct dans le Compartement des Animaux et de l'Homme" (G. Grasse, ed.), pp. 387–452. Masson, Paris.

Schneirla, T. G. 1965. Aspects of stimulation and organization in approach–withdrawal processes underlying vertebrate behavioral development. *Adv. Study Behav.* **1**, 1–74.

Schwartz, B. 1974. On going back to nature. A review of Seligman and Hager's Biological Boundaries of Learning. *J. Exp. Anal. Behav.* **21**, 183–198.

Seligman, M. E. P. 1970. On the generality of the laws of learning. *Psychol. Rev.* **77**, 406–418.

Shettleworth, S. J. 1972. Constraints on learning. *Adv. Study Behav.* **4**, 1–68.

Simpson, G. G. 1967. "The Meaning of Evolution." Yale Univ. Press, New Haven, Connecticut.

Skinner, B. F. 1966. The phylogeny and ontogeny of behavior. *Science* **153**, 1205–1213.

Slobodkin, L. B., and Rapoport, A. 1974. An optimal strategy of evolution. *Q. Rev. Biol.* **49**, 181–200.

Smith, J. L. B. 1962. "Old Four-Legs." Longmans, London.

Staddon, J. E. W. 1975. Learning as adaptation. *In* "Handbook of Learning and Cognitive Processes" (W. K. Estes, ed.), Vol. 2, pp. 37–98. Lawrence Erlbaum Assoc., Hillsdale, New Jersey.

Staddon, J. E. R., and Simmelhag, V. L. 1971. The superstition experiment. A re-examination of its implications for the principle of adaptive behavior. *Psychol. Rev.* **78**, 3–43.

Thoday, J. M. 1953. Components of fitness. *Symp. Soc. Exp. Biol.* **7**, 96–113.

Thoday, J. M. 1964. Genetics and the integration of reproductive systems. *Symp. R. Entomol. Soc. London* **2**, 108–119.

Thoday, J. M. 1975. Non-Darwinian "evolution" and biological progress. *Nature (London)* **255**, 675–677.

Tinbergen, N. 1951. "The Study of Instinct." Clarendon Press, Oxford.

Waddington, C. H. 1957. "The Strategy of the Genes." Allen & Unwin, London.

Waddington, C. H. 1959. Evolutionary adaptation. *In* "Evolution after Darwin" (S. Jax, ed.), pp. 381–402. Univ. of Chicago Press, Chicago, Illinois.

Waddington, C. H. 1969. Paradigm for an evolutionary process. *In* "Towards a Theoretical Biology 2: Sketches" (C. H. Waddington, ed.), pp. 106–128. Edinburgh Univ. Press, Edinburgh.

Waddington, C. H. 1975. "The Evolution of an Evolutionist." Edinburgh Univ. Press, Edinburgh.

Whyte, L. L. 1965. "Internal Factors in Evolution." Tavistock, London.

Williams, G. C. 1966. "Adaptation and Natural Selection: A Critique of Some Current Evolutionary Thought." Princeton Univ. Press, Princeton, New Jersey.

Wilson, E. O. 1971. "The Insect Societies." Belknap Press, Cambridge, Massachusetts.

Wilson, E. O. 1975. "Sociobiology: The New Synthesis." Belknap Press, Cambridge, Massachusetts.

Wright, S. 1931. Evolution in Mendelian populations. *Genetics* **16,** 97–159.

Wright, S. 1949. Population structure and evolution. *Proc. Am. Philos. Soc.* **93,** 471–478.

Social Behavior, Group Structure, and the Control of Sex Reversal in Hermaphroditic Fish

Douglas Y. Shapiro

DEPARTMENT OF MARINE SCIENCES
UNIVERSITY OF PUERTO RICO
MAYAGUEZ, PUERTO RICO

I. Introduction

Hermaphroditism is widespread in marine fish, thus far having been found in more than 100 species belonging to 15 families (Choat, 1969; Reinboth, 1970) and is thought to have evolved independently on at least ten occasions (Smith, 1975). Hermaphroditism can be defined as the presence of male and female tissues in the same individual (Yamamoto, 1969). It is regarded as functional if the individual functions reproductively both as a male and as a female during its life, and it is called "normal" in a species if it occurs in many or all of the species' members (Atz, 1964). In fishes hermaphroditism may take the form of simultaneous function of testis and ovary, even with self-fertilization as in *Serranus subligarius* (Clark, 1959), of protandry where a male fish reverses sex and becomes female (e.g., Reinboth, 1962a), or of protogyny where a female

Copyright © 1979 by Academic Press, Inc.
All rights of reproduction in any form reserved.
ISBN 0-12-004510-9

changes functionally into a male (e.g., Chan, 1970). Since, in protandrous and protogynous species, an individual progresses from one sex to the other, such species are called successive hermaphrodites.

For most protogynous and protandrous species the evidence for sex reversal consists of two observations: (1) one sex predominates in a larger size range than the other sex, often not being found at all in the small size ranges, and (2) gonadal histology changes dramatically with an increase in the size of individual fish (e.g., Liem, 1963; Lodi, 1967; Longhurst, 1965; Zei, 1950). This evidence has led to an often stated hypothesis that sex reversal is internally initiated in an individual when it attains a given size or age (e.g., Choat, 1969; McErlean and Smith, 1964; Roede, 1972; Warner et al., 1975).

In 1972 it was reported, for the cleaner wrasse, *Labroides dimidiatus*, that sex reversal in a female could be initiated by the removal of a male from a social group containing one male and 9–16 females (Robertson, 1972). A similar finding was very briefly reported for another species, *Anthias squamipinnis* (Fishelson, 1970). These were exciting observations. If it could be clearly established, under properly controlled conditions, that sex reversal is initiated by the removal of a male from a group, then social groups of these fish would be excellent systems for the causal analysis of the influence of social factors on the physiological status of the individual, and of various features of social interactions and group structure. The study of these problems could have wide application to other highly social, group-living species. The male removal studies produced two additional hypotheses on the control of sex reversal: (1) females have a physiological tendency to change sex, which is inhibited by the presence of a male, and (2) males inhibit females from changing sex by aggression.

This article has three purposes: (1) to describe the phenomena of protogynous sex reversal, (2) to discuss the mechanisms that control its initiation and completion, and (3) to relate these processes to the spatial structure and composition of social groups. The social and developmental hypotheses mentioned above for the control of sex reversal will be critically examined in light of available evidence, and a new hypothesis will be proposed: females are stimulated to change sex by changes of a specific behavioral measure of a critical magnitude. While a large literature exists documenting the occurrence of hermaphroditism in fish (see reviews in Atz, 1964; Chan, 1970; Reinboth, 1970), recent work on the control of sex reversal and its relation to social systems has been confined to a few species. This article will concentrate on one of those, *Anthias squamipinnis* (Peters).

A. squamipinnis is a small, sexually dimorphic fish that occupies stable, sedentary social groups in the shallow water around coral reefs. During the day, these fish hover in the midwater column above their home site and feed on zooplankton swept toward them by the current. Individuals spawn in pairs throughout the year, but show a pronounced increase in spawning frequency during January, February, and March in the northern tropics (Fishelson, 1975). The social groups are of variable size and composition, ranging from small, single-male groups to groups containing 35 males and more than 350 females. Some groups are subdivided into two or three spatially separated subgroups (Shapiro, 1977a). As will be discussed later, the process by which a group becomes subdivided is closely related to the control of sex reversal.

II. The Phenomena of Sex Reversal

A. Population Structure

Where sample sizes are large and all individuals of small–medium size are sexually mature and of the same sex, it is difficult to avoid concluding that large individuals of the opposite sex have developed by sex reversal. Sex-separated size distributions are such a standard feature of the populations of species with simple protogyny (e.g., Bruslé and Bruslé, 1975, 1976; Bullough, 1947; Choat and Robertson, 1975; Lavenda, 1949; Liem, 1963; Liu, 1944; McErlean and Smith, 1964; Quignard, 1966; Warner, 1975) that several authors have listed this as a criterion for judging the protogynous nature of a new species (Reinboth, 1970; Smith, 1959). In simple cases little size overlap occurs between the sexes and this interpretation is straightforward. In species in which size overlap is large, the diagnosis of sex reversal must reside more with histological evidence than with the size structure of the population.

Size overlap can be a valid feature of a population, as in *Sparisoma viride* (Robertson and Warner, 1978) or can result from inadvertent pooling of data from populations of different size structure. The size–sex distributions of two populations of *A. squamipinnis* from separate collecting sites 800 m apart on the same Aldabra Island reef are shown in Fig. 1. Little overlap occurred at either site. In fact, each population was composed of 5–6 independent social groups. In each group males were larger than all females (Shapiro, 1977b). Considerable overlap would occur if the data from both sites were pooled. Studies of protogyny have tended to present a single size–sex distribution as being typical of a species (see the references in the paragraph above for examples) without specifying the size or nature of the collecting site or the justification for believing that the data were derived from a homogeneous population. Since one of the most widely discussed hypotheses for the control of sex reversal, that females change sex upon reaching a certain size, would become immediately unattractive if the size at which males appear in the population varied widely from site to site, the reason for size overlap in a particular set of data assumes importance. In this

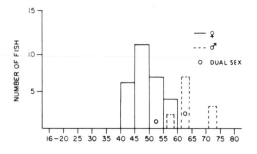

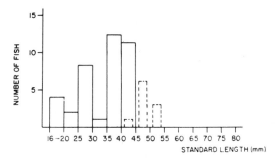

FIG. 1. Size distribution of male and female *A. squamipinnis* at Aldabra. Top: Site 1, $N = 43$. Bottom: Site 2, $N = 48$. (From Shapiro, 1980a.)

regard, geographic variation in population structure within a species is also possible (Choat, 1969; Liem, 1968; Shapiro and Lubbock, 1980).

Another source of overlap occurs in diandric species (Reinboth, 1967), that is, species with two ontogenetically distinct types of males: those that mature initially as males (primary males) and those that are sex-reversed females (secondary males). In diandric populations of some Labridae and Scaridae primary males may occur in all size categories, while secondary males tend to be restricted to the larger size ranges (Choat, 1969; Choat and Robertson, 1975; Warner and Robertson, 1978). In species in which all males are derived ontogenetically from females, that is, monandric species, this source of overlap is not present.

Monandric species tend to have female-biased sex ratios ranging from 1.04 in *Sparisoma radians* to 8.5 in *Bodianus rufus* (Robertson and Warner, 1978) and 8.4 in *A. squamipinnis* (Shapiro, 1977a). Individual groups of *A. squamipinnis* have sex ratios as high as 36 adult females per male. Populations of diandric species also tend to have more females than males, but sex ratios seldom exceed 4.5 (Robertson and Warner, 1978).

B. INDIVIDUAL SEX REVERSAL

When a female changes sex, depending on the species, she may undergo striking changes in external coloration and morphology, general and reproductive behavior, and a thorough reorganization of the gonad. Since direct observation of behavior and color change is relatively new, gonadal histology figures prominently in the literature and will be discussed first.

1. Gonadal Structure

The basic structure of the gonad is similar for most protogynous species. In both sexes the gonad is bilobate, with both lobes fused posteriorly to form, in the ovary, a common oviduct. In the ovary each lobe is a hollow tube with a membrane-lined central cavity and an encapsulating wall of smooth muscle and connective tissue. Most of the interior surface of the ovarian wall is lined with germinal epithelium, which folds inward to form ovarian lamellae.

In all protogynous species the gonad of a sex-reversing individual shows widespread degeneration of the ovarian tissue and active proliferation of testicular tissue (e.g., Chan and Phillips, 1967a, b; Okada, 1965a, b; Reinboth, 1962a, 1963, 1964; Sordi, 1962). Details of these changes vary somewhat with the structure of the ovary prior to sex reversal. The serranids show two types of gonadal arrangement (Smith, 1965), which exemplify the variations seen in other species. In the *Rypticus* type the ovary contains two narrow bands of testicular tissue along the ventrolateral surface of the ovarian wall. During sex reversal spermatogenic tissue proliferates from these pre-existing testicular areas. This arrangement also occurs in some labrids and sparids (Reinboth, 1962a) and in some anthiids (Reinboth, 1963, 1964). In the *Epinephelus* type small crypts of testicular cells are intermixed with the predominant ovarian tissue in the germinal epithelium of the ovarian lamellae. When sex reversal begins the spermatogenic crypts proliferate and spread throughout the gonad. This is also seen in other species, such as *Monopterus albus* (Chan, 1970) and, in a slightly altered form, in *Thalassoma bifasciatum* (Reinboth, 1970). In *A. squamipinnis* (Gundermann, 1972; Shapiro, 1977b) and some other anthiids (Shapiro and Lubbock, in preparation) no spermatogenic cells can be found in the ovaries of most females and testicular tissue proliferates during sex reversal from isolated regions in the ovarian wall (Fishelson, 1975).

When protogynous sex reversal has been completed, the gonad is a testis that has retained the general organizational features of the ovary—a lamellar structure and a central cavity—while developing a new duct system for the transport of sperm, either in the tissue surrounding the ovarian cavity (Reinboth, 1962a) or by a hollowing of the entire central region of the testis as the gonad fills with motile sperm (Shapiro, 1977b). Other ovarian remnants to be found in the testis include atretic follicles (Hoar, 1969; Mehl, 1973; Moe, 1969) and first- or second-stage oocytes. The ovarian features of the protogynous testis are sufficiently distinct that in many cases a species can be diagnosed as protogynous largely on the basis of gonadal histology. In diandric species primary males can be distinguished from secondary males histologically by the lack of ovarian features in the primary male and by differences in the structure of the vas deferens (Reinboth, 1962a).

Limited observations have been made on the duration required for gonadal sex change. The most reliable estimates have been made in species in which sex reversal has been experimentally initiated by removing a male from a social group. *A. squamipinnis* required 14 days or less to undergo gonadal transformation from the day of male removal (Shapiro, 1977b). Sex-changing *Labroides dimidiatus* showed a gonadal activity level (that is, the number of spermatogenic crypts divided by the number of oocytes per unit area of gonad) well above the female and into the male range by the fifth day after male removal. Sperm appeared in the sperm ducts by Day 14 (Robertson, 1974).

In species for which the initiation of sex change is less well understood few workers have followed individuals through their sex change with any study technique. Chan and his colleagues have performed serial biopsies on the gonads of identified individual *Monopterus albus*. They report variously that sex reversal in this species required 3–5 months (Chan, 1970) or less than 30 weeks (Chan *et al.*, 1972b). A male *Amphiprion bicinctus* changed to a female and laid eggs 26 days after being separated from its female mate (Fricke and Fricke, 1977). Of 40 female *Halichoeres poecilopterus* kept in a tank, 30 changed sex in a 40-day period (Okada, 1962), but no definite onset time was reported for these sex reversals. Sordi (1962) felt that protogynous sex reversal occupied about a year in the life of female *Labrus turdus* and *L. merula*, but he had no direct evidence. Colombo *et al.* (1972) examined the changes in steroid biosynthesis of *in vitro* preparations of gonads from the protandric *Sparus auratus*, but could put no time dimension on the changes observed. Reinboth (1962a) removed one lobe of the gonad of a drab-colored female *Coris julis*. After five and a half months the color had changed to the gaudy male form, and the remaining portion of gonad showed predominantly active testicular tissue with some degenerative ovarian tissue. The females used had already begun their transformation, however, at the start of the study. With the exception of the work with *A. squamipinnis*, *Amphiprion*, and *L. dimidiatus*, all of these studies have been hampered by the inability to place a definite marker on the onset of sex reversal. The behavioral control of sex reversal thus provides a valuable tool permitting one to date sex reversal from the moment of male removal.

2. Coloration, External Morphology, and Size

Knowledge of the coloration changes of sex-reversing species is spotty, and few generalizations can yet be made. In the Serranidae (Bohlke and Chaplin,

1968; Fourmanoir and Laboute, 1976; Randall, 1968) and Labridae (Choat, 1969) many species are monochromatic and show no coloration change associated with sex reversal. In sexually dichromatic Scaridae and Labridae (Choat, 1969; Robertson and Warner, 1978) and Anthiinae (Popper and Fishelson, 1973; Shapiro, 1977b; Shapiro and Lubbock, in preparation) coloration change tends to be associated temporally with gonadal sex change. However, individuals have been found that have changed color but not gonadal sex (e.g., Gundermann, 1972; Robertson and Warner, 1978; Roede, 1972, 1975) and vice versa (e.g., Buckman and Ogden, 1973; Quignard, 1966; Randall and Randall, 1963; Warner and Robertson, 1978), and in few instances has the full sequence of events been closely examined in individual fish.

Perhaps the best studied case concerns *A. squamipinnis* in which females are a uniform orange-gold color with an orange-red stripe running from just beneath the eye to the mid-pectoral base. Males are red to brown-violet on the head and on the dorsal, caudal, and pelvic fins, pale red to yellow-brown on the flanks, with a red to orange-red stripe running from the posterior margin of the orbit to the center of the pectoral fin base (Fig. 2). The pectoral fins are pale pink or yellow with a large round orange-red or violet spot on the upper fin. In seven laboratory cases color change and sex reversal was initiated by the removal of a male from a social group. After male removal the day of appearance of the first change of coloration was noted for five body regions on the sex-reversing fish. A record was also made of the day when the basic changes in each region were considered to be complete. The onset and completion days for the color changes of sex-reversing fish in natural groups in the Red Sea and on Aldabra Island were also studied (Shapiro, 1977b).

In 75% of the 44 observed sex reversals color changes followed a typical sequence characterized by an early onset after male removal, a short interim period between onset and completion, and a short total time for completion (Table I). In the laboratory, male coloration appeared first on the head–nape region 3–6 days after male removal. These changes were followed, in order, by changes on the pelvic fins, dorsal fin, pectoral fins, and caudal fin. On the pelvic fins, dark coloration appeared first on the distal margins and then spread proximally. The spot on the superior portion of the pectoral fins developed from 4 to 8 parallel, horizontal black lines. The lines forming the center of the spot appeared first, followed by lines above and below them. Once the coloration changes of the typical sequence had begun, 2–11 days were required for their completion. The full male color pattern was recognizable by the 7–16th day after male removal, but more subtle changes in hue continued for at least two months. Pigment appearing in the first two weeks was predominately black. In the next two months this was replaced increasingly by red pigment.

Some typical sequences in the field were slower to begin than all typical sequences in the laboratory (see Table I). In addition, two types of atypical

FIG. 2. Normal female (a) and male (b) *A. squamipinnis.*

sequence were also observed. The first type, which was seen in 18% of 44 observed sex changes, differed from the typical sequence by having a much greater onset time following male removal and/or a longer interim period (Table II). In the second type, representing 7% of 44 sex changes, the order of appearance of male coloration on different body parts was completely different from that of the typical sequence. Both types of atypical sequence and the slower typical sequences seen in the field tended to occur in socially unusual groups, that is, in groups containing 3 females or fewer, and in groups that had lost a male during the first week of sex reversal of another group member (Shapiro, 1977b; see Section IV,A for a description of group composition). When three males were removed simultaneously from a multimale field group, one female showed an early onset time of color change, a second female showed the first color changes 2–3 days after the first female, and a third female delayed 2–3 days after the second. This was observed in three instances of multimale removal. It would appear that the coloration sequence of a sex-reversing fish may be influenced by the composition of its social group and by the timing of the male removal that initiated the sequence.

In spite of the indication given by these results that the sequence of color changes may depend on intragroup circumstances, by sex reversal standards *A. squamipinnis* is a very simple case. Every color change in these studies was accompanied by gonadal and behavioral sex reversal. In diandric species most primary males have the same "initial phase" color pattern as most females (for a discussion of terminology, see Reinboth, 1975a; Warner and Robertson, 1978).

TABLE I

ONSET DAY, INTERIM TIME, AND COMPLETION DAY FOR LABORATORY AND FIELD CASES OF THE TYPICAL COLORATION SEQUENCE IN *Anthias squamipinnis* [a,b]

	Day of onset	Interim time (days)	Day of completion
Laboratory cases			
Median	4	6	10
Range	3–6	2–11	7–16
$N = 7$			
Field cases			
Median	5	8	14
Range	4–9	5–11	8–19
$N = 18$			
Mann–Whitney U test			
p Values, two-tailed	0.02	NS	0.02

[a] Days are counted from the day of male removal.
[b] From Shapiro (1980b).

TABLE II

Onset Day, Interim Time, and Completion Day for Atypical Coloration Sequences in
A. squamipinnis [a,b]

Group	Fish number	Day of onset	Interim time (days)	Day of completion
Laboratory				
1974 LT[c]	4	4	30	34
1974 LT	2	19	34	53
1975 MT	3	28	6	34
Field				
Aldabra, Site 2		20	6	26
Red Sea				
29A		20	17	37
29A		25	12	37
23		37	?	?
22		?	>18	?
Median:		20	17	35.5
Range:		4–37	6–34	26–53

[a] Days are counted from the day of male removal.
[b] From Shapiro (1980b).
[c] LT and MT refer to two aquaria in which these studies were conducted.

Color change from the initial phase to the "terminal phase," the pattern shown by large secondary males, can involve either a female, which may or may not change gonadal sex synchronously with the color change, or a primary male. Although initial-phase *Scarus croicensis* are said to require as little as ten days to complete a coloration change (Robertson and Warner, 1978), the sequence and timing of natural color change have not been carefully examined in any of these diandric species, and it is not known what factors initiate and control either color or gonadal change.

In *A. squamipinnis* the early appearance of dark pigment followed in a later stage by a change to violet and red is similar to the sequence of chromatophore appearance during larval and juvenile development (Goodrich and Greene, 1959; Hama and Hasegawa, 1967; Matsumato, 1965; Shapiro, 1977b) and during regeneration of destroyed fin regions in other fish (Goodrich and Nichols, 1931; Goodrich *et al.*, 1954). The hormonal basis of these changes in *A. squamipinnis* is not fully known, but in several dichromatic, protogynous species treatment of females with testosterone has resulted in the appearance of male coloration. This has been found in *Coris julis* (Reinboth, 1957, 1962b), *Thalassoma bifasciatum* (Reinboth, 1962b, 1975b; Roede, 1972; Stoll, 1955), *Scarus croicensis* (Ogden and Buckman, 1973), *A. squamipinnis* (Fishelson, 1975), and *Halichoeres poecilopterus* (Okada, 1962), but was not found in *H. bivittatus* or *H. garnoti*

(Reinboth, 1975b; Roede, 1972), the difference perhaps being related to dose (cf. Okada, 1962, 1964). Dose-related phenomena may also occur in natural populations. Primary males of several labrid and scarid species, although possessing larger testes than those of brightly colored secondary males (Roede, 1966, 1972; Choat, 1969), retain a drab, initial phase coloration but become a bright and gaudy, terminal phase color when injected with testosterone (Reinboth, 1962b; Stoll, 1955). Primary males may thus be secreting less androgen than secondary males. This view is supported by the observation that large, terminal phase males of *Halichoeres poecilopterus* contain large numbers of testicular interstitial cells, which in *Monopterus albus* are the source of testosterone (Tang *et al.*, 1975), while the testes of small, drab colored, initial phase males rarely contain these cells (Kinoshita, 1934).

In summary, it can be said that the color phenomena of sex reversal depend largely on the structure of the male population and the species studied. In the one relatively simple protogynous species that has been closely studied, coloration changes accompanied gonadal change and appeared in regular, temporal sequences probably influenced by the composition and circumstances of the social group in which they occurred.

In addition to gonadal and coloration changes, sex reversal may be accompanied by changes in external morphology and size. Transformation of fin shape and size has been reported in some labrids (Quignard, 1966; Roede, 1975). In *A. squamipinnis* the third dorsal spine and the caudal fin streamers were often greatly elongated in males and only occasionally so in females. When females with a short third dorsal spine changed sex, the third dorsal spine elongated in the first two weeks following initiation of the sex reversal (Shapiro, 1977b). While changes in body size during sex reversal have not been carefully documented for any species, a growth spurt almost certainly accompanies sex reversal in *A. squamipinnis* and in protandrous *Amphiprion* species (Allen, 1972; Fricke, 1974; Fricke and Fricke, 1977). A growth spurt is also suggested by the tendency for the size distribution of protogynous populations to form bimodal peaks with relatively few individuals occupying the size ranges separating male and female peaks (e.g., figures in Liem, 1963, 1968; Robertson and Warner, 1978; Warner and Robertson, 1978). Certain Labridae show a decrease in growth rate with increasing size. At body lengths at which color changes often occur the growth rate is close to zero. After the period of color change, growth rate increases again (Roede, 1972). Sex reversal and changes of growth rate may well be closely related physiologically.

3. Behavior

Qualitative male–female differences exist in all protogynous species, at least during courtship and spawning. Hence, sex-reversing females must alter their behavior as they become males. This process has been examined in two cases.

In the group-living cleaner wrasse, *L. dimidiatus,* males differ from females (1) by being the sole performers of "flutter runs," (2) by having cleaning relationships with individuals from a greater number of status classes in the group, (3) by using qualitatively different, less intensely threatening, and more ritualized movement patterns during aggression, (4) by being less likely to be aggressive toward a subordinate and more tolerant of close proximity to another individual but more likely to attack in response to another individual's initiation of interaction, (5) by having a larger territory and a higher rate of movement around the territory, (6) by responding more positively to a female's sexual approach, and (7) by adopting the superior position in a spawning rush (Robertson, 1974; Robertson and Hoffmann, 1977). Robertson concluded that this imposing list of behavioral sex differences represented merely a difference of degree and that the change to a male behavioral role required only a few relatively simple qualitative behavioral adjustments and an increase in the "amount of social activity." The most important new behavior that developed during sex reversal was the "flutter run," and this could appear for the first time in as little as 2 hr after removal of the group's male. According to Robertson (1974), behavioral sex changes could be "virtually completed" in 4–5 hr; however, the time course for the change in most of the behavior patterns listed above was never documented.

The behavioral changes of sex reversal in *A. squamipinnis* have been studied in laboratory social groups in which sex reversal was initiated by male removal (Shapiro, 1977b). The effects of male removal are multiple, however, and concern all individuals remaining in the group. To describe the behavioral changes of the sex-reversing individual in isolation from the changes shown by other group members would artifically remove it from the nexus of social relationships within the group and obscure phenomena highly relevant to the control of sex reversal. The behavioral changes of *A. squamipinnis* therefore will be described in detail as part of the discussion of the control of sex reversal. It is sufficient here to say that the sex-reversing individual shows thorough qualitative and quantitative changes in general social behavior as well as in spawning activity, and the changes require 3–4 weeks for completion.

III. The Control of Sex Reversal

Controlling influences on sex reversal can be divided conveniently into two categories: (1) factors required to initiate sex reversal, and (2) factors that tend to maintain the sex reversal process or to take the process to completion once it has been initiated. In Section III,A, five hypotheses will be discussed on the initiation of sex reversal; two are developmental in nature and three involve social or behavioral factors. In Section III,B, on the continuance and completion of sex reversal, genetic and hormonal influences will be discussed.

A. INITIATION OF SEX REVERSAL

1. Developmental Hypotheses

a. Size. In the absence of any clear environmental factors that initate sex reversal, evidence of the segregation of sexes into different size groups has been interpreted in many species to indicate that sex reversal is a developmental process that begins when a female attains a threshold size (e.g., Choat, 1969; Fishelson, 1975; Liem, 1963; Liu, 1944; Roede, 1972; Smith, 1967). This hypothesis has gained support through its use in models explaining the evolution of sex reversal (Ghiselin, 1969, 1974; Leigh *et al.*, 1976; Warner, 1975; Warner *et al.*, 1975). The evidence of *A. squamipinnis* from Aldabra is not consistent with such an interpretation (see Fig. 1). If sex reversal occurred whenever a female reached a threshold size, the maximum size of females (or the minimum size of males) should be the same at both collecting sites. This was not the case. The maximum female–minimum male size range at Site 1 was at least 10 mm larger than that at Site 2. With these data it would be impossible to construct a single size threshold beyond which females change sex that would apply equally to the two Aldabra sites.

The same conclusion can be drawn from other data on *A. squamipinnis*. Popper and Fishelson (1973) showed a sample of fish in which no male occurred below a total length of 100 mm and no female occurred above a length of 105 mm, and concluded that "... this morphological change (of sex reversal) occurs in fish which achieve at least 100 mm TL ..." (p. 415). Subsequently, Fishelson (1975) presented a different set of data in which the male–female size division occurred at 75–85 mm total length. There he concluded that "... sex reversal normally begins after the fish, first functioning as a female, attains a length of ±65 mm TL ..." (Fishelson, 1975, pp. 288). Discussing the same data, Gundermann (1972) claimed, "... it is possible to hypothesize that some of the females cease to grow before the process of sex change occurs or that all females change sex before they attain the critical size" (p. 37). It is apparent, however, that for the species as a whole, there is no critical size.

No study has yet been done on any species to demonstrate that individual females raised in isolation change sex at a uniform size. Consequently, the size hypothesis rests entirely at present on the sex-separated size distributions of protogynous populations. In the few instances in which these distributions have been examined for separate populations of the same species on the same or neighboring reefs, the size at sex reversal has been found to vary widely (Shapiro, 1977b; suggested by data in Warner and Robertson, 1978, and in Robertson and Warner, 1978). If neighboring adult populations of a protogynous species were genetically distinct, differences in size at sex reversal might be genetically determined and the size hypothesis could be maintained. However, early in development individuals of most protogynous species undergo a pelagic phase in which larvae and young juveniles have been reported to drift as far as 120 miles from the nearest spawning site (e.g., Fourmanoir, 1969, 1971, 1976) before drifting over and settling out on a shallow reef. It is thus improbable that variations in sex reversal size between neighboring populations of the same species are genetically determined. The existing facts, then, provide at the very most weak, circumstantial evidence that sex reversal is initiated by the attainment of a critical size.

One additional word must be said about the size hypothesis. The growth rate of fish is often not uniform throughout the year (Brown, 1946, 1957). In periods in which favorable environmental conditions obtain, fish will grow rapidly and may attain a major proportion of the year's growth in a short time. When favorable conditions cease, growth may slow or halt altogether (Lowe-McConnell, 1975). If sex reversal depended both on the attainment of a critical size and on the occurrence of suitable environmental conditions, it would be possible for conditions to favor rapid growth but not be sufficient to permit sex reversal. For a female living in this situation, continued growth would carry her beyond the threshold size without her changing sex. The size hypothesis can thus be rescued by wedding it to a second hypothesis, which together would explain the differences in maximum female–minimum male size in separate populations. To date there is no evidence to support such a complicated explanation.

b. Stage of Development. An alternative to the size hypothesis argues that sex reversal is initiated, not by the attainment of a critical size, but by reaching a critical age or a threshold level of development. In the literature there is evidence of developmental influences on sex reversal only in the Cyprinodont fish, *Rivulus marmoratus*. When kept in isolation in aquaria water of moderate temperature, all hatchlings of this species became simultaneous hermaphrodites (Harrington, 1961, 1968) and remained so throughout their lives. If fish were early reared at a high temperature, the individuals matured as simultaneous hermaphrodites and later changed into pure males in response to short day length (Harrington, 1971). No work of this kind has been done on a protogynous species, nor, indeed, upon any successive hermaphrodite.

2. Social Hypotheses

a. Male Removal. (1) *The Inhibition Hypothesis.* Social initiation of sex reversal has been shown clearly in *L. dimidiatus* (Robertson, 1972), *A. squamipinnis* (Fishelson, 1970; Shapiro, 1977b), and in two protandric *Amphiprion* species (Fricke and Fricke, 1977), and has been claimed for *Thalassoma bifasciatum* (Warner *et al.,* 1975), *Amphiprion melanopus* (Ross, 1978), two *Paragobiodon* species (Lassig, 1977), and *Centropyge interruptus* (Moyer and

Nakazono, 1978). In each protogynous case the evidence consists of observations that removal of a male from a social group results in the sex reversal of a single female. Since the removed animal was invariably the largest and most dominant individual of his sex in the group, two hypotheses were immediately proposed to explain these results. The first was that mature females have a continual physiological tendency to change sex and that this tendency is inhibited by the presence of a male (Fishelson, 1970; Robertson, 1972, 1973; Robertson and Warner, 1978). The second was an extension of the first: inhibition of sex reversal is achieved directly by aggression (Robertson, 1973) or aggressive dominance (Fricke and Fricke, 1977; Robertson and Warner, 1978). While each of these hypotheses is possible, in neither study for which control by aggressive dominance was claimed to have been shown was evidence presented that teased apart dominance or aggression from even one of the multiple other features characterizing the relationship between the individual fish of these social groups (Shapiro, 1980a). In *A. squamipinnis*, when the behavioral interactions of all group members, including the sex-reversing fish, were examined before and after male removal (Shapiro, 1977b), the results suggested several new hypotheses that did not depend on male inhibition or directly upon dominance per se. It is worth looking at this evidence in more detail.

In the laboratory, male removal studies were performed on four social groups, each in a separate aquarium and each consisting of one male and six females that had lived together undisturbed for two months before onset of the study. Each group was used in a succession of experimental rounds in which 2½ weeks of baseline observations were followed either by male removal or by a control manipulation (the male was netted as though to remove him but he was then released), and another 2½ weeks of observation. In each round two groups served as control groups and two as experimental groups. In the initial round, control and experimental groups were selected at random. The results showed clearly that male removal led to the sex reversal of one female in the group, generally the largest female, while control manipulations produced little change (Table III).

TABLE III

LABORATORY MALE REMOVAL STUDY ON *A. squamipinnis*:
NUMBER OF INSTANCES IN WHICH A FEMALE CHANGED SEX
AFTER MALE REMOVAL (TOP ROW, MARKED EXPERIMENTAL)
AND AFTER A CONTROL MANIPULATION (BOTTOM ROW,
MARKED CONTROL)[a]

	Sex change	No change
Experimental	7	1
Control	1	6

[a] From Shapiro (1980a).

Small, single-male, bisexual groups were commonly found on coral reefs (Shapiro, 1977a, b), where single-male removals produced similar results (Table IV). In the field, however, large multimale, bisexual groups also existed. Removing a single male from such groups left multiple males remaining in the group but also produced a sex reversal (Table IV). Furthermore, in the field all-female groups were found in which no sex reversal occurred for as long as they were observed (that is, for three months). Similar groups could be artificially created in the laboratory. Thus, in all-female groups sex reversal did not occur in spite of a complete absence of males, while in multimale groups sex reversal did occur in the continued presence of many males. While the small number of all-female groups thus far observed (two in the laboratory and seven in the field) render any conclusion preliminary, one may tentatively conclude that sex reversal in this species is not a matter of simple presence or absence of a male (cf. Fishelson, 1970; Robertson, 1974), nor of a simple inhibition by males of females. It is triggered rather by something specific about male removal.

In theory, any aspect of the male communicable to females could be responsible, when removed, for sex reversal. In a small number of preliminary studies the male of a single-male group was placed within a plastic cylinder in the center of the group's aquarium. The cylinder was either: (1) opaque, allowing acoustic and chemical, but not visual or direct, interactive communication between the isolated male and the group, or (2) transparent, allowing acoustic, chemical, and visual, but not direct, interactive communication. In both cases sex reversal tended to occur, suggesting the involvement of direct behavioral interaction in the initiation process. These experiments are currently being repeated in a systematic manner.

TABLE IV

SUMMARY OF *A. squamipinnis* SINGLE MALE REMOVAL STUDIES IN THE FIELD[a]

No. of males in group	No. of groups studied	Total no. males removed	Total no. sex changes
1	7	7	8
2	2[b]	5	3
3	1	1	1
4	1	3	3
10	1	3	3
Totals:	12	19	18

[a] From Shapiro (1980a.)

[b] In these groups males were removed successively. When a male removal resulted in a female changing sex, then the new male was removed. In this way, more males were removed, in total, than the number initially present in the two groups studied.

The first seventeen days of the complete male removal study in the laboratory was called the *initial control period* for all four groups. The period of approximately two weeks immediately following the removal of a male from a group was called an *experimental period* for that group, and the period following a control manipulation in a group was called a *control period*. Since each social group was used for successive experimental rounds, females were present during successive *control periods* and *experimental periods*. To distinguish the *control* and *experimental periods* associated with a particular female's sex reversal from prior *control* and *experimental periods* not associated with her sex reversal, the following designations were made. For each particular female that changed sex the period immediately preceding the male removal which resulted in her sex reversal was called her *C period* and the period immediately following that male removal was called her *E period*. Each day the groups to be observed were randomized for observation order. All individuals within a group then were observed, in random order, each for 10 min. During that time a record was made of the actor, the recipient, and the type of interaction, for all interactions performed or received by the focal animal (Altmann, 1974; Shapiro and Altham, 1978).

During the *initial control period* males performed four of seven measured movement patterns at significantly higher mean rates than females (Shapiro, 1977b). Following male removal, sex reversing females increased their performance rates of three of the behaviors (Table V): (1) rushes (an aggressive movement; (2) U swims (a behavior performed exclusively by males to females and used both in general social and in spawning contexts); and (3) third dorsal erections (a rapid erection of the elongate third dorsal spine). Mean rates during the first two weeks after male removal, while significantly above female levels, still had not attained the full male level (Table V). These behavioral changes by sex-reversing individuals did not occur after control manipulations and were significantly greater in incidence and in magnitude than changes shown by any other individuals in the group (Shapiro, 1977b). The declines in mean rates of nose bump/cross and bent approach shown by *E period* females in Table V were in the "male" direction but were not statistically significant.

Occasionally several females in a group showed a small, transient rise in the performance rate of rushes in the first two days after male removal. These rushes seldom were directed to the sex-reversing fish, and these females showed no change in the performance rate of other behaviors. No female other than a sex-reversing fish showed even the earliest indication of developing U swims. From the first day of his disappearance, then, the male's removal affected the largest female of each group differently, in kind and in degree, from the smaller females in the group. There is no evidence that male removal initiated a competition among the remaining females, with the "winner" changing sex. It is more likely that the individuals in each group had previously interacted with one

TABLE V

MEAN RATES (FREQUENCY PER 10 MIN) OF GIVING BEHAVIORS BY SEX-REVERSING *A. squamipinnis*
BEFORE AND AFTER MALE REMOVAL AND BY CONTROL PERIOD MALES[a]

	Sex-reversing females		
	C period: $N = 7$	E period: $N = 7$	Males Initial control period: $N = 4$
Rushes			
Range	0–5.4	2.4–8.4	5.4–11.3
Median	1.8	5.3	9.2
U swims			
Range	0	2.0–8.3	8.5–28.9
Median	0	5.4	13.0
3rd dorsal erection			
Range	0.1–2.2	0.5–9.1	10.4–29.8
Median	0.2	6.5	26.5
LDD[b]			
Range	0.2–5.1	0–6.7	0.6–3.3
Median	0.3	1.6	1.2
NB/CR[b]			
Range	0–0.5	0–0.2	0–0.1
Median	0.2	0.1	0.05
BA[b]			
Range	0–2.1	0–0.6	0–0.21
Median	1.4	0	0.09
M to M[b]			
Range	0–0.2	0	0.4–1.0
Median	0	0	0.8

[a] From Shapiro (1980c).

[b] LDD, lateral dorsal display; NB/CR, nose bump/cross; BA, bent approach; M to M, mouth-to-mouth fight.

another in such a way as to have left only one individual able to change sex in the existing social context.

Males also differed from females initially in the mean rates at which three movement patterns were received from other group members (Shapiro, 1977b). In the two weeks after male removal the reception rates for these behaviors were significantly different, for the sex-reversing females, from the normal female level and were well on the way to being typical male reception rates (Table VI). The rate at which the sex-reversing fish receives a behavior from other group members is a summary measure of how all other individuals treat it. These results indicate, then, that male removal affects the behavior both of the individual that changes sex and of other group members as well, at least in their relationship with the sex-reversing fish.

(2) Distinguishing Causes. The changes that sex-reversing females showed in the performance and reception of behaviors were not instantaneous. Each movement pattern increased or decreased in its own daily progression after male removal. Figure 3 illustrates the daily changes in performance rate for three movement patterns. The sex-reversing females showed an initial increase in the performance of rushes on the second day after male removal. With various ups and downs this increase was sustained thereafter. In those instances in which sex-reversing fish were observed for longer than 2½ weeks following male removal the sex-reversing individual continued to increase the performance rate of rushes until, by the third or fourth week, the normal male level was attained.

TABLE VI

MEAN RATES (FREQUENCY PER 10 MIN) OF RECEIVING BEHAVIORS BY SEX-REVERSING
A. squamipinnis BEFORE AND AFTER MALE REMOVAL AND BY CONTROL PERIOD MALES[a]

	Sex-reversing females		Males Initial control period: $N = 4$
	C period: $N = 7$	E period: $N = 7$	
Rushes			
Range	0.1–2.9	0–0.3	0–0.1
Median	2.2	0.1	0
U swims			
Range	0–4.4	0–0.5[c]	0
Median	1.5	0	0
LDD[b]			
Range	0.3–6.3	0.3–2.2	0.2–3.0
Median	0.8	1.0	0.2
BA[b]			
Range	0–0.4	0.4–2.1	0.7–2.8
Median	0.1	0.6	1.3

[a] From Shapiro (1980c).

[b] LDD, lateral dorsal display; BA, bent approach.

[c] During one *control period* in one laboratory group a female changed sex while the male was still present (see Table III). The group then contained two males. When one of these males was later removed the remaining male performed four U swims, during eight observation periods (hence, a mean rate of 0.5) to the female which subsequently changed sex. For a complete discussion of this unusual occurrence see Shapiro (1980a,d).

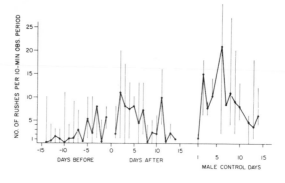

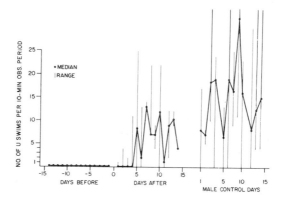

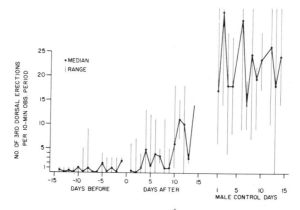

FIG. 3. The daily rate of rushes (top), U swims (center), and third dorsal erections (bottom) by sex-reversing *A. squamipinnis* before and after male removal (left portion of

In contrast to the performance rate of rushes, which showed an early increase, the performance rate of U swims and third dorsal erections showed time lags of 4–5 days and 10–11 days, respectively, before beginning to rise (see Fig. 3). For both behaviors more than 14 days were required for the performance rate to reach the full male level. The reception rate of U swims and aggressive rushes by the sex-reversing fish, in all cases, fell virtually to zero on the first day after male removal and remained at zero thereafter. This reflects the fact that the males were the only performers of U swims and that the sex-reversing female, being the largest female in the group, received very few rushes from any female smaller than herself. The reception rate of bent approaches (possibly a conciliatory gesture) showed a 3–4-day lag before it increased (Shapiro, 1977b).

It has always been difficult to distinguish cause from effect in social groups, including pairs (e.g., Simpson, 1968, 1973; Slater, 1973), when several behaviors by more than one individual are occurring synchronously. In the development of male performance rates by sex-reversing females the separate time courses for separate movements permit this distinction. This is illustrated by the relation of bent approaches to rushes. Bent approaches were directed primarily by the two largest females in each group to the male. When a female underwent sex reversal, she began to receive an increased rate of bent approaches from the females beneath her in size (see Table VI) and she also increased the rate of directing rushes to them (see Table V). The increased performance of rushes toward the smaller females could be the cause of the smaller females' increased performance of bent approaches, or vice versa. Since bent approaches were not received at an increased rate by the sex-reversing fish until two days after she initially increased her performance of rushes (Shapiro, 1977b), it is unlikely that the change in her rushing behavior was caused by the changed behavior of the other females toward her. On the contrary, it is quite possible that the remaining females began to show bent approaches to the sex-reversing fish as the result of her increased rushing behavior. A similar argument makes it unlikely that changes in third dorsal erection or in U-swim performance by the sex-reversing fish influenced the bent approaches directed to her by the other group members.

The time lags before the initial upsurge in performance rates of different movement patterns also suggest that the casual factors influencing performance rates are not the same for all behaviors. Rushes showed an immediate response to male removal and are therefore likely to have been influenced directly by the alteration of group composition. The four-day and ten-day lags of U swims and

---

each graph) and by normal males during the initial control period (right portion of each graph). The median and range plotted on each day represent only the sample of individuals observed on that day. The sample size for females varied from one to seven, with a median of four individuals, and for males, from two to four with a median of four individuals. (From Shapiro, 1980c.)

third dorsal erections, respectively, suggest that these behaviors responded to influences that were slower in appearing. These may have been subtle social changes, but it is tempting to suppose they resulted from hormonal changes within the sex-reversing fish. In the protogynous rice-field eel, *Monopterus albus,* sex reversal was accompanied by a gradual decrease in circulating estrogens. Testosterone levels increased rapidly to a peak midway through the sex reversal and then declined to a moderate level, which was still substantially above its level in the female (Chan *et al.,* 1975). The gonadal and coloration changes of sex reversal in *A. squamipinnis* were completed, in these experiments, in 7–16 days. Midway through sex reversal would be 4–8 days, which is just the interval during which U swims and third dorsal erections showed their initial increase. A mid-sex-reversal peak in testosterone initiating an increase in the performance of U swims and third dorsal erections would explain these findings nicely.

(3) Behavior Profiles. In addition to affecting performance and reception rates, male removal also affected the relative proportions of behaviors given and received. Behaviors given profiles were constructed by finding the proportion of the total number of behavioral acts given by an individual during a particular period, which were rushes, lateral dorsal displays, bent approaches, mouth-to-mouth fights, nose bump/cross, and U swims. Such a profile shows the relative tendency for an individual to perform one behavior (for example, lateral dorsal display) as opposed to other behaviors (for example, U swim or bent approach). Profiles were also constructed for the behaviors an individual received from group members.

During the initial control period, the four males showed a characteristic profile for behaviors given (Fig. 4). U swims were performed about twice as frequently as rushes. Other movement patterns were relatively infrequent in the male profile. Three of the four males showed this profile. One male deviated slightly. The summary male profile of Fig. 4 was consistently presented by the male to each female in the group (Shapiro, 1977b). The only slight exception to this was the profile presented to the highest-ranking female, in which U swims fell slightly beneath rushes in relative frequency. Females showed two profiles for behavior given, both distinctly different from the male profile (see Fig. 4). Half of the females (*n* = 11) showed the rush profile, in which rushes were relatively more frequent than lateral dorsal displays, with other behaviors generally relatively rare. The remaining half of the females (*n* = 13) showed a display profile, in which lateral dorsal displays were greater in frequency than rushes, again with other behaviors generally much less common. The individual profiles of most females fell neatly and obviously into one of these two categories. These two behavioral profiles represented bimodal frequency distributions among the females of these groups and each female consistently maintained only one of these profiles through time despite the upsets that successive male removals

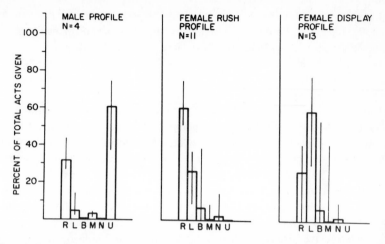

FIG. 4. Behavior given profiles of *A. squamipinnis* males and females during the initial control period. Six behaviors are listed along the horizontal axis: rushes (R), lateral dorsal displays (L), bent approaches (B), mouth to mouth (M), nose bump/cross (N), and U swims (U). The height of the bar is the median and the thin vertical line is the range for each behavior. (From Shapiro, 1980d.)

brought to their groups (Shapiro, 1977b). Behaviorally, then, there were two types of female in these groups.

In some protogynous labrids (Quignard, 1966; Sordi, 1962) and serranids (Moe, 1969), the fact that females occasionally grew as large as the largest male has been taken to indicate the presence of two types of female, one type capable

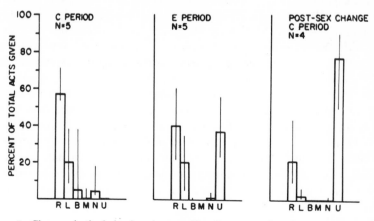

FIG. 5. Changes in the behavior given profile of sex-reversing *A. squamipinnis* showing the rush profile during *C. period*. Legend as in Fig. 4. (From Shapiro, 1980d.)

of changing sex and a second type incapable of sex change (see discussion in Reinboth, 1970). In these species no means of placing small- or medium-sized individual females into each of these two categories has yet been proposed. In *A. squamipinnis,* females could be separated by behavior. However, both types of female were capable of changing sex and when they did their behavioral profiles assumed the characteristic male profile (for example, Fig. 5).

Males showed a characteristic profile for behaviors received (Fig. 6) in which the predominant behavior was the bent approach. The male received more than several behavioral acts from only the two largest and top-ranking females in each group. The male's profiles for behavior received from each of them were identical to Fig. 6. All females showed very much the same behavior received profile (Fig. 6), which was quite distinct from that of the male. They received primarily rushes with somewhat fewer U swims and lateral dorsal displays. Individual females received behavioral acts both from other females and from the male. When acts received from the male were excluded from consideration, the behavior received profile distinguished individual females clearly on the basis of dominance rank (Fig. 7).* With decreasing dominance there was a progressive increase in the relative frequency of rushes and a progressive decrease in the relative frequency of lateral dorsal displays. The farther apart two individuals were in rank, the greater was the difference between their behavior received profiles. When acts received from the male were included, the behavior received profile ceased to distinguish nicely females of different rank (Fig. 8).

It would appear that females interacted with each other in such a way as to create a clear and regularly progressing difference between all females on the basis of the profile of behaviors each received from the others. The male interacted with the females in such a way as to obscure this orderly difference. If one of the physiological requirements determining that only the largest female changes sex after male removal is for females to be behaviorally different from one another, then the consequence of female–female interactions is to create those differences, while the consequence of male–female interactions is to obscure them. At first glance this appears to suggest that the male, by reducing the differences between females, was inhibiting all of them from changing sex. But the male did not act in the same way toward all females. In each group, the male directed specific behaviors to different females at different absolute rates (this is not, of course, inconsistent with the male directing the same behavioral profile to each female, since the behavioral profile is a measure of relative rate). The similarity in the females' behavior received profiles, when interactions with the male were included, resulted from the counterbalancing effect, on the profile of each individual, of behaviors received from him and from other females.

---

*In each group a linear dominance rank order was constructed from the matrix of rushes directed by each individual to each other individual in the group.

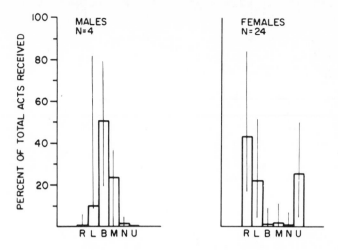

FIG. 6. Behavior received profiles of *A. squamipinnis* males and females during the initial control period. Legend as in Fig. 4. (From Shapiro, 1980d.)

Combined with the observations that male removal results in the sex reversal of only one female and that no sex reversal occurred in up to 90 days in all-female groups, these data support a view that sex reversal is not a matter of simple inhibition of females by the male.

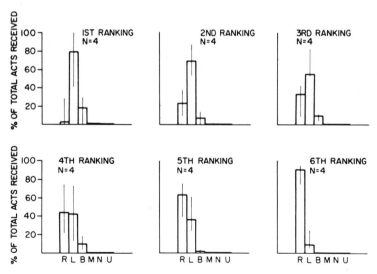

FIG. 7. Profiles for *A. squamipinnis* females of different dominance rank of behavior received only from females. Legend as in Fig. 4. (From Shapiro, 1980d.)

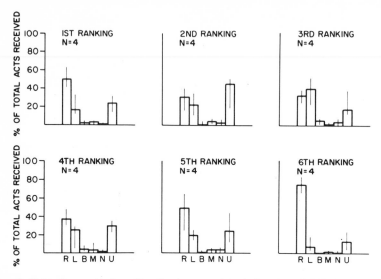

FIG. 8. Behavior received profiles for *A. squamipinnis* females of different dominance rank. The profiles include behaviors received from the male and from females. Legend as in Fig. 4. (From Shapiro, 1980d.)

When females reversed sex, the relative frequency of the various behaviors that they received altered from a female to a male profile, that is, the remaining females in the group changed the way they behaved toward the sex-reversing fish and began to treat her as though she were a male (Fig. 9). This finding raises the possibility that a female might be induced to change sex, not directly as the result

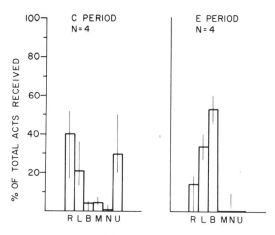

FIG. 9. Changes in the behavior received profile of four sex-reversing *A. squamipinnis*. Legend as in Fig. 4. (From Shapiro, 1980d.)

of removing the male, but in response suddenly to being treated like a male by the other members of the group.

Two pieces of evidence argue that suddenly being treated like a male is insufficient to bring about a sex reversal. Both observations derive from studies of all-female groups. Two all-female groups were artificially established in the laboratory by putting females together immediately after their arrival from commercial sources. For as long as these groups remained intact, a minimum of 90 days, no female in either group showed external evidence, in coloration or behavior, of sex change.

Quantitative behavioral data were taken on both groups, which contained seven and eight females, respectively. During an initial 15-day period the largest female in Group 3LT behaved like a normal rush profile female whose only suggestion of male behavior were rare attempts at U swims (Fig. 10, top left profile; cf. Fig. 4). Her color and her gonadal histology were typically female. However, her behavior received profile showed that she was treated by the other group members exactly as though she were a male (Fig. 10, top right profile; cf. Figs. 6 and 7). During this period the second largest female showed typical female profiles both for behaviors given and for behaviors received.

When the largest female was removed from this group, the second largest female continued to show a female rush profile for behaviors given (Fig. 10,

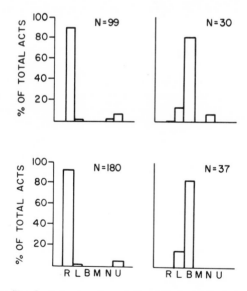

FIG. 10. The profiles for behavior given (left column) and for behavior received (right column) of the largest female (top row) and the second largest female (bottom row) in an all-female *A. squamipinnis* group. See text for additional explanation. (From Shapiro, (1980d.)

bottom left profile), but her behavior received profile changed to one that was typically male (Fig. 10, bottom right profile). The other group members, then, began to treat this female as though she were a male, but her coloration, behavior, and gonadal histology remained entirely female. The second all-female group showed similar phenomena for the largest female. These observations suggest that being treated like a male could not be the sole cause of sex reversal.

(4) Factors Controlling Sex Reversal. If being treated like a male and simple absence of a male are both usually insufficient to induce a female to change sex, then precisely what factor does stimulate her to change sex? A promising behavioral candidate would be one that sets one female apart from the others and changes dramatically for the sex-reversing fish after the male is removed. Such a candidate is a measure called the percentage of rushes given. As applied to each individual, this is the proportion over a particular time period of the number of rushes given divided by the sum of the number given plus the number received. When the percentage of rushes given was plotted against the dominance rank of each individual in the four social groups during the first 15 days of the study, the resulting curves were quite similar (Fig. 11). In each case the dominant female showed a lower value than the male dominant to her and at least two females subordinate to her. When the male was removed, the dominant female immediately jumped to a value close to 100%. This represented a greater proportional change than that of any other female.

This behavioral measure may be closely paralleled by internal physiological changes. In a group of brown trout in an aquarium containing 12 liters per fish (Brown, 1946) the specific growth rate of individuals depended on the individuals' size rank in a way (Fig. 12) that closely paralleled the curves of Fig. 11. Generally, in fish, dominance rank and relative size are very highly correlated

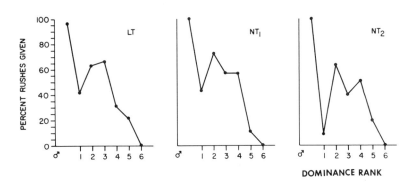

Fig. 11. The percentage of rushes given for individuals of different dominance rank in three *A. squamipinnis* groups during the initial control period. The horizontal axis lists, from left to right, the male and six females in order of decreasing dominance. (From Shapiro, 1980d.)

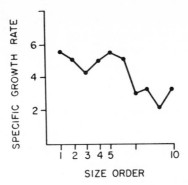

FIG.. 12. The growth rate of individual trout of different size rank. (After Brown, M. E. 1946. The growth of brown trout. II. Growth of two-year-old trout at constant temperature 17.5°C. *J. Exp. Biol.* **22**, 130–144. By permission of Cambridge University Press.)

(e.g., Barlow and Ballin, 1976; Henderson and Chiszar, 1977), so that speaking of one is practically synonymous with speaking of the other. The main result of Brown's study was that the growth rate of an individual could be manipulated by changing its relative position in the size order of the group. Individuals suddenly propelled to the top of the size order by removing all of the larger fish underwent an immediate increase in growth rate.

In the *Anthias* experiments the sex-changing female appeared to undergo a similar growth spurt following the removal of a male (Shapiro, 1977b). This has also been observed in wrasses at the time they change sex (Roede, 1972) and in juvenile *Amphiprion* after the removal of adults (Allen, 1972; Fricke and Fricke, 1977). Could it not be that sex reversal and growth are closely allied physiologically (see Mittwoch, 1975, p. 440), as they seem to be in prawns (Allen, 1959), and that, while growth depends on the absolute value of the percentage of rushes given, sex reversal requires a very large increase in this value, which occurs only in the dominant female following male removal?

The percentage of rushes given curves could be generated in several ways. A low value on this measure for an individual could be the result either of an unusually high rate of receiving rushes or of an unusually low rate of giving them, or both. When the rate of giving rushes was calculated for each individual and plotted against dominance rank, the resulting curves for all four groups closely reproduced, in shape, the curves of Fig. 11. The curves for rates of rushes received plotted against dominance were, in all four groups, the inverse shape to those of Fig. 11. Thus, individuals which gave rushes at a high rate received them at a low rate and vice versa. Since the rate of rushes given and the rate of rushes received changed with decreasing dominance in the same or in the inverse manner as the percentage of rushes given, all three measures are potential candidates for a behavioral factor influencing sex reversal.

In *Labroides dimidiatus* (Robertson and Warner, 1978) and *Amphiprion* (Fricke and Fricke, 1977), dominance has been suggested as the behavioral factor controlling sex reversal. In *A. squamipinnis* this is unlikely. The existence of all-female groups in which the dominant female does not change sex shows that the controlling factor could not be the absolute dominance rank of an individual. Since the removal of a male from a one-male group results in an increase in dominance by one rank of *all* remaining females, but in sex reversal of *only one* female, the controlling factor could also not be an increase in dominance rank. This latter argument also applies to *Amphiprion* and *L. dimidiatus,* in both of which only one fish changes sex although all increase in dominance rank after removal of the largest fish. It would seem that dominance per se, being a largely qualitative aspect of relationships (cf. Hinde, 1974, pp. 342–343), does not, in *A. squamipinnis* at least, serve as the factor controlling sex reversal. One must find a more quantitative behavioral measure capable of revealing finer distinctions between the changes shown by individuals in these groups if sex reversal is to be understood.

The percentage of rushes given and the associated rates of giving and receiving rushes fit this description. So does the profile of behaviors received. Figures 7 and 8 suggest that male removal would result in an immediate shift in the behavior received profile of each individual, from that of Fig. 8 to that of Fig. 7. To prove this, of course, would require a detailed comparison of behavior received profiles from one day to the next following male removal. Nevertheless, if such a shift occurred and if sex reversal depended on an alteration in this profile of a certain magnitude, then the largest and most dominant female would show a change of greater magnitude than the change of any other female. This could explain the finding that only the largest female changed sex in each group. Since all females in the group should show some change in the profile after male removal, with the magnitude of change being directly related to dominance rank (see Figs. 7 and 8), the second- and perhaps the third-ranking female should also be capable of sex change under the proper circumstances. However, if the control factor were the percentage of rushes given only the most dominant female should be capable of sex change at any given time, because only she is in a behavioral position for male removal to increase her percentage of rushes given a sufficient amount to induce sex change (see Fig. 11).

It thus becomes possible to perform an experiment to choose between the percentage of rushes given (and its associated measures) and the behavior received profile as the control factor inducing sex reversal. The male and the dominant female can be removed simultaneously from a one-male group. If the percentage of rushes given is the control factor, no sex reversal should occur, but if the behavior received profile is critical, then the second ranked female should change sex. Such an experiment is currently in progress.

(5) The Priming Hypothesis. Hypothesizing a critical role for the percentage of rushes given or the behavior received profile carries with it a very different set of physiological implications from a hypothesis of inhibition. Inhibition implies that females have a continual tendency to change sex, which is constrained by male presence. As Robertson (1972) puts it, "The male in each harem suppresses the tendency of the females to change sex by actively dominating them. Death of the male releases this suppression and the dominant female of the harem changes sex immediately" (p. 1007). The theory based on the behavioral measures of *A. squamipinnis* implies neither that females have an internal tendency to change sex nor that males inhibit this tendency. It suggests, rather, that a female is actively stimulated to change sex by the occurrence of a particular set of behavioral changes of a critical magnitude. In the absence of these changes sex reversal would not occur, whether a male was present or not. Since sex reversal, by this hypothesis, would depend on a critical change in a behavioral measure, producing this change requires both a final behavioral state and an initial or preexisting behavioral state. For example, if the percentage of rushes given controlled sex reversal, the dominant female would show the critical increase after male removal, only because she showed an unusually low value prior to male removal. Showing this low value requires the presence of the male. This can be seen by examining all-female groups in which the percentage of rushes given-versus-dominance curves were straight lines, with no fish showing the peculiarly low value typical of the dominant female in a one-male group (Shapiro, 1977b). Similarly, if the behavior received profile is critical, the largest female would be in a behavioral position to change sex only because interactions with the male resulted in a profile sufficiently different from the profile received exclusively from females that his removal would lead to the critical change. The male's presence is thus actively required to create the behavioral preconditions enabling a female to change sex after his removal. In this sense, under this theory, the male not only does not inhibit females but he forms a necessary part of the positive stimulation which eventually results in sex reversal.

For convenience, a female in the behavioral precondition to change sex may be said to be "primed," and this theory referred to as the priming hypothesis. The priming hypothesis makes a number of testable predictions; for example, (1) only a limited number of females in each group are primed at a time; (2) after all primed females have been removed from a group, a minimal time interval must elapse for another female to become primed; (3) the critical change in behavior that produces sex reversal in a primed female might occur very rapidly, as it does following male removal, or fairly slowly, and hence (a) sex reversal might be initiated in some way other than by male removal, and (b) the morphological changes of such a sex reversal should follow a more prolonged time course. Work is currently being done to investigate these predictions.

b. Sex Ratio Threshold. Inititiating sex reversals by male removal results in the replacement of males lost from a group, but fails to explain how the number of males in a group may increase. Diandric species have two means of producing males: by the initial maturation of juveniles as males, and by the sex reversal of females (e.g., Choat and Robertson, 1975; Reinboth, 1957, 1975b; Robertson and Choat, 1974; Robertson and Warner, 1978). Monandric species, such as many anthiines (Gundermann, 1972; Shapiro and Lubbock, in preparation) and some labrids (Robertson and Warner, 1978) do not produce primary males. If monandric species occupy large social groups containing multiple males, as well as small, single-male groups, as many anthiines do, and if groups grow only by the addition of females, as suggested for *A. squamipinnis*, there must be an additional mechanism for initiating sex reversal.

Shapiro and Lubbock (1980) suggested that whenever the ratio of adult females to males within a group exceeded a threshold value, a female would change sex even though no male had been removed. They constructed a model that assumed a sex ratio threshold. The model, in turn, made certain predictions concerning the sex ratio of groups and the size of individual fish at the moment of sex reversal, which were compared with data on *A. squamipinnis* collected in the Red Sea and the Indian Ocean. The data conformed reasonably closely to the model's predictions, and Shapiro and Lubbock concluded that a sex ratio threshold was likely to exist.

In brief, the model made three basic assumptions: a single female changes sex (1) whenever a male is removed from the group and (2) whenever a group's adult female-to-male sex ratio equals or exceeds a threshold value r; and (3) juveniles mature into the adult female pool of a group at a constant recruitment rate of k females per day. It was then shown that the time T required to produce n sex reversals in a group could be expressed as

$$T = \frac{n(r + 1)}{(k - d_f + d_m r)}, \tag{1}$$

where d_f and d_m are constant female and male mortality rates, respectively. It followed that the average time interval Δt between successive sex reversals would be

$$\Delta t = \frac{(r + 1)}{(k - d_f + d_m r)}. \tag{2}$$

This expression states that the average time required to produce successive sex reversals is a function of the mortality rates of males and females, the recruitment rate of new adult females into a group, and the value of the sex ratio threshold. Furthermore, the expression tells us that, in terms of the time required to produce a sex reversal, the loss of a male is equivalent to the addition of r new adult females to a group.

When a social group of fish is subject to the parameters of this model, the sex ratio of the group will change in a regular and predictable fashion. Figure 13 shows a hypothetical example of a simple case in which male and female mortality rates are negligibly small. As juveniles mature to adult females, the sex ratio of the group rises until the sex ratio threshold is reached, when one female changes sex. If sex reversal is completed very rapidly, there is an immediate fall in the female-to-male sex ratio. Thereafter, the sex ratio rises gradually again.

Female mortality does not greatly affect the shape of this curve but male mortality does, provided that the sex reversal which results from loss of a male requires a substantial number of days for completion (as in *A. squamipinnis*). Figure 14 illustrates the same hypothetical case as Fig. 13 with the addition of a constant, evenly spaced male mortality and a completion time of ten days for sex reversal following a male loss. The loss of a male (point a in Fig. 14) suddenly increases the sex ratio of the group and initiates a sex reversal. During the ten days required for its completion two additional females are recruited into the group and the sex ratio rises (segment a–b in Fig. 14). After ten days the new male is present (point b in Fig. 14), and the sex ratio falls. The resulting curve is a series of overlapping, broken lines.

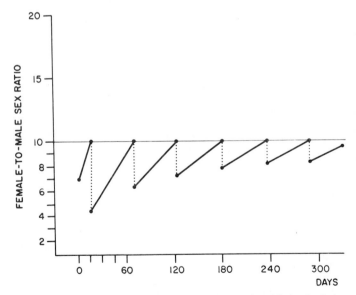

FIG. 13. The life history curve (i.e., the change of sex ratio with time) of a hypothetical *A. squamipinnis* group with a very rapid completion time (less than one day) for sex reversal. The horizontal line represents the sex ratio threshold r. $k = 0.2$ females per day; $r = 10$ females-to-male threshold; $d_m = d_f = 0$. (From Shapiro and Lubbock, 1980.)

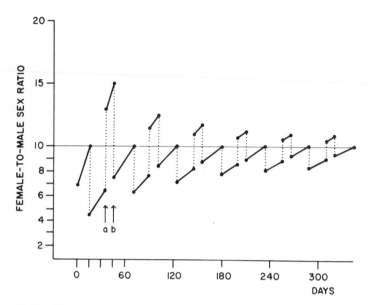

FIG. 14. The life history curve of a hypothetical *A. squamipinnis* group with a constant, evenly spaced male mortality, a moderate completion time (10 days) for sex reversals resulting from male mortality, and a rapid completion time (less than one day) for sex reversals resulting from the attainment of the sex ratio threshold. $k = 0.2$ females per day; $r = 10$ females-to-male threshold; $d_m = 0.018$ males per day. See text for additional explanation. (From Shapiro and Lubbock, 1980.)

If this model is a reasonable approximation of events on a coral reef, then a sample of naturally occuring groups of different size should represent different stages of a common process by which the sex ratio of a group changes as the group increases in size. The form of the curve of Fig. 14 then should approximate the curve produced by plotting sex ratio versus size for the groups of the sample. This curve for 36 groups of *A. squamipinnis* studied at a single locality in the Sudanese Red Sea is shown in Fig. 15. The solid oblique lines connect groups containing the same number of males. Regression curves have been fitted to the maxima and minima to show that the maxima tend to decline and the minima to rise, and both approximately approach a common value just as in Fig. 14.

The model also generated an algebraic expression relating the size at which individual fish change sex to the growth rate, sex ratio threshold, female recruitment rate, and the male mortality rate (Shapiro and Lubbock, 1980). This expression predicts that in a well-localized population, that is, where groups are found in a small, compact area in which demographic conditions are likely to be homogeneous for all groups, the size at sex reversal will be virtually identical for all groups. If separate populations from demographically distinct areas were compared, the size at sex reversal should be different in each population. That is exactly what was found in ecologically distinct populations of *A. squamipinnis*, as Fig. 1 shows, and of *A. randalli* (Shapiro and Lubbock, 1980).

Definitive evidence that a sex ratio threshold exists in any particular species would require two kinds of study: (1) long-term observation of identified social groups and measurement of recruitment, mortality, and the time between sex reversals, and (2) an experimental study in which females are added successively to a social group and the sex ratio at which sex reversal occurred is monitored. Few observations in either category have yet been made. In six social groups of *L. dimidiatus* a female changed sex without prior male removal. In each case, prior to sex reversal, either the female moved progessively farther away from the core area of the resident male's territory or the male concentrated his activity in a new part of his territory and ignored one female (Robertson, 1974). If such occurrences prove to be a regular feature of oversize groups (in *L. dimidiatus*, groups have never been seen to contain more than one male, so sex ratio changes directly with group size), they would constitute evidence for a sex ratio threshold.

It is not necessary that individual fish recognize the sex ratio of their group in order for the sex ratio threshold method of producing sex reversal to work. All that is required is for some factor, which can affect individuals, to change with the sex ratio. As the *Labroides* case illustrates, an obvious candidate would be a

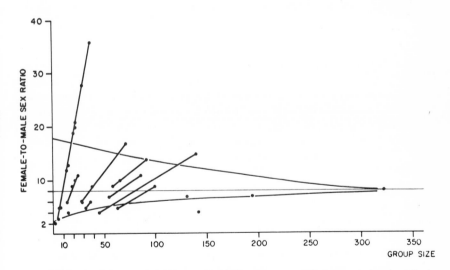

FIG. 15. Sex ratio versus group size of 36 groups of *A. squamipinnis*. Solid, oblique lines connect groups containing the same number of males. The horizontal line defines the hypothetical sex ratio threshold of eight females per male, estimated by the sex ratio of the largest group in the sample. The thin, curving lines are regression curves fitted to the maxima and minima of the solid, oblique lines. (From Shapiro and Lubbock, 1980.)

feature of behavior. If, for example, a critical increase in the percentage of rushes given induced sex reversal, the entry of new adult females into a group might stimulate the group's dominant female to direct a large number of rushes to them, and/or might draw attention of the male away from the dominant female so that the rate at which he rushed the dominant female declined. In either circumstance the dominant female's percentage of rushes given would rise. When sufficient new females had entered that the critical increase in the percentage of rushes given were achieved, the dominant female would reverse sex. Unlike the case of male removal in which the percentage of rushes given changes immediately, the arrival of new females would produce the same change gradually. Under this hypothesis, then, sex reversal would be controlled by a single behavioral factor but could be triggered either by male removal or by increasing the number of females in the group to or beyond a threshold sex ratio.

B.   CONTINUANCE AND COMPLETION OF SEX REVERSAL

Sex reversal, as a process, extends over a more or less prolonged period of time. Once initiated it follows a regular sequence of events (Chan *et al.,* 1975; Shapiro, 1977b) and finally reaches completion. While it is conceivable that the entire sequence of behavioral, morphological, and physiological changes that constitute sex reversal could be initiated, in all-or-none fashion, by a single set of eliciting stimuli, it is more likely that continuing series of events play casual roles in the completion of the sequence:

1.   Ongoing hormonal and genetic influences certainly direct, at least in part, the development of secondary sex characteristics.

2.   Social factors influence the timing of alterations of external coloration and gonadal structure, as was suggested for *A. squamipinnis* by the data on coloration. Some of these factors are likely to be important after sex reversal has been initiated. For instance, in *L. dimidiatus,* in eight cases a dominant female reversed sex when isolated from all members of its group. However, the gonadal activity level of the sex-changing fish took up to eight times longer to reach the male range than it would have if the dominant female were left with smaller females still in the group (Robertson, 1974). A similar finding was reported for two *Amphiprion* species (Fricke and Fricke, 1977).

3.   In the discussion of the initiation of sex reversal an implicit distinction was made between behavioral changes, such as in the percentage of rushes given, which seem to initiate sex reversal and other behavioral changes requiring 3–4 weeks to be completed. Social and hormonal events occurring after initiation are likely to mediate between the two. Of the many factors that may affect the continuance and completion of sex reversal, only two will be discussed: hormonal and genetic influences.

1. Endocrinology of Sex Reversal

A variety of workers have attempted to induce sex reversal in individuals of normally hermaphroditic species by treating them with estrogens or androgens. In some cases a single treatment was given; in others treatment was repeated. In most cases no distinction was made between treatment initiating sex reversal and treatment necessary for its completion. The results of these studies have not all been consistent, but most evidence suggests that females of protogynous species can be induced to change sex by injections of androgen. This has been found unequivocally in *Spicara maena* and *Spondyliosoma canthurus* (Reinboth, 1962a), in *Thalassoma bifasciatum* (Reinboth, 1972, 1975b; Roede, 1972; Stoll, 1955), and in *Epinephelus tauvina* (Chen *et al.*, 1977). Androgens have un-equivocally not produced or enhanced sex reversal in female *Monopterus albus* (Tang *et al.*, 1974a). The color and gonadal structure of some female *Coris julis* completely changed following androgen administration (Reinboth, 1962b), while in other females only the color changed, their gonads showing degeneration of oocytes, which regenerated after seven months (Reinboth, 1962a, b). Two female *Halichoeres poecilopterus* changed sex fully after receiving testosterone (Okada, 1964), but others changed only in coloration (Okada, 1962, 1964). After androgen injections, females changed sex gonadally but not in external coloration in *Halichoeres bivittatus* (Reinboth, 1975b; Roede, 1972) and *H. garnoti* (Roede, 1972). Testosterone added to aquaria water induced spermatogenesis, oocyte degeneration, and external changes in female *A. squamipinnis* (Fishelson, 1975).

In several protogynous species, estrogen has had a degenerating effect on the spermatocytes and spermatogonia of males (Okada, 1964; Reinboth, 1970). Estrogen suppressed gonadal development generally in sexually undifferentiated juveniles of the protandric *Sparus auratus* (Reinboth, 1962a). In the simultaneous hermaphrodites *Serranus cabrilla, S. scriba,* and *S. hepatus,* neither estrogen nor testosterone had a specific masculinizing or feminizing effect on the gonads (Reinboth, 1962a).

Even where exogenous hormones have a clear effect on the gonads, the conclusion does not immediately follow that hormones initiate sex change normally (see Okada, 1964; Reinboth, 1970, 1972). While several workers have reported briefly on the gonadal steroid metabolism of hermaphroditic species (Columbo *et al.*, 1972; Idler *et al.*, 1976; Lupo di Prisco and Chieffi, 1965; Reinboth, 1974, 1975b; Reinboth *et al.*, 1966), the most illuminating work on the hormonal processes of sex reversal concerns the freshwater, rice-field eel, *Monopterus albus.* In this species, as female sex reversal progressed, (1) ovarian tissue degenerated, spermatogenic tissue proliferated, and there was a gradual development of interstitial Leydig cells (Chan and Phillips, 1967a, b; Chan *et al.*, 1967, 1972b); (2) increasing amounts of androgen were synthesized by the testicular interstitium (Chan and Phillips, 1967c, 1969; Tang *et al.*, 1974c 1975);

(3) circulating testosterone levels increased, reaching a peak midway through sex reversal; and (4) circulating estradiol-17β levels decreased (Chan *et al.,* 1975). In spite of the large output of androgen during sex reversal, a variety of androgens injected into females at different dosages failed to elicit sex reversal (Chan *et al.,* 1972a), even with simultaneous cyanoketone blockage of estrogen synthesis (Tang *et al.,* 1974a). Mammalian prolactin, however, initiated sex reversals in some specimens (Chan, 1977) and mammalian LH stimulated injected females to change sex fully (Chan *et al.,* 1975; Tang *et al.,* 1974b). As measured by biological assays, pituitary extracts of female *M. albus* had 7–8 times as much National Institutes of Health (NIH)–luteinizing hormone (LH) activity as the extracts of males or individuals midway through sex reversal (Chan, 1975). It is not yet clear whether the high pituitary content of LH in females represents storage of LH with little release or a high level both of synthesis and release. While this work clearly establishes the occurrence of hormonal changes throughout the course of sex reversal, Chan cautions against interpreting these findings to mean that sex reversal necessarily is internally initiated. Presumably a large LH increase from the pituitary must be triggered by something and the initiating factor could lie in the social or nonsocial external environment (Chan, 1975). This warning is equally applicable to a consideration of the possible hormonal basis for the completion of sex reversal.

2. *Genetics of Sex Determination*

Knowledge of the genetics of normally hermaphroditic fish species is practically nonexistent. The only work of which I am aware concerns the self-fertilizing, synchronous hermaphrodite *Rivulus marmoratus* (Harrington, 1961, 1963; Harrington and Rivas, 1958). In aquaria 27 uniparental generations were produced by individuals kept in lifelong isolation from the egg stage (Harrington, 1975; Kallman and Harrington, 1964). Tissue grafts between parent and offspring, and between siblings, were histocompatible, a result interpreted as evidence for homozygosity (Harrington and Kallman, 1968). Although Harrington never found males in natural populations in Florida, adult hermaphrodites in captivity tended to lose the ovarian portion of the gonad with increasing age. The timing of the change to the male form was influenced by seasonal changes in day length (Harrington, 1971) and by the ambient water temperature at critical stages of early development (Harrington, 1967, 1968). Differences between individuals in the timing of the change to the male form were thought to depend on differences in genotype (Harrington, 1971, 1975).

The genetics of sex determination in fish generally have been thoroughly studied only in a few species of freshwater Cyprinodont (see reviews by Atz, 1964; Beatty, 1964; Dodd, 1960; Mittwoch, 1967, pp. 89–100, 1975; Yamamoto, 1969). These species are usually either male heterogametic (for example, *Oryzias latipes* and *Poecilia reticulata*) or female heterogametic (for example, *Xiphophorus maculatus*), but populations of *X. maculatus* could be either (Yamamoto, 1969, p. 134). In several normally male heterogametic species, unusual heterogametic females have been found. Successive breeding of these individuals with normal, XY males and their offspring have produced XX males, YY males, and XY females. In several species, treatment of individuals of known sexual genotype with heterologous sex hormones prior to sexual differentiation has induced the individuals to develop as the opposite sex (Yamamoto, 1953, 1958, 1969). This "genetic sex reversal" should be clearly distinguished from the adult sex change of normally protogynous and protandrous hermaphrodites.

In summary, then, it is generally held that sex determination in fishes occurs on the genic rather than the chromosomal level, and that the chromosomes bearing these genes need not to be different from one another or from any pair of autosomes (e.g., Chan, 1970; Mittwoch, 1967; Yamamoto, 1969). The genetic influence on sex determination in successive hermaphrodites remains an open question.

Two technical barriers have impeded genetic research on sex determination in normally hermaphroditic fish: the difficulty of maintaining breeding stocks of tropical marine fish in the laboratory and the lack of knowledge of the nutritional requirements of young fry necessary for successful rearing of marine fish in captivity. Recent advances in both areas make such work currently feasible (Houde, 1973, 1974, 1975; Houde and Palko, 1970; Houde *et al.*, 1976; Spotte, 1970). Furthermore, the possibility that histocompatibility-Y (H–Y) antigen directs differentiation of the gonad of the heterogametic sex (Silvers and Wachtel, 1977; Wachtel, 1977) makes available a new approach to the genetics of sex reversal which is not dependent on breeding fish in captivity (P. Pechan, personal communication). Histocompatibility-Y antigen is a cell-surface component found on the cells of males of mammalian species (Wachtel *et al.*, 1973), including man (Wachtel *et al.*, 1974, 1975b), and is responsible for the rejection of male tissue transplanted to females in isogenic strains (Wachtel, 1977). A variety of mammalian genetic sex anomalies—including XX mice, which are phenotypic males (Bennett *et al.*, 1977); XY wood lemmings, which are phenotypic females (Wachtel *et al.*, 1976b), bovine freemartins (Ohno *et al.*, 1976), mouse (Bennett *et al.*, 1975) and human (Koo *et al.*, 1977) testicular feminization syndromes, and human XYY males (Wachtel *et al.*, 1975b) and XX males and XX true hermaphrodites (Wachtel *et al.*, 1976a)—have been explained by demonstrating that, regardless of sexual genotype, gonadal tissue has differentiated into testes when H-Y antigen is present and into ovaries when H-Y antigen is absent. Histocompatibility-Y antigen has also been found in all nonmammalian species thus far tested, that is, two species of frog and the white leghorn chicken (Wachtel *et al.*, 1975a). In the South African clawed frog, *Xenopus laevis,* and the white leghorn chicken, *Gallus domesticus,* the female is the heterogametic sex and it is the cells of females that display H-Y antigen.

Thus, the prevailing hypothesis is that H-Y antigen is a chromosomal product that adheres to the cell membrane of most cells and induces embryologic gonadal cell clusters to differentiate as the heterogametic sex. To date, H-Y antigen in fish has been looked for and found only in the heterogametic males of *Oryzias latipes* (Wachtel, personal communication), but rejection of male-to-female tissue grafts in the 32nd generation of brother sister inbreeding in the platyfish has also been demonstrated (Miller, 1962). These results suggest that H-Y antigen may be found in teleosts generally. If H-Y antigen directs gonadal differentiation in successively hermaphroditic fish, as it appears to do in mammals (Wachtel *et al.*, 1975c), then H-Y antigen might first appear in a protogynous individual at or immediately after the initiation of sex reversal and may be responsible for the completion of gonadal reorganization. A search for H-Y antigen in these fishes may thus prove rewarding (P. Pechan, personal communication).

IV. Sex Reversal and Group Structure

The range of group composition and spatiobehavioral structures available to a species may be restricted by the social control of sex reversal to certain specific combinations. Particular forms of group organization may not be compatible with particular methods of controlling sex reversal. For example, suppose that a female were induced to change sex whenever a male ceased interacting with her during 6 consecutive daylight hours. Then it would not be possible to have a stable, one-male group consisting of two, spatially separated subgroups in which the male resided in one subgroup for the first 8 hr of each day and in the second subgroup for the remaining 4 daylight hours. With the male dividing his time in such a manner, one female in the second subgroup would change sex and the group would no longer be a one-male group. Similarly, if a female relied on the continual visual presence of a male to determine whether to change sex or not, then a single-male harem would be incompatible with a group structure in which the male and the females inhabited opposite sides of an opaque coral knoll, or in which the male moved so far away from the females that he could no longer be seen, or in which there were so many females separating the male on one side of the group from the females on the opposite side of the group that not all females could see the male. Careful scrutiny of the composition and spatial structure of social groups in conjunction with a causal study of the initiation of sex reversal may thus be expected to reveal principles by which the form of a particular group structure is maintained over time within constraining bounds. While we are currently unable to specify these principles in detail, it is nevertheless worth examining the structure of social groups of hermaphroditic species with this view in mind.

A. Group Composition and Spatial Structure

Protandric *Amphiprion bicinctus* form monogamous pairs, each of which occupies a territory in a sea anemone (Fricke, 1975). Both members of the pair defend their territory against intruders (Fricke, 1973). When a pair is split and the male left alone in its anemone, the male changes into a female. When two males are paired, one changes sex and the two fish form a mated pair. When two females are placed together, one female kills or wounds the other (Fricke and Fricke, 1977). The lack of plasticity of the form and size of the social group in this species thus appears to be due primarily to the aggressive and territorial behavior of individuals. Since placing two males or a male and a female together will both result in a bisexual, mated pair, the social control of sex reversal in this species appears to introduce plasticity to the social system rather than to restrict it.

Another protandric species, *Amphiprion akallopisos*, shows greater variability in the size of groups. An adult male and female pair may be joined by up to eight juveniles (Fricke and Fricke, 1977). In this species, as in *Dascyllus aruanus* (Fricke, 1974), the size of the group and the spatial distribution of individuals within it are probably restricted both by the size of the occupied coelenterate host (Sale, 1972a) and by intragroup aggressive behavior (Sale, 1972b), which may also be important in the initiation of sex reversal (Fricke and Fricke, 1977). The restriction of group composition to two adults in *A. akallopisos* appears related to (1) the tendency of the adults to exclude intruding adults, and (2) the failure of juvenile group members to grow (Allen, 1972) or to mature sexually (Fricke and Fricke, 1977) as long as both adults are present, and/or to their failure to remain in the group after reaching adulthood.

Unlike that of *Amphiprion* or *D. aruanus*, the territory of the cleaner wrasse, *Labroides dimidiatus*, is not restricted to an area occupied by a host coelenterate. A single male defends a 25–1100-m² territory containing the much smaller territories and cleaning stations of 5–16 females (Robertson, 1974). The male frequently swims to all parts of his territory and interacts successively with each female in the harem. Territorial defense by the male effectively eliminates the possibility of multimale groups. Since the precise interactional factors controlling sex reversal are not known in this species, it is not yet clear how spatial and behavioral subgroups are prevented from forming. However, some restrictions are clearly implied by the tendency for a female to change sex and establish its own harem whenever the male concentrates its activity in areas away from that female or substantially reduces its frequency of visiting the female in her area (Robertson, 1972). It would appear, then, that the only variability in composition or spatial structure available to these groups is in the number of females present within the harem, and this number is restricted by the ability of the male to match the behavioral requirements controlling sex reversal.

Several species of protogynous labrid (Robertson and Choat, 1974; Warner and Robertson, 1978) and scarid (Choat and Robertson, 1975; Robertson and Warner, 1978) are said to form permanent harems resembling those of *L. dimidiatus*. Others form temporary feeding aggregations of ten to hundreds of individuals, with some individuals defending territories while spawning. *Scarus croicensis* is a protogynous, diandric species in which small primary males and all females have a striped coloration pattern. Large individuals have gaudy, terminal phase coloration and may be primary males that have changed color or secondary males (that is, females that have changed both color and sex). Populations of this parrotfish consist of three categories of social grouping: foraging groups, nonterritorial stationary groups, and small territorial groups. One population of 1771 adults and 1472 juveniles contained two foraging groups and multiple territorial and stationary groups. One foraging group ranged in size from 53 to 275 individuals over 9 observation days (Ogden and Buckman, 1973). The color composition of this group ranged from 8 to 25 striped fish for each terminal phase fish.

With data provided by Ogden and Buckman (1973) the color composition of the entire population can be estimated and compared with the color composition of this foraging group. The median values for the number of striped phase fish migrating off the reef each night along the only two migratory routes used by the adults of the population were summed and divided by the sum of the median values for the number of terminal phase fish counted migrating along the same routes during 2-hr intervals on different days (Ogden and Buckman, 1973, Table 6). The resulting ratio of striped to terminal phase individuals is 4.3. This ratio is significantly lower (χ^2 test, $p < 0.01$, two-tailed) than the corresponding ratio for the foraging group on seven of the eight days for which data are available (Ogden and Buckman, 1973, Table 5). In spite of its variability in size and color composition the foraging group is thus not a simple, randomly selected subdivision of the entire population but represents a uniquely composed social grouping.

Territorial groups contained one terminal phase male and 1–3 striped females occupying an area of 10–12 m² in depths of 1–2 m (Buckman and Ogden, 1973). Thirty-three territorial groups contained 68 striped individuals and 33 terminal phase males, an overall ratio of 2.1 striped to terminal phase fish. This value is significantly lower ($\chi^2 = 11.49$, $p < 0.001$, two-tailed) than the color ratio provided above for the population as a whole. Territorial groups, then, also represent aggregations of nonrandomly selected individuals. Comparable data are not available for stationary groups.

Having examined the size, sex, and color composition of the three group types of *S. croicensis*, Warner and Downs (1977) hypothesized that each group type is important to individuals at particular stages in their life history. Small females would begin their mature life in stationary groups and in shallow foraging groups. At a larger size they would enter territorial groups where they spawn.

When a female fish changed sex and color, it would return to a stationary or a foraging group and eventually, when very large, would find its way back to a territorial group, where it would remain. Primary males would follow a similar course, moving from stationary and foraging groups to territorial groups after they changed to terminal phase coloration. The size at which individuals, change color compared with the size of the members of each type of group suggests that color and sex change occur primarily within territorial or shallow foraging groups.

In the protogynous diandric wrasse, *Halichoeres maculipinna*, several foraging groups, containing initial phase coloration males and females, are included in the territory of each terminal phase male. This male ". . . regularly patrols his entire territory, moving rapidly from aggregate to aggregate and feeding briefly within each" (Thresher, personal communication). Both in this species and in *Scarus croicensis* the means for initiating sex reversal and color change remain completely unknown.

Groups of *Anthias squamipinnis* show greater variability of size, composition, and spatial structure than any of the monogamous or harem-like, territorial species described above, but show a considerably greater degree of intragroup organization than has been reported thus far for the other labrids and scarids. In one population intensively studied in the Sudanese Red Sea 45 groups varied in size from 1 to 370 fish, with a median group size of 31 fish (Shapiro, 1977a, b). Groups contained as many as 35 males and 287 adult females, with a median of 2 males and 21 adult females. The water volume occupied by a group was directly related to group size. As a result, fish density was relatively constant for all groups, that is, 41 of the 45 groups had a fish density less than 4 fish per cubic meter. Since sex reversal can be initiated in this species by the removal of a male, it was expected that all groups would contain a similar proportion of females to males. Contrary to expectation, the adult female-to-male sex ratio ranged between 2 and 41 females per male. Small groups showed greater variation in the sex ratio than large groups. In groups of 1–50 individuals the highest sex ratio was 36. In larger groups, the sex ratios did not exceed 17. Since females change sex in response to male removal, it is unlikely that the number of females in a group is independent of the number of males. These variations in sex ratio therefore cannot be satisfactorily explained as a result of chance sampling errors from a binomial distribution, but can be explained by the sex ratio threshold model of sex reversal (see Section III,A,2,b).

Of the 45 groups surveyed, 22 consisted of a single, spatially homogeneous aggregation of fish, that is, they were "unitary" groups. The remaining 23 groups were subdivided spatially into two or three subgroups and were called "bipartite" and "tripartite," respectively. If more than one fish moved between neighboring aggregations within a 15-min observation period, the aggregations were considered to be subgroups belonging to the same group. Otherwise, the

TABLE VII

THE SIZE OF UNITARY, BIPARTITE, AND TRIPARTITE GROUPS
OF *A. squamipinnis*[a]

	Unitary	Bipartite	Tripartite
Median	19.5	33	68
Range	1–370	14–199	38–156
N	22	16	7

[a] From Shapiro (1980a).

aggregations were taken to be separate groups. By this criterion, 16 groups were bipartite and 7 were tripartite.

The frequency distribution of unitary, bipartite, and tripartite groups, the spatial relation of groups to the substrate, the size of subgroups, the distance between subgroups of the same group, and the total linear extent of a group (the sum of the three linear dimensions of all subgroups within a group plus the linear distance separating those subgroups) each varied regularly with increasing group size. For example, unitary groups were smaller than bipartite groups which were smaller than tripartite groups (Table VII; Kruskal–Wallis one-way analysis of variance, $p < 0.01$; Median test or Mann–Whitney U test, $p < 0.01$). The only measure that did not vary with group size was the fish density of groups and of subgroups. In addition, the largest of the three subgroups of tripartite groups resembled the larger of the two subgroups of bipartite groups in several ways. Both showed significantly higher sex ratios than the smaller subgroups of their respective groups. Both showed a common spatial relationship to the coral substrate that was identical with that of unitary groups but different from that of the smaller subgroups in their respective groups. Similarly, the smaller subgroups of bipartite groups resembled the smaller subgroups of tripartite groups (Shapiro, 1977a, b).

These regularities in the data suggested that groups undergo a common developmental process initiated by increases in the size of each group (larval recruitment tends to increase group size; see Fishelson, 1975; Fourmanoir, 1976; Shapiro, 1977b; Suzuki *et al.,* 1974). Under this hypothesis, when a group has grown to a size of about 33 fish, it would tend to divide into two subgroups, a small subgroup containing 1–2 males and 5–10 females, which would depart and permanently establish itself several meters away from the original group, and a larger subgroup (the original group) with about 12 females per male, which would remain at the same location and retain its original spatial relation to the substrate. As the size of the group increased, each subgroup would expand its occupied water volume, thereby (1) increasing the total linear extent of the group, (2) maintaining a constant fish density within each subgroup and within the group as a whole, and (3) reducing the distance between the peripheral boundaries of the subgroups. The data showed that all three of these measures did change with an increase in a group size in the expected manner. At a group size of about 85 fish, a third subgroup would break from the larger subgroup of the bipartite group and establish itself at a permanent site several meters from the larger subgroup from which it originated.

Above a group size of about 100 fish the trends relating various measures to an increase in group size tended to reverse themselves. For example, below 100 fish an increase in group size was associated with a decrease in the distance separating the subgroups of tripartite groups. Above 100 fish, an additional increase in group size accompanied an increase in subgroup separation (Fig. 16). Similar reversals affected the size–frequency distribution of unitary, bipartite, and tripartite groups and the total linear extent of groups. These reversals of trends are explained (Shapiro, 1977a) by hypothesizing that at a group size of about 100 fish some subgroups, whose boundaries approach each other as subgroups expand in size, finally make contact. These groups are no longer recognized as being subdivided. The groups that remain subdivided are those whose subgroups were initially very far apart. These subgroups eventually would become new, independent groups of their own. The data, then, suggest a developmental process in two phases. In the first phase, small groups tend to subdivide increasingly as they grow. In the second phase, subgroups tend either to break completely

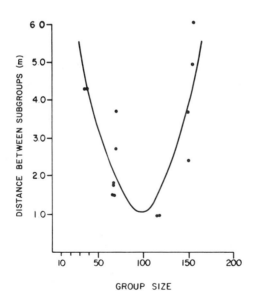

FIG. 16. The subgroup separation of tripartite *A. squamipinnis* groups of different size. $N = 14$. (From Shapiro, 1980e.)

from their group and to become independent groups, or to reunite as their increasing expansion into surrounding water brings them closer to neighboring subgroups. The largest groups, then, would be expected to be unitary, which they were in this study.

Aspects of this hypothesized process of group development have been observed: new juveniles have been seen to swell the size of existing groups (Shapiro, 1977b; Suzuki *et al.*, 1974); individual males with small clusters of females have been seen to depart from large groups (Fishelson, 1975). Alternative hypotheses for these data, such as the hypothesis that the size and spatial structure of groups are determined solely by the quantity and distribution of critical resources on the reef, do not satisfactorily explain the striking coincidence of regular trends between so many spatial and compositional measures and the size of the group. For a detailed discussion of alternative hypotheses, see Shapiro (1977b).

· B. THE SOCIAL UNIT OF SEX REVERSAL

The division of *A. squamipinnis* groups into spatially disparate subgroups represents a complexity of intragroup organization not known in the monogamous pomacentrids or in the single-male groups of labrids and scarids. Before the role of sex reversal in the maintenance of this more complex group structure can be ascertained, additional information is required. How stable is the composition and location of a subgroup? What is the relationship between individual members of different subgroups in the same group? Does the removal of a male from one subgroup affect the composition of other subgroups in the group and/or the relationship between members of separate subgroups? If a male is removed from one subgroup, will the female that subsequently reverses sex derive from the same or a different subgroup? Stated differently, this last question asks whether the social unit for the control of sex reversal is the individual subgroup or the group as a whole. Finally, how might group size affect the relationship between subgroups or the social unit for sex reversal?

These problems have been studied in two subdivided groups in the field (Shapiro, 1977b). The 2 adult males and the 12 adult females of a bipartite group in the Sudanese Red Sea were individually identified. The substrate over which the group lived was divided into three blocks, called Positions 1, 2, and 3. During 951 observation minutes over a 25-day period, the position of each group member was recorded immediately after any individual in the group moved from one position to another. The time of the move was recorded to the nearest minute. When two or more fish were present at the same position for more than 30 or less than 90 sec, they were said to be together at that position for 1 min. The data showed that Male M_0 and Females 4 and 5 spent more than 80% of the total observation time together, that is, at the same position (Table VIII), but

TABLE VIII

PERCENT OF TOTAL OBSERVATION TIME (951 MIN) EACH INDIVIDUAL *A. squamipinnis* SPENDS WITH EACH OTHER INDIVIDUAL[a]

Fish identity	M_0	4	5	8	M_1	2	3	6	7	9	10	11	12	13
M_0		89	80	46	4	2	1	1	1	4	1	1	1	1
4			81	47	4	7	7	7	7	9	7	7	7	7
5				62	16	16	15	16	16	18	16	16	16	16
8					51	53	53	53	53	55	53	53	53	53
M_1						97	96	96	96	94	96	96	96	96
2							99	99	99	97	99	99	99	99
3								100	100	98	100	100	100	100
6									100	98	100	100	100	100
7										98	100	100	100	100
9											98	98	98	98
10												100	100	100
11													100	100
12														100
13														

[a] From Shapiro (1980f).

very little time together with any other fish except Female 8. These three individuals formed the small subgroup. Male M_1 and all remaining individuals except Female 8 were together for 94% or more of the time, but were seldom together with individuals of the small subgroup (see Table VIII). These 10 fish constituted the large subgroup. Female 8 was exceptional. She spend about half of the time together with members of the small subgroup and the other half with members of the large subgroup.

Of the total time each individual of the large subgroup spent together with each other individual of the large subgroup, 99% was spent at Position 1. This position then was the location of the large subgroup. Members of the small subgroup were almost never together at Position 1. Of the total time they were together, more than 89% of the time was at Position 2. Members of the large subgroup were never together at this position. Position 2 was therefore the site of the small subgroup. The composition and spatial structure of each subgroup, as defined by these data, were quite stable throughout the 25 days of the study and varied very little from day to day.

Movement between the two subgroups was asymmetrical. When members of the small subgroup were together with members of the large subgroup, it was almost always at the large subgroup's site and rarely at the small subgroup's site. Thus, although small subgroup members occasionally moved to the large subgroup, large subgroup members rarely entered the small subgroup. At night, all members of the small subgroup returned to Position 1, the site of the large subgroup, where they entered crevices in the rocks and slept. At morning they resumed their station at Position 2. The converse was never observed. In the sense that the large subgroup served as a home base for the small subgroup, the small subgroup was relatively dependent on the other.

At the conclusion of these observations, Male M_1 was removed from the large subgroup at Position 1. If each subgroup functioned as an independent social unit for the control of sex reversal, then the only consequence of this removal should have been that the largest female from within the large subgroup, namely, Female 2, would begin to change sex. Instead, the entire small subgroup returned immediately (that is, within one day) and permanently (that is, did not change for the remaining five weeks of observation) to Position 1. The female that subsequently changed sex was Female 4, a member of the small subgroup and the largest female in the entire group. The social unit for sex reversal was therefore the group as a whole. The two subgroups of this group did not represent, in terms of the control of sex reversal, a division of the group into two, single-male harems functioning independently of one another. As the data from the next field study show, in larger groups whose subgroups are relatively independent of each other behaviorally, each subgroup does function as its own social unit for the control of sex reversal.

In a study of the effect of male removal from one subgroup on the movements of individuals back and forth between subgroups, a tripartite group containing 4 males, 35 adult females, and 12 juveniles was observed 33 times, for 25–35 min each (median 30 min), over a 44-day period on Aldabra Island (Shapiro, 1977b). The males and one or two large females were individually identified in each subgroup. Subgroups 2 and 3 contained one male each, and Subgroup 1 contained two males. The boundaries between subgroups were marked. Each of three observers was assigned a station opposite one subgroup. At a common signal, each observer recorded time zero and the number and identity (where possible) of males and females present within his subgroup. Thereafter, every time one or more individuals entered or left the boundaries of the subgroup the time, to the nearest half-minute, was recorded along with the number of fish entering or leaving and the resulting number of males and females present in the subgroup. The observation period was ended with another common signal.

For each observation period the mean number of fish movements per minute was calculated separately for each subgroup. A movement was defined as the entry into or exit from a subgroup of a single fish. A fish exiting from one subgroup generally entered a neighboring subgroup directly and immediately. Since fish frequently moved in and out of subgroups, the problem arose as to how to estimate the general size of each subgroup. A size–duration histogram was prepared for each period showing the total duration each subgroup spent at each size (that is, containing 8, 9, 10, or 11, etc. individuals). The median value of this distribution was taken to indicate the size of the subgroup for that period. Baseline observations were made for 13 days. The male then was removed from Subgroup 3. Three days later the male from Subgroup 2 was removed. Twenty-two days later the new male in Subgroup 2 (one female changed sex following the first removal) was removed to repeat the male removal performed earlier. The group then was observed for an additional seven days.

Prior to any male removal, Subgroup 3 was the smallest subgroup (Table IX). The removal of the male from Subgroup 3 was expected to diminish the median size of that subgroup by one fish. Instead, the size decreased by two fish (see Table IX). Since neither Subgroup 1 nor Subgroup 2 showed an increase in median size, the extra female lost from Subgroup 3 appears to have left the group entirely. Eighteen days later the largest female in Subgroup 3 showed the first external evidence of sex reversal (an atypical delay; see Section II,B,2). She was an identified female, was known to have spent most of the time within the subgroup, and was not the largest female in the entire group. She completed the coloration changes of her sex reversal seven days after it began. The social unit for sex reversal in this case was the subgroup itself.

The removal of the male from Subgroup 3 did not affect the median size of the other two subgroups. At the end of Period B (in Table IX) the male was removed from Subgroup 2. The median size of this subgroup was expected subsequently to decrease by one, but in fact it decreased by three (see Table IX). At the same time Subgroup 1 increased in size by one fish. Thus, after male removal, two females left Subgroup 2 and one entered Subgroup 1. The other female seemed to have left the group entirely.

On the fourth day after the male was removed from Subgroup 2, one of the identified females known to have spent most of her time in this subgroup showed the first coloration signs of sex reversal. Her sex reversal was subsequently

TABLE IX

MEDIAN SIZE OF THE THREE SUBGROUPS OF *A. squamipinnis* GROUP B DURING FIVE TREATMENT PERIODS[a]

Subgroup	Treatment periods[b]				
	A	B	C	D	E
1	13	13	14	9[c]	13[c]
2	16	16	13[c]	18[c]	13[c]
3	10	8[c]	8	8	7

[a] From Shapiro (1980f).

[b] Treatment periods: A: before male removals. B: after removal of male DL from Subgroup 3. C: after removal of male RC from Subgroup 2. D: after completion of sex change in Subgroup 2. E: after removal of male WD from Subgroup 2.

[c] $p < 0.001$ on median test comparing the starred value with the immediately preceding value.

completed within ten days. In the period following completion (that is, Period D) Subgroup 2 increased in median size by five females. All five came from Subgroup 1 (see Table IX). There was no change in the size of Subgroup 3 at this time.

At the end of Period D the new male in Subgroup 2 was removed. Again Subgroup 2 lost females, this time four, which entered Subgroup 1 (see Table IX). There was also a loss of one female from Subgroup 3. Of the five females that left subgroups, one left the group completely. Four days after this male was removed, a large, identified female, known to have spent most of her time in Subgroup 2, showed the early coloration signs of impending sex reversal. By the end of the study four days later she was well on her way to completion.

Thus, in each case in which a male was removed from either Subgroup 2 or 3, the subgroup itself served as the social unit for sex reversal. After each male removal, one or more females left that subgroup, generally to enter a neighboring subgroup containing a male, but occasionally to leave the group altogether. The fate of these females is not known. Since the three subgroups were linearly arrayed in space with no fish moving directly between Subgroups 1 and 3, the females that left the centrally placed Subgroup 2 had a choice between entering Subgroups 1 or 3. While Subgroup 1 was closer, it also contained more males (two) than Subgroup 3 (none and one on two occasions). The females thus went to the closer subgroup with more males. Following the appearance by sex reversal of a new male, females returned to Subgroup 2. Changes in the partitioning of females between subgroups thus were influenced by changes in the distribution of males.

Before any male removal, individuals moved into and out of Subgroup 1 at a median rate of 0.43 movements per minute (range: 0.17–1.00). In the first three days following the removal of the male from Subgroup 3 (Period B in Table IX), the movement rate of Subgroup 1 increased significantly ($p < 0.002$, Mann–Whitney U test, two-tailed) to 1.81 (range: 1.17–2.00). The maximum daily rate increased further after the removal of the male from Subgroup 2. Following the completion of sex reversal by a fish in Subgroup 2, the movement rate of Subgroup 1 fell significantly ($p < 0.002$) from its high rate following male removals to a low median of 0.20 movements per minute (Shapiro, 1977b). Thus, the appearance of a new male resulted in a return to the low levels of fish movement seen prior to the male removals. After the new male in Subgroup 2 was removed the movement rate of Subgroup 1 increased significantly again ($p < 0.03$).

Since movements into and out of Subgroup 1 primarily involved an interchange with Subgroup 2, and vice versa, the movement rates of both subgroups changed in precisely the same way and to the same extent in response to male removals. These results show that, even when subgroups functioned as their own social unit for sex reversal, the removal of a male from one subgroup affected the members of neighboring subgroups. The removal of a male seems to have had a destabilizing effect on the group as a whole, while the appearance of a new male to replace the one lost had a calming effect, at least in terms of individual movements between subgroups.

Subgroup 3 was clearly distinguished from Subgroups 1 and 2 by its rate of movement. During the baseline period prior to any male removal (Period A in Table IX), the median rate of movement of Subgroup 3 was 0.03 (range: 0.00–0.15) fish per minute, an order of magnitude lower than the comparable rates for Subgroups 1 and 2. So, from the beginning, Subgroup 3 was considerably more isolated and independent than the other subgroups. Indeed, the median movement rate for Subgroup 3 for the first eight days of the study (that is, 0.09 movements per minute or 1.5 movements per 15 min) just barely exceeded the criterion of 1 movement per 15 min adopted in the Red Sea census to distinguish two neighboring independent groups from two subgroups of a bipartite group.

The removal of a male from Subgroup 3 further lowered the movement rate of this subgroup ($p < 0.01$, median test, two-tailed). Thus, in contrast to its effects on Subgroups 1 and 2, the removal of a male from this subgroup resulted in the subgroup turning in on itself, that is, reducing its interchange with Subgroup 2. Furthermore, neither the removal of one male from, nor the appearance of another male in Subgroup 2 affected the movements into and out of Subgroup 3. This result strengthens the impression that Subgroup 3 was more independent than Subgroups 1 and 2, both of which responded vigorously to events in other subgroups.

In spite of the different degrees of independence of Subgroups 1 and 2, and Subgroup 3, each subgroup from which a male was removed served as its own social unit of sex reversal and each retained its daytime position at night. This group contrasts strongly with the smaller, bipartite group studied in the Red Sea, in which one subgroup was relatively dependent on the other, and the group as a whole was the social unit of sex reversal. These contrasting results suggest that: (1) the social unit of sex reversal may be determined by the degree of independence of the subgroups composing a group; (2) the independence of a subgroup may be, in part, a function of the size of the group (that is, it may be a reflection of the group's stage of development), and therefore (3) the size and developmental stage of a group may influence the social unit of sex reversal.

V. Conclusion

Ample field and laboratory evidence has been presented to conclude that sex reversal may be initiated by changes of group composition in protandrous and protogynous species. Details of the manifestations of sex reversal vary with the degree of sexual dimorphism, the population structure, and the social structure of the species, but the events initiating sex reversal are at least superficially similar in all socially controlled species. In the protogynous species, removal of a male from a social group leads to the sex reversal of a large female. In protandrous species, a male changes sex when his mated female is removed.

The dependence of sex reversal on social events influences the composition and structure of social groups viable for these species. While in *Amphiprion bicinctus* sex reversal permits new combinations of individuals to form mated pairs, in species living in more elaborately structured groups, such as *A. squamipinnis,* sex reversal constrains group structure and composition to a finite range, which, in principle, should be fully discoverable.

Field studies on *A. squamipinnis* suggest that groups undergo a regular sequence of spatial and compositional changes as they increase in size. The group tends to subdivide into spatially separate subgroups, which maintain contact with each other by a movement of individuals swimming back and forth between neighboring subgroups. The partitioning of females and the rate of movement between subgroups depend on changes in the partitioning of males. For example, the removal of a male from one subgroup affects the movement rate and the size of the specific subgroup from which the male was removed and of neighboring subgroups as well. Thus, it is not only the sex-reversing female that is affected by male removal; all group members are affected. This result was also found when the behavioral interactions of all group members were studied before and after male removal in controlled laboratory groups. In those studies, the largest female changed behavior dramatically, according to several measures, and began to behave like a male toward other members of the group. At the same time, other group members altered their behavior toward the sex-reversing fish and began to treat it as though it were a male. The behavioral changes shown by the sex-reversing fish were different, in kind or in degree, from the changes displayed by other members of the group. However, the fact that all group members were affected to some extent suggests that sex reversal may be an extreme or unusually dramatic form of behavioral and physiological response to an event common to them all. Some of the changes demonstrated by the members of these fish groups may prove to be similar to the intragroup changes seen in many non-sex-reversing animal species following the disappearance of a conspicuous or highly significant member of a group. The validity of this view can only be ascertained by a comparative and detailed causal analysis of the precise behavioral events preceding and following individual removals in sex-reversing fish and in other group-living animals.

The sex-separated size distributions typical of many protogynous species usually has been explained by postulating that sex reversal occurs whenever an individual attains a critical size. Considering that the hermaphroditic species presently known to change sex under social control also have sex-separated size distributions, it would not be surprising if this aspect of population structure is eventually seen to result from an entirely different cause. The model presented by

Shapiro and Lubbock (1980), based on the assumption of a sex ratio threshold, predicts a sex-separated size distribution without requiring that sex reversal be initiated by the attainment of a critical size. Once it is clear that sex reversal is socially initiated, then it may be seen that male mortality and/or female recruitment will affect the rate of occurrence of sex reversal, and a particular sex-separated size distribution will be produced for each separate population experiencing homogeneous demographic conditions. The prevalence of sex-separated size distributions might then be taken to suggest that the sex reversal of large numbers of protogynous species will prove to be under social control.

The superficial resemblance between *L. dimidiatus, A. squamipinnis,* and the two *Amphiprion* species in the events initiating sex reversal should be regarded with care. It could well be, for example, that in *L. dimidiatus* male removal disinhibits an internally driven sequence of physiological changes, as predicted by the inhibition hypothesis, whereas male removal in *A. squamipinnis* actively stimulates a particular female to change sex, as predicted by the priming hypothesis. The physiological changes underlying sex reversal would be quite different, under these two hypotheses, even though sex reversal was initiated in both cases by male removal.

Distinguishing the specific behavioral feature that initiates sex reversal from other behavioral features characterizing the relationships between group-living individuals is critical: (1) to understand precisely how the behavioral and physiological status of the individual is affected by alterations in the social group, and (2) to specify the constraints that social initiation of sex reversal places on group structure. The long-range hope is that, in the process of resolving these issues, certain basic principles by which individuals relate to each other and to the spatiobehavioral structure of their social group will be revealed.

Acknowledgments

Aspects of this work have been supported by a studentship and a Paton-Taylor Travelling Fellowship from Gonville and Caius College, Cambridge, the Durham Fund, King's College, Cambridge, the Cambridge Philosophical Society, a grant from the British Science Research Council, and in various ways by the University Sub-Department of Animal Behaviour and the Zoology Department of Cambridge University, the Cambridge Coral Starfish Research Group, and the Royal Society. The author is grateful to the following people for help and advice at various stages of this work: the late T. Weis-Fogh, D. A. Parry, H. W. Lissmann, M. J. A. Simpson, R. A. Hinde, P. P. G. Bateson, G. F. Potts, R. Sankey, R. Lubbock, P. Colin, K. Leighton, D. Allen, J. Rodford, M. E. Leighton-Shapiro, and R. Boulon.

References

Allen, G. R. 1972. "Anemone Fishes." T. F. H. Publ. Neptune City, New Jersey.
Allen, J. A. 1959. On the biology of *Pandalus borealis* Kroyer, with reference to a population off the Northumberland Coast. *J. Mar. Biol. Assoc. U.K.* **38,** 189–220.

Altmann, J. 1974. Observational study of behavior sampling methods. *Behavior* **49**, 227–267.

Atz, J. W. 1964. Intersexuality in fishes. *In* "Intersexuality in Vertebrates Including Man" (C. N. Armstrong and A. J. Marshall, eds.), pp. 145–232. Academic Press, New York.

Barlow, G. W., and Ballin, P. J. 1976. Predicting and assessing dominance from size and coloration in the polychromatic midas cichlid. *Anim. Behav.* **24**, 793–813.

Beatty, R. A. 1964. Chromosome deviations and sex in vertebrates. *In* "Intersexuality in Vertebrates Including Man" (C. N. Armstrong and A. J. Marshall, eds.), pp. 15–144. Academic Press, New York.

Bennett, D., Boyse, E. A., Lyon, M. F., Mathieson, B. J., Scheid, M., and Yanagisawa, K. 1975. Expression of H-Y (male) antigen in phenotypically female Tfm/y mice. *Nature (London)* **257**, 236–238.

Bennett, D., Mathieson, B. J., Scheid, M., Yanagisawa, K., Boyse, E. A., Wachtel, S., and Cattanach, B. M. 1977. Serological evidence for H-Y antigen in Sxr, XX sex-reversed phenotypic males. *Nature (London)* **265**, 255–257.

Bohlke, J. E., and Chaplin, C. C. G. 1968. "Fishes of the Bahamas and Adjacent Tropical Waters." Livingston, Wynnewood, Pennsylvania.

Brown, M. E. 1946. The growth of brown trout. II. Growth of two-year old trout at constant temperature 17.5°C. *J. Exp. Biol.* **22**, 130–144.

Brown, M. E. 1957. Experimental studies on growth. *In* "The Physiology of Fishes" (M. E. Brown, ed.), pp. 361–400. Academic Press, New York.

Bruslé, J., and Bruslé, S. 1975. Ovarian and testicular intersexuality in two protogynous Mediterranean groupers, *Epinephelus aeneus* and *Epinephelus guaza*. *In* "Intersexuality in the Animal Kingdom" (R. Reinboth, ed.), pp. 222–227. Springer-Verlag, Berlin and New York.

Bruslé, J., and Bruslé, S. 1976. Contribution a l'etude de la reproduction de deux esèces de Merous (*Epinephelus aeneus* et *Ep. guaza*) des côtes de Tunisie. *Rapp. Comm. Int. Mer Mediterr.* **23**(8), 49–50.

Buckman, N. S., and Ogden, J. C. 1973. Territorial behavior of the striped Parrotfish *Scarus croicensis* Bloch (Scaridae). *Ecology* **54**, 1377–1382.

Bullough, W. S. 1947. Hermaphroditism in the lower vertebrates. *Nature (London)* **160**, 9–11.

Chan, S. T. H. 1970. Natural sex reversal in vertebrates. *Philos. Trans. R. Soc. London* **259**, 59–71.

Chan, S. T. H. 1975. *Endocrinol. Sex Reversal, Semin., Physiol. Dep., Cambridge Univ., Cambridge, Eng.*

Chan, S. T. H. 1977. Spontaneous sex reversal in fishes. *In* "Handbook of Sexology" (J. Money and H. Musaph, eds.), pp. 91–105. Elsevier/North Holland Biomed. Press, Amsterdam.

Chan, S. T. H., and Phillips, J. G. 1967a. The structure of the gonad during natural sex reversal in *Monopterus albus*. *J. Zool.* **151**, 129–141.

Chan, S. T. H., and Phillips, J. G. 1967b. Some aspects of the structure and function of the gonad during the process of natural sex reversal in *Monopterus albus*. *Proc. Asia Oceania Congr. Endocrinol, 3rd,* Manila, Philippines.

Chan, S. T. H., and Phillips, J. G. 1967c. Seasonal changes in the distribution of gonadal lipids and spermatogenic tissue in the male phase of *Monopterus albus* (Pisces: Teleostei). *J. Zool.* **152**, 31–41.

Chan, S. T. H., and Phillips, J. G. 1969. The biosynthesis of steroids by gonads of the rice-field eel *Monopterus albus* at various phases during natural sex reversal. *Gen. Comp. Endocrinol.* **12**, 619–636.

Chan, S. T. H., Wright, A., and Phillips, J. G. 1967. The atretic structures in the gonad of the rice-field eel (*Monopterus albus*) during natural sex reversal. *J. Zool.* **153**, 527–539.

Chan, S. T. H., Tang, F., and Lofts, B. 1972a. The role of sex steroids on natural sex reversal in *Monopterus albus*. *Int. Congr. Endocrinol., 4th, Washington, D.C., Excerpta Med. Found. Int. Congr. Ser.* No. 256, p. 348.

Chan, S. T. H., Wai-sum O, Tang, F., and Lofts, B. 1972b. Biopsy studies on the natural sex reversal in *Monopterus albus* (Pisces: Teleostei). *J. Zool.* **167**, 415–421.

Chan, S. T. H., Wai-sum O, and Hui, W. D. 1975. The gonadal and adenohypophysial functions of natural sex reversal. *In* "Intersexuality in the Animal Kingdom" (R. Reinboth, ed.), pp. 201–221. Springer-Verlag, Berlin and New York.

Chen, F. Y., Chow, M., Chao, T. M., and Lim, R. 1977. Artificial spawning and larval rearing of the grouper, *Epinephelus tauvina* (Forskal) in Singapore. *Singapore J. Pri. Ind.* **5**(1), 1–21.

Choat, J. H. 1969. Studies on the Biology of Labroid Fishes (Labridae and Scaridae) at Heron Island, 1969, Great Barrier Reef. Ph.D. Thesis, Univ. of Queensland, Brisbane, Australia.

Choat, J. H., and Robertson, D. R. 1975. Protogynous hermaphroditism in fishes of the family Scaridae. *In* "Intersexuality in the Animal Kingdom" (R. Reinboth, ed.), pp. 263–283. Springer-Verlag, Berlin and New York.

Clark, E. 1959. Functional hermaphroditism and self-fertilization in a serranid fish. *Science* **129**, 215–216.

Colombo, L., del Conte, E., and Clemenze, P. 1972. Steroid biosynthesis *in vitro* by the gonad of *Sparus auratus* L. (Teleostei) at different stages during natural sex reversal. *Gen. Comp. Endocrinol.* **19**, 26–36.

Dodd, J. M. 1960. Genetic and environmental aspects of sex determination in cold-blooded vertebrates. *Mem. Soc. Endocrinol.* **7**, 17–44.

Fishelson, L. 1970. Protogynous sex reversal in the fish *Anthias squamipinnis* (Teleostei, Anthiidae) regulated by the presence or absence of a male fish. *Nature (London)* **227**, 90–91.

Fishelson, L. 1975. Ecology and physiology of sex reversal in *Anthias squamipinnis* (Peters) (Teleostei: Anthiidae). *In* "Intersexuality in the Animal Kingdom" (R. Reinboth, ed.), pp. 284–294. Springer-Verlag, Berlin and New York.

Fourmanoir, P. 1969. Contenus stomacaux d'Alepisaurus (Poissons) dans le sud-ouest Pacifique. *Cah. O.R.S.T.O.M., Sér. Océanogr.* **7**, 51–60.

Fourmanoir, P. 1971. Liste des espèces de poissons contenus dans les estomacs de thons jaunes, *Thunnus albacores* (Bonnaterre) 1788 et de thons blancs, *Thunnus alalunga* (Bonnaterre) 1788. *Cah. O.R.S.T.O.M., Sér. Océanogr.* **9**, 109–118.

Fourmanoir, P. 1976. Formes post-larvaires et juvéniles de poissons côtiers pris au chalut pélagique dans le sud-ouest Pacifique. *Cah. Pac.* **19**, 47–88.

Fourmanoir, P., and Laboute, P. 1976. "Poissons de Nouvelle Calédonie et des Nouvelles Hébrides." Éditions Pacifique, Papeete, Tahiti.

Fricke, H. W. 1973. Individual partner recognition in fish: Field studies on *Amphiprion bicinctus*. *Naturwissenschaften* **60**, 204–205.

Fricke, H. W. 1974. Öko-ethologie des monogamen Anemonenfisches *Amphiprion bicinctus* (Freiwasseruntersuchung aus dem Roten Meer). *Z. Tierpsychol.* **36**, 429–512.

Fricke, H. W. 1975. Evolution of social systems through site attachment in fish. *Z. Tierpsychol.* **39**, 206–211.

Fricke, H. W., and Fricke, S. 1977. Monogamy and sex change by aggressive dominance in coral reef fish. *Nature (London)* **266**, 830–832.

Ghiselin, M. T. 1969. The evolution of hermaphroditism among animals. *Q. Rev. Biol.* **44**, 189–208.

Ghiselin, M. T. 1974. "The Economy of Nature and the Evolution of Sex." Univ. of California Press, Berkeley.

Goodrich, H. B., and Greene, J. M. 1959. An experimental analysis of the development of a color pattern in the fish *Brachydanio albolineatus* Blyth. *J. Exp. Zool.* **141**, 15–46.

Goodrich, H. B., and Nichols, R. 1931. The development and regeneration of the color pattern in *Brachydanio rerio*. *J. Morphol.* **52**, 513–523.

Goodrich, H. B., Marzullo, C. M., and Bronson, W. R. 1954. An analysis of the formation of color patterns in two freshwater fish. *J. Exp. Zool.* **125**, 487–506.

Gundermann, N. 1972. The Reproductive Cycle and Sex Inversion of *Anthias squamipinnis* (Peters). M. S. Thesis, Dept. Zool., Tel-Aviv Univ., (In Hebrew.)

Hama, T., and Hasegawa, H. 1967. Studies on the chromatophores of *Oryzias latipes* (Teleostean: Fish): behavior of the pteridine, fat, and carotenoid during xanthophore differentiation in the color varieties. *Proc. Jpn. Acad.* **43**, 901-906.

Harrington, R. W., Jr. 1961. Oviparous hermaphroditic fish with internal self-fertilization. *Science* **134**, 1749-1750.

Harrington, R. W., Jr. 1963. Twenty-four hour rhythms of internal self-fertilization and of oviposition by heramphrodites of *Rivulus marmoratus*. *Physiol. Zool.* **36**, 325-341.

Harrington, R. W., Jr. 1967. Environmentally controlled induction of primary male gonochorists from eggs of the self-fertilizing hermaphroditic fish *Rivulus marmoratus* Poey. *Biol. Bull. (Woods Hole, Mass.)* **132**, 174-199.

Harrington, R. W., Jr. 1968. Delimitation of the thermolabile phenocritical period of sex determination and differentiation in the ontogeny of the normally hermaphroditic fish *Rivulus marmoratus* Poey. *Physiol. Zool.* **41**, 447-460.

Harrington, R. W., Jr. 1971. How ecological and genetic factors interact to determine when self-fertilizing hermaphrodites of *Rivulus marmoratus* change into functional secondary males with a reappraisal of the modes of intersexuality among fishes. *Copeia* pp. 389-432.

Harrington, R. W., Jr. 1975. Sex determination and differentiation among the uniparental homozygotes of the hermaphroditic fish *Rivulus marmoratus*. *In* "Intersexuality in the Animal Kingdom" (R. Reinboth, ed.), pp. 249-262. Springer-Verlag, Berlin and New York.

Harrington, R. W., Jr., and Kallman, K. D. 1968. The homozygosity of clones of the self-fertilizing hermaphroditic fish *Rivulus marmoratus* Poey (Cyprinodontidae, Atheriniformes). *Am. Nat.* **102**, 337-343.

Harrington, R. W., Jr., and Rivas, L. R. 1958. The discovery in Florida of the Cyprinodont fish, *Rivulus marmoratus*, with a redescription and ecological notes. *Copeia* pp. 125-130.

Henderson, D. L., and Chiszar, D. A. 1977. Analysis of aggressive behavior in the bluegill sunfish *Lepomis macrochirus* Rafinesque: Effects of sex and size. *Anim. Behav.* **25**, 122-130.

Hinde, R. A. 1974. "Biological Bases of Human Social Behaviour." McGraw-Hill, New York.

Hoar, W. S. 1969. Reproduction. *In* "Fish Physiology" (W. S. Hoar and D. J. Randall, eds.), Vol. 3, pp. 1-72. Academic Press, New York.

Houde, E. D. 1973. Some recent advances and unsolved problems in the culture of marine fish larvae. *Proc. Wildl. Maricult. Soc.* **3**, 83-112.

Houde, E. D. 1974. Effects of temperature and delayed feeding on growth and survival of larvae of three species of subtropical marine fishes. *Mar. Biol.* **26**, 271-285.

Houde, E. D. 1975. Effects of stocking density and food density on survival, growth and yield of laboratory-reared larvae of sea bream *Archosargus rhomboidalis* (L.) (Sparidae). *J. Fish Biol.* **7**, 115-127.

Houde, E. D., and Palko, B. J. 1970. Laboratory rearing of the clupeid fish *Harengula pensacolae* from fertilized eggs. *Mar. Biol.* **5**, 354-358.

Houde, E. D., Berkeley, S. A., Klinovsky, J. J., and Schekter, R. C. 1976. Culture of larvae of the white mullet, *Mugil curema* Valenciennes. *Aquaculture* **8**, 365-370.

Idler, D. R., Reinboth, R., Walsh, J. M., and Truscott, B. 1976. A comparison of 11-hydroxytestosterone and 11-ketotestosterone in blood of ambisexual and gonochoristic teleosts. *Gen. Comp. Endocrinol.* **30**, 517-521.

Kallman, K. D., and Harrington, R. W., Jr. 1964. Evidence for the existence of homozygous clones in the self-fertilizing hermaphroditic teleost *Rivulus marmoratus* (Poey). *Biol. Bull. (Woods Hole, Mass.)* **126**, 101-114.

Kinoshita, Y. 1934. On the differentiation of the male color patterns and the sex ratio in *Halichoeres poecilopterus*. *J. Sci. Hiroshima Univ., Ser. B, Div. 1* **3**, 65-76.

Koo, G. C., Wachtel, S. S., Saenger, P., New, M. I., Dobik, H., Amarose, A. P., Dorus, E., and Ventruto, V. 1977. H-Y antigen: Expression in human subjects with the testicular feminization syndrome. *Science* **196**, 655–656.

Lassig, B. R. 1977. Socioecological strategies adopted by obligate coral dwelling fishes. *Proc. Int. Coral Reef Symp., 3rd, Rosenstiel Sch. Mar. Atmos. Sci., Univ. Miami* pp. 565–570.

Lavenda, N. 1949. Sexual differences and normal protogynous hermaphroditism in the Atlantic sea bass, *Centropristes striatus. Copeia* pp. 185–194.

Leigh, E. G., Jr., Charnov, E. L., and Warner, R. R. 1976. Sex ratio, sex change, and natural selection. *Proc. Natl. Acad. Sci. U.S.A.* **73**, 3656–3660.

Liem, K. F. 1963. Sex reversal as a natural process in the synbranchiform fish *Monopterus albus. Copeia* pp. 303–312.

Liem, K. F. 1968. Geographical and taxonomic variation in the pattern of natural sex reversal in the teleost fish order Synbranchiformes. *J. Zool.* **156**, 225–238.

Liu, C. K. 1944. Rudimentary hermaphroditism in the synbranchoid eel *Monopterus javanensis. Sinensia* **15**, 1–8.

Lodi, E. 1967. Sex reversal of *Cobitis taenia* L. (Osteichthyes, Fam. Cobitidae). *Experientia* **23**, 446–447.

Longhurst, A. R. 1965. The biology of West African polynemid fishes. *J. Cons., Cons. Perm. Int. Explor. Mar* **30**, 58–74.

Lowe-McConnell, R. H. 1975. "Fish Communities in Tropical Freshwaters: Their Distribution, Ecology, and Evolution." Longmans, London.

Lupo di Prisco, C., and Chieffi, G. 1965. Identification of steroid hormones in the gonadal extract of the synchronous hermaphrodite teleost, *Serranus scriba. Gen. Comp. Endocrinol.* **5**, 689–699. (Abstr.)

McErlean, A. J., and Smith, C. L. 1964. The age of sexual succession in the protogynous hermaphrodite, *Mycteroperca microlepis. Trans. Am. Fish. Soc.* **93**(3), 301–302.

Matsumoto, J. 1965. Role of pteridines in the pigmentation of chromatophores in cyprinid fish. *Jpn. J. Zool.* **14**, 45–94.

Mehl, J. A. P. 1973. Ecology, osmoregulation and reproductive biology of the white Steenbras *Lithognathus lithognathus* (Teleostei: Sparidae). *Zool. Afr.* **8**(2), 157–230.

Miller, I. 1962. The Eichwald–Silmer phenomenon in an inbred strain of platyfish. *Transplant Bull.* **30**, 147–149.

Mittwoch, U. 1967. "Sex Chromosomes." Academic Press, New York.

Mittwoch, U. 1975. Chromosomes and sex differentiation. *In* "Intersexuality in the Animal Kingdom" (R. Reinboth, ed.), pp. 438–446. Springer-Verlag, Berlin and New York.

Moe, M. A., Jr. 1969. Biology of the red grouper *Epinephelus morio* (Valenciennes) from the eastern Gulf of Mexico. *Fla. Dep. Nat. Resour. Mar. Res. Lab. Prof. Pap. Ser.* No. 10.

Moyer, J. T., and Nakazono, A. 1978. Population structure, reproductive behavior and protogynous hermaphroditism in the angelfish *Centropyge interruptus* at Miyake-jima, Japan. *Jpn. J. Ichthyol.* **25**(1), 25–39.

Ogden, J. C., and Buckman, N. S. 1973. Movements, foraging groups and diurnal migrations of the striped parrotfish *Scarus croicensis* Bloch (Scaridae). *Ecology* **54**, 589–596.

Ohno, S., Christian, L. C. Wachtel, S. S., and Koo, G. C. 1976. Hormone-like role of H-Y antigen in bovine freemartin gonad. *Nature (London)* **261**, 597–599.

Okada, Y. K. 1962. Sex reversal in the Japanese wrasse *Halichoeres poecilopterus. Proc. Jpn. Acad.* **38**, 508–513.

Okada, Y. K. 1964. Effects of androgen and estrogen on sex reversal in the wrasse, *Halichoeres poecilopterus. Proc. Jpn. Acad.* **40**, 541–544.

Okada, Y. K. 1965a. Sex reversal in the serranid fish *Sacura margaritacea.* I. Sex characters and changes in gonads during reversal. *Proc. Jpn. Acad.* **41**(8), 737–740.

Okada, Y. K. 1965b. Sex reversal in the serranid fish *Sacura margaritacea*. II. Seasonal variations in gonads in relation to sex reversal. *Proc. Jpn. Acad.* **41**(8), 741–745.

Popper, D., and Fishelson, L. 1973. Ecology and behavior of *Anthias squamipinnis* (Peters, 1855) (Anthiidae, Teleostei) in the coral habitat of Eilat (Red Sea). *J. Exp. Zool.* **184**, 409–424.

Quignard, J.-P. 1966. Recherches sur les Labridae (poissons téléostéens perciformes) des côtes européenes: Systématique et biologie. *Nat. monspel. Ser. Zool.* **5**, 247.

Randall, J. E. 1968. "Caribbean Reef Fishes." T. F. H. Publ., Jersey City, New Jersey.

Randall, J. E., and Randall, H. A. 1963. The spawning and early development of the Atlantic parrotfish, *Sparisoma rubripinne*, with notes on other scarid and labrid fishes. *Zoologica (N.Y.)* **48**(2), 49–59.

Reinboth, R. 1957. Sur la sexualité du Téléostéen *Coris julis* (L.) *C. R. Acad. Sci.* **245**, 1662–1665.

Reinboth, R. 1962a. Morphologische und Funktionelle Zweigeschlechtigkeit bei marinen Teleostiern (Serranidae, Sparidae, Centracanthidae, Labridae). *Zool. Jahrb., Abt. Allg. Zool. Physiol. Tiere* **69**, 405–480.

Reinboth, R. 1962b. The effects of testosterone on female *Coris julis* (L.), a wrasse with spontaneous sex inversion. *Gen. Comp. Endocrinol.* **2**, 629. (Abstr.)

Reinboth, R. 1963. Natürlicher Geschlechteswechsel bei *Sacura margaritacea* (Hilgendorf) (Serranidae). *Annot. Zool. Jpn.* **36**, 173.

Reinboth, R. 1964. Inversion du sexe chez *Anthias anthias* (Serranidae). *Vie Milieu, Suppl.* **17**, 499–503.

Reinboth, R. 1967. Biandric teleost species. *Gen. Comp. Endocrinol.* **9**, 486.

Reinboth, R. 1970. Intersexuality in fishes. *Mem. Soc. Endocrinol.* **18**, 515–543.

Reinboth, R. 1972. Hormonal control of the teleost ovary. *Am. Zool.* **12**, 307–324.

Reinboth, R. 1974. Studies on gonadal steroid metabolism in ambisexual fish. *Gen. Comp. Endocrinol.* **22**, 341.

Reinboth, R. (ed.) 1975a. "Intersexuality in the Animal Kingdom." Springer-Verlag, Berlin and New York.

Reinboth, R. 1975b. Spontaneous and hormone-induced sex-inversion in wrasses (Labridae). *Pubbl. Stn. Zool. Napoli* **39**, Suppl., 550–573.

Reinboth, R., Callard, I. P., and Leathem, J. H. 1966. *In vitro* steroid synthesis by the ovaries of the teleost fish, *Centropristes striatus* (L.). *Gen. Comp. Endocrinol.* **7**, 326–328.

Robertson, D. R. 1972. Social control of sex-reversal in a coral reef fish. *Science* **177**, 1007–1009.

Robertson, D. R. 1973. Sex changes under the waves. *New Sci.* **58**(848), 538–540.

Robertson, D. R. 1974. A Study of the Ethology and Reproductive Biology of the Labrid Fish, *Labroides dimidiatus* at Heron Island, Great Barrier Reef. Ph.D. Thesis, Queensland Univ., Brisbane, Australia.

Robertson, D. R., and Choat, J. H. 1974. Protogynous hermaphroditism and social systems in labrid fish. *Proc. Int. Coral Reef Symp., 2nd, Great Barrier Reef Comm., Brisbane, Aust.* pp. 217–225.

Robertson, D. R., and Hoffman, S. G. 1977. The roles of female mate choice and predation in the mating systems of some tropical labroid fishes. *Z. Tierpsychol.* **45**, 298–320.

Robertson, D. R., and Warner, R. R. 1978. Sexual patterns in the labroid fishes of the Western Caribbean: II. The parrotfishes (Scaridae). *Smithson. Contrib. Zool.* **255**, 1–26.

Roede, M. J. 1966. Notes on the labrid fish *Coris julis* (Linnaeus 1758) with emphasis on dicromatism and sex. *Vie Milieu* **17**, 1317–1333.

Roede, M. J. 1972. Color as related to size, sex, and behavior in seven Caribbean labrid fish species (genera *Thalassoma, Halichoeres, Hemipteronotus*). *Stud. Fauna Curaçao Carib. Isl.* **42**, 1–166.

Roede, M. J. 1975. Reversal of sex in several labrid fish species. *Pubbl. Stn. Zool. Napoli* **39**, Suppl., 595–617.

Ross, R. M. 1978. Reproductive behavior of the anemonefish *Amphiprion melanopus* on Guam. *Copeia* pp. 103–107.

Sale, P. F. 1972a. Effect of cover on agonistic behavior of a reef fish: A possible spacing mechanism. *Ecology* **53**, 753–758.

Sale, P. F. 1972b. Influence of corals in the dispersion of the pomacentrid fish, *Dascyllus aruanus*. *Ecology* **53**, 741–744.

Shapiro, D. Y. 1977a. The structure and growth of social groups of the hermaphroditic fish *Anthias squamipinnis* (Peters). *Proc. Int. Coral Reef Symp., 3rd, Rosenstiel Sch. Mar. Atmos. Sci., Univ. Miami* pp. 571–578.

Shapiro, D. Y. 1977b. Social Organization and Sex Reversal of the Coral Reef Fish, *Anthias squamipinnis* (Peters). Ph.D. Thesis, Univ. of Cambridge, Cambridge, England.

Shapiro, D. Y. 1980a. Size, maturation and the social control of sex reversal in the coral reef fish, *Anthias squamipinnis* (Peters). *J. Zool.* (in press).

Shapiro, D. Y. 1980b. The sequence of coloration changes during sex reversal in the tropical marine fish, *Anthias squamipinnis* (Peters). *Bull. Mar. Sci.* (in press).

Shapiro, D. Y. 1980c. The behavioral changes of protogynous sex reversal in a coral reef fish. *Anim. Behav.* (in press).

Shapiro, D. Y. 1980d. Intra-group behavioral changes and the initiation of sex reversal in a coral reef fish. *Anim. Behav.* (in press).

Shapiro, D. Y. 1980e. The composition, spatial structure and development of social groups of a sex-reversing coral reef fish. *Ecology* (in press).

Shapiro, D. Y. 1980f. The stability and independence of sub-groups and the social unit of sex reversal in the coral reef fish, *Anthias squamipinnis* (Peters). *Anim. Behav.* (in press).

Shapiro, D. Y., and Altham, P. M. E. 1978. Testing assumptions of data selection in focal animal sampling. *Behaviour* **67**, 115–133.

Shapiro, D. Y., and Lubbock, R. 1980. Group sex ratio and sex reversal in coral reef fish. *J. Theor. Biol.* (in press).

Silvers, W. K., and Wachtel, S. S. 1977. H-Y antigen: Behavior and function. *Science* **195**, 956–960.

Simpson, M. J. A. 1968. The display of the Siamese fighting fish, *Betta splendens*. *Anim. Behav. Monogr.* **1**, 1–73.

Simpson, M. J. A. 1973. The social grooming of male chimpanzees. *In* "The Comparative Ecology and Behavior of Primates" (J. H. Crook and R. P. Michael, eds.), pp. 411–505. Academic Press, New York.

Slater, P. J. B. 1973. Describing sequences of behavior. *In* "Perspectives in Ethology" (P. P. G. Bateson and P. H. Klopfer, eds.), pp. 131–153. Plenum, New York.

Smith, C. L. 1959. Hermaphroditism in some serranid fishes from Bermuda. *Pap. Mich. Acad. Sci. Arts Lett.* **44**, 111–118.

Smith, C. L. 1965. The patterns of sexuality and classification of serranid fishes. *Am. Mus. Novit.* No. 2207, 1–20.

Smith, C. L. 1967. Contributions to a theory of hermaphroditism. *J. Theor. Biol.* **17**, 76–90.

Smith, C. L. 1975. The evolution of hermaphroditism in fishes. *In* "Intersexuality in the Animal Kingdom" (R. Reinboth, ed.), pp. 295–310. Springer-Verlag, Berlin and New York.

Sordi, M. 1962. Ermafroditismo proteroginico in *Labrus turdus* L. e in *L. merula* L. *Monit. Zool. Ital.* **69**, 69–89.

Spotte, S. 1970. "Fish and Invertebrate Culture: Water Management in Closed Systems." Wiley (Interscience), New York.

Stoll, L. M. 1955. Hormonal control of the sexually dimorphic pigmentation of *Thalassoma bifasciatum*. *Zoologica (N.Y.)* **40**(3), 125–132.

Suzuki, K., Kobayashi, K., Hioki, S., and Sakamoto, T. 1974. Ecological studies of the Anthiine fish *Sacura margaritacea* in Suruga Bay, Japan. *Jpn. J. Icthyol.* **21**, 21-33.

Tang, F., Chan, S. T. H., and Lofts, B. 1974a. Effect of steroid hormones on the process of natural sex reversal in the rice-field eel, *Monopterus albus. Gen. Comp. Endocrinol* **24**, 227-241.

Tang, F., Chan, S. T. H., and Lofts, B. 1974b. Effect of mammalian luteinizing hormone on the natural sex reversal of the rice-field eel, *Monopterus albus. Gen. Comp. Endocrinol.* **24**, 242-248.

Tang, F., Lofts, B., and Chan, S. T. H. 1974c. $\Delta^5 3\beta$-hydroxysteroid dehydrogenase activities in the ovary of the rice-field eel *Monopterus albus* (Zuiew). *Experientia* **30**, 316-317.

Tang, F., Chan, S. T. H., and Lofts, B. 1975. A study of the 3β- and 17β-hydroxysteroid dehydrogenase activities in the gonads of *Monopterus albus* (Pisces: Teleostei) at various sexual phases during natural sex reversal. *J. Zool.* **175**, 571-580.

Wachtel, S. S. 1977. H-Y antigen: Genetics and serology. *Immunol. Rev.* **33**, 33-58.

Wachtel, S. S., Gasser, D. L., and Silvers, W. K. 1973. Male-specific antigen: Modification of potency by the H-2 locus in mice. *Science* **181**, 862-863.

Wachtel, S. S., Koo, G. C., Zuckerman, E. E., Hammerling, V., Scheid, M. P., and Boyse, E. A. 1974. Serological cross-reactivity between H-Y (male) antigens of mouse and man. *Proc. Natl. Acad. Sci. U.S.A.* **71**(4), 1215-1218.

Wachtel, S. S., Koo, G. C., and Boyse, E. A. 1975a. Evolutionary conservation of H-Y ('male') antigen. *Nature (London)* **254**, 270-272.

Wachtel, S. S., Koo, G. C., Breg, W. R., Elias, S., Boyse, E. A., and Miller, O. J. 1975b. Expression of H-Y antigen in human males with two Y chromosomès. *N. Engl. J. Med.* **293**, 1070-1072.

Wachtel, S. S., Ohno, S., Koo, G. C., and Boyse, E. A. 1975c. Possible role for H-Y antigen in the primary determination of sex. *Nature (London)* **257**, 235-236.

Wachtel, S. S., Koo, G. C., Breg, W. R., Thaler, H. T., Dillard, G. M., Rosenthal, J. M., Dosik, H., Gerald, P. S., Saenger, P., New, M., Lieber, E., and Miller, O. J. 1976a. Serologic detection of a Y-linked gene in XX males and XX true hermaphrodites. *N. Engl. J. Med.* **295**, 750-754.

Wachtel, S. S., Koo, G. C., Ohno, S., Gropp, A., Dev, V. G., Tantravahi, R., Miller, D. A., and Miller, O. J. 1976b. H-Y antigen and the origin of XY female wood lemmings (*Myopus schisticolor*). *Nature (London)* **264**, 638-639.

Warner, R. R. 1975. The adaptive significance of sequential hermaphroditism in animals. *Am. Nat.* **109**, 61-82.

Warner, R. R., and Downs, I. F. 1977. Comparative life histories: Growth *vs* reproduction in normal males and sex-changing hermaphrodites of the striped parrotfish, *Scarus croicensis. Proc. Int. Coral Reef Symp., 3rd, Rosenstiel Sch. Mar. Atmos. Sci., Univ. Miami*, pp. 275-282.

Warner, R. R., and Robertson, D. R. 1978. Sexual patterns in the labroid fishes of the Western Caribbean: I. The wrasses (Labridae). *Smithson. Contrib. Zool.* **254**, 1-27.

Warner, R. R., Robertson, D. R., and Leigh, E. G. 1975. Sex change and sexual selection. *Science* **190**, 633-638.

Yamamoto, T. 1953. Artificially induced sex-reversal in genotypic males of the Medaka (*Oryzias latipes*). *J. Exp. Zool.* **123**, 571-594.

Yamamoto, T. 1958. Artificial induction of functional sex reversal in genotypic females of the Medaka (*Oryzias latipes*). *J. Exp. Zool.* **137**, 227-264.

Yamamoto, T. 1969. Sex differentiation. *In* "Fish Physiology" (W. S. Hoar and D. J. Randall, eds.), Vol. 3, pp. 117-175. Academic Press, New York.

Zei, M. 1950. Typical sex reversal in teleosts. *Proc. Zool. Soc. London* **119**, 917-920.

Mammalian Social Odors:
A Critical Review

RICHARD E. BROWN

DEPARTMENT OF PSYCHOLOGY
DALHOUSIE UNIVERSITY
HALIFAX, NOVA SCOTIA, CANADA

Copyright © 1979 by Academic Press, Inc.
All rights of reproduction in any form reserved.
ISBN 0-12-004510-9

I. ODOR CLASSIFICATION

In this review the information contained in the odorous secretions of mammals is examined and a classification system based on behavioral and chemical analyses is provided. This classification divides social odors into two groups: identifier and emotive, and subdivides each of these groups as shown in Table I. Identifier odors are defined as those produced through the regular metabolic processes of the animal, without specific stimulation. The emotive odors are those produced as the result of some transient emotional state or external stimulus. Odors that appear to be merely identifiers may provide the basis for emotional attachments. In addition to their sensory function, the olfactory bulbs are involved in the nonsensory modulation of aggressive and sexual arousal through their limbic system connections (see e.g., Pribram and Kruger, 1954; Alberts, 1974; Cain, 1974).

There have been a number of earlier systems for classifying the odorous secretions of mammals. Bethe (1932) distinguished between *endohormones* or hormones that are secreted within an animal's own body and *ectohormones,*

TABLE I

A CLASSIFICATION OF MAMMALIAN SOCIAL ODORS

Identifier Odors: those produced by the body's normal metabolic processes that are stable for long periods of time

   Individual: odors that are unique to each individual animal

   Colony: odors that are characteristic to all members of a colony, nest, den, deme, or living group; in some cases family odors

   Species typical: odors that are characteristic of all members of a species or subspecies

   Age specific: odors that distinguish animals of different age classes; such as preweanling (infant), postweanling (juvenile), and adult

   Sex specific: odors that distinguish males from females

Emotive odors: those produced or released only in special circumstances

   Rut: odors distinguishing animals in breeding condition from those that are not

   Social status: odors that distinguish subordinate from dominant animals, usually males

   Stress: odors released in response to fear or alarm inducing stimuli; also odors released in frustration situations

   Maternal: odors distinguishing lactating from nonlactating females, and possibly pregnant from nonpregnant females which are secreted from the exterior of the body. The ectohormones were divided into two subgroups, depending on whether their action was intra-specific (homoiohormones) or interspecific (alloiohormones). Karlson and Luscher (1959) coined the term *pheromone* for those substances termed homoiohormones by Bethe, and Brown (1968) coined the term *allomone* for the alloiohormones. Because of its completeness, Kirschenblatt's (1962) terminology is worth noting, but its complexity and the fact that many secretions may not fall exclusively into one category have kept this system from being used to describe mammalian social odors. At present the terms pheromone and allomone are the most widely used (see Eisenberg and Kleiman 1972; Wilson, 1975, p. 321; Thiessen and Rice, 1976).

The present review is concerned with those substances termed pheromones, and suggests that in mammalian research, such secretions should be termed "social odors" until it can be shown that they meet all of the criteria of pheromones. These criteria are as follows: a single chemical compound is released through a gland to the outside of the body, which is received by another animal of the same species and alters a specific behavior or physiological process in the recipient (Karlson and Luscher, 1959). Only the sex-specific odors of mammals appear to meet these criteria. In the majority of cases both the chemical nature of the secretions and the behavioral responses are quite complex (see Bronson, 1971, 1976; Beauchamp *et al.*, 1976).

Odoriferous secretions are produced in the urine, feces, and specialized skin glands. Extensive reviews on the scent-producing organs of different species of mammals are given by Pocock (1910, 1916), Schaffer (1940), Ortmann (1960), Fiedler (1964), Altmann (1969), and Thiessen and Rice (1976). In the following review an attempt is made to integrate data from behavioral, chemical, and microbiological approaches to the analysis of the role of odors in mammalian social communication.

A. BEHAVIORAL RESPONSES

Two primary techniques have been used to demonstrate the ability of mammals to discriminate between conspecifics of various types by their odors alone: preference and reinforcement-training methods. In the preference method, a test animal is presented either simultaneously or successively with an odor from animal A and an odor from a second animal B. A preference for one odor is indicated by a significant difference in the time spent investigating or scent-marking each odor.

A criticism of the preference test method is as follows: because the motivation for preferring one odor to another is intrinsic to the test animal, a preference for one odor indicates an ability to discriminate between the two odors (Irwin, 1958), but a lack of preference does not indicate inability to discriminate. No preference may indicate simply a lack of motivation for choosing between odors or that the two odors are equally attractive. A summary and further critique of preference test methods has been given by Doty (1975).

The reinforcement-training procedures consist of presenting an animal with two odors, A and B, and rewarding the choice of one. A criterion is then set for "learning" the discrimination, and the trials to reach this criterion are recorded. The reinforcement-training procedure introduces a number of extraneous variables such as the nature of the reward, the drive or deprivation level, the apparatus, the training procedure, and the criterion set to indicate learning. It is also necessary to ensure that the reinforcement used is appropriate for the response to be learned (Glickman, 1973). As long as these variables have been carefully controlled, the reinforcement-training procedure can be used to demonstrate that an animal is able to discriminate between two odors. A summary and critique of reinforcement-training procedures used in olfactory research is given by Stevens (1975).

B.   EXPERIENTAL DETERMINANTS OF RESPONSES TO SOCIAL ODORS

Animal responses to social odors range from simple approach and withdrawal to sexual, aggressive, and maternal behavior and may include long-term hormonal and physiological changes. These responses appear to be stereotyped only when they involve physiological changes, and even here experience may alter the response (Lott and Hopwood, 1972). Variables such as age, sex, hormonal state, dominance status, and social and sexual experience of an animal may influence its responses to social odors.

C.   CHEMICAL AND MICROBIOLOGICAL BASES OF MAMMALIAN SOCIAL ODORS

Glandular secretions and urine presumably contain specific chemical compounds that carry the chemical messages underlying the social odors summarized in Table I. Often these chemical secretions take on a characteristic odor only after being metabolized by bacteria that colonize the skin and glandular pouches (Albone et al., 1977). Therefore, a distinctive odor may be caused by a combination of three factors: the internal metabolism and hormonal state, the chemistry of the scent gland, and the action of external microorganisms.

Early research on the chemistry of mammalian scent gland secretions showed numerous species differences. Lederer (1949) lists 40 components found in the anal glands of the Canadian beaver, and Lederer (1950) lists the components of the anal glands of the musk deer, the civet, the musk rat, and the skunk as well as the urine of pregnant mares. Early reports such as these, however, were oriented toward perfumery and not zoology, and contain no bioassay of the identified components.

Chemical analysis of scent-gland secretions involves three principle techniques: gas–liquid chromatography (GLC), thin-layer chromatography (TLC), and mass spectrometry (MS). Chromatographic methods are used to separate samples into bands, each containing different chemical components in the sample. Spectrometric methods record the molecular weights of compounds and thus aid in the determination of their chemical structure. Claesson and Silverstein (1977) have reviewed the chemical methodology used in collecting, isolating, identifying, and synthesizing mammalian social odors.

Once the chemical constituents of a secretion have been determined by GLC and TLC, a behavioral bioassay must be performed to determine the "active component" of the secretion. Before this bioassay is done, however, the responses of animals to whole-gland secretion must be known. The "active" component is the chemical or group of chemicals to which the animal responds as if it were the entire secretion. Problems in the development of bioassay procedures have been discussed by Müller-Schwarze (1977), O'Connel (1977), and Thiessen (1977).

II. INDIVIDUAL ODORS

A. BEHAVIORAL RESPONSES

Does each individual animal have its own "chemical fingerprint" (Epple, 1976) or "olfactory signature" (Wilson, 1975, p. 205), and, if so, can other animals learn to distinguish them by this odor? Animals presented with body odor from different individuals give some evidence that they use odor for individual recognition. The source of the odor may not be known, however. Harrington (1976a) showed that *Lemur rufus* and *Lemur fulvus* could discriminate between the body odor of two conspecifics, and Mertl (1975) found in addition that, for *Lemur catta,* antibrachial (wrist) gland odor could be used for individual recognition. Using general body odor, Bowers and Alexander (1967) trained male and female mice in a Y maze to discriminate between two male mice of the same strain.

In species with several scent glands, some odors are more readily used for individual discrimination than others. Gerbils (*Meriones unguiculatus)* use cage shavings, ventral gland secretions, and urine for individual recognition, but do not use fecal pellets (Dagg and Windsor, 1971; Halpin, 1974; see Fig. 1). The African dwarf mongoose (*Helogale undulata rufula)* can discriminate between two unfamiliar males on the basis of anal gland odor, but not cheek gland, odor (Rasa, 1973). Dalesbred sheep can learn to discriminate between two individuals

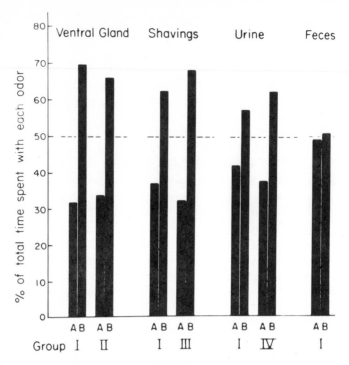

FIG. 1. Times spent by adult male Mongolian gerbils in investigating odors from two male conspecifics, A and B. Group I was tested with ventral gland secretions, cage shavings, urine, and feces. Groups II, III, and IV are replications. (From Halpin, 1974.)

of the same sex using odors from urine, saliva, feces, and wool samples, and the secretions from the infraorbital pouches, the interdigital glands, and the inguinal glands. Differences in odors from urine are easier to discriminate than those from the infraorbital glands, whereas odors from interdigital glands and wool samples are more difficult (Baldwin and Meese, 1977).

Anecdotal reports (Ellis, 1905) suggest that humans possess individual odors and can use these for individual recognition:

> The case has been recorded of a man who with bandaged eyes could recognise his acquaintances, at the distance of several paces, the moment they entered the room. In another case a deaf and blind mute woman in Massachusetts knew all her acquaintances by smell and could sort linen after it came from the wash by odor alone. Governesses have been known to be able, when blindfolded, to recognise the ownership of their pupil's garments by smell (p. 61).

Humans, like other mammals, have a variety of odor sources. The most important of these are: (1) general skin odor, which is a faint, but agreeable, fragrance often detected on the skin even after washing; (2) the hair and scalp; (3) the breath; (4) the armpits; (5) the feet; (6) the perineum; (7) in men, the preputial smegma; (8) in women, the mons veneris, the vulvar smegma, vaginal mucus, and menstrual fluid (Ellis, 1905, p. 62). Kalmus (1955) notes that regional differences in body odors of the armpits, the palms of the hand, and the soles of the feet of an individual, which may smell different to humans, do not prevent dogs from recognizing that they all belong to the same individual. Whether humans are able to recognize this does not seem to have been explored.

Human infants are able to discriminate between their own mother and other lactating mothers by the odor of their breast and/or milk odors (Macfarlane, 1975; Russell, 1976).

Dogs are able to discriminate between the odors of two unrelated people and also between two people of the same family, but have difficulty in discriminating between the odors of identical twins (Kalmus, 1955; see also McCartney, 1968). Human subjects are able to discriminate their own axillary odor from that of other individuals of the same age and sex (Russell, 1976; Hold and Schleidt, 1977). Wallace (1977) found that humans could discriminate between the hand odors of two individuals of the same sex and that the odors from the unrelated people of the same sex were easier to discriminate than odors from siblings or twins.

Animals may use odors to distinguish between familiar and unfamiliar individuals. Male saddle-backed tamarins (*Saguinus fuscicollis*) investigate the mixed circumgenital gland and urine odor of an animal with which they have just completed a social encounter more than the odor of an alien of the same sex (Epple, 1970, 1973). Male and female rats, however, investigate the general body odor of a novel rat of the opposite sex more than the odor of a rat with whom they have just completed a sexual encounter (Carr *et al.*, 1970a). Male gerbils contact unfamiliar males socially (sniffing, boxing, fighting) more than familiar cagemates (Halpin, 1976). The fact that anosmia produced by zinc sulfate intranasal infusion reduces the social behavior they show toward unfamiliar as well as familiar males may be based upon their inability to discriminate between them or, perhaps, a general reduction in social behavior.

The ability of one animal to recognize another by its odor may lead to social interaction, or the avoidance of such interaction. In mammals, many of the introductory acts between animals, when they meet for the first time, are to smell or lick the nose or anal and genital regions (see Schloeth, 1956). It is possible that there are various kinds of behavior performed by mammals to enhance their olfactory identity. For example, self-marking increases the odorous area of an animal's body by which it can be identified, and behavior such as tail wagging, "presenting" the anal–genital area, or penile displays (including erection) may act to enhance the area of odor around the animal (Wickler, 1967).

Male guinea pigs (*Cavia porcellus*), rats, and mice (*Mus*) are able to discriminate between their own odor and that of other males (Beauchamp, 1973; Brown,

1975, Carr *et al.*, 1976). By using their odors as scent marks they can be used for place recognition as well as individual recognition. A well-marked personal space could enable an animal and its family to move swiftly around on known paths, provide a sense of security, or deter conspecific rivals from entering the territory (Martin, 1968). In addition to a dependence on the animal recognizing its own odor as distinct from those of other conspecifics, these functions also suggest some "emotional" attachment to the marked locations. Some of the conditions affecting place marking have been incorporated into the "odor-field" theory proposed by Eisenberg and Kleiman (1972):

> One way in which we can approach the motivation of scent-marking is to consider that for each individual—depending upon his age, sex, mood, and reproductive state—there is an optimum odor field which will provide the optimum level of security. This odor field is composed of a combination of olfactory stimuli emanating from the individual, the environment, the conspecifics. If a disturbance in the odor field occurs (a) through a change in physiological state which alters the sensitivity of the individual and his perception of the optimum odor field, (b) through the introduction of a foreign odor, or (c) through a change in the odor field caused by the dissipation of scent previously deposited, then the individual will attempt to restore the previous balance by the release and deposition of scent. The changed odor field may also arouse the individual, depending upon the nature of the change, and other behavior patterns may be initiated (p. 26).

This odor-field theory may help to explain why animals mark "clean" cages, or objects having "strange" odors. Mongooses, for example, show arousal when they encounter alien odors and cover these with their own scent. An animal's own scent mark does not cause symptoms of disturbance (Rasa, 1973). A male rabbits' own odor on a novel neutral ground stimulates the confidence of both the male and his female partner (Mykytowycz *et al.*, 1976).

Much of the literature on scent marking contains the assumption that this behavior has a "territorial" function (see review by Johnson, 1973), but the results of preference experiments indicate that conspecific odors are usually attractive rather than repulsive. Animals may be deterred only from areas where they perceive the *fresh* mark of a *known* conspecific. Schaller (1967) suggests that scent marks might prevent encounters between conspecifics by relaying different messages depending on the age of the mark:

> Fresh mark = section closed, going on implies the danger of a hostile encounter. Less fresh mark = proceed with caution. Old mark = go ahead. The individual, before passing such a mark, regularly covers it with its own thus "closing the section" (p. 253).

A distinction might be drawn between marking trails, paths, and other locations throughout the home range and marking boundaries around a home range. Extensive boundary marking may occur because, and only because, of frequent encounters between the territory resident and aliens, or their odors at the boundary. Thus, scent marking at boundaries may be a "threat" display, or a "displacement'' activity, arising from competing approach–withdrawal responses at the boundary. The amount of boundary marking may be correlated with population density and the reactions of different animals to these marks may form part of a natural population density regulatory system (Rogers and Beauchamp, 1976b).

B. EFFECTS OF EXPERIENCE

There are many situations in which a preference for one individual over another could be important. Three of these are mother–young relationships, mating or pair-bond relationships, and agonistic encounters.

Domestic goats require as little as 2–5 min of contact with their kids after birth to be able to recognize them (Klopfer *et al.*, 1964). Maternal rabbits are agonistic to alien young, but not to their own (Mykytowycz and Dudzinski, 1966), and rats and gerbils prefer to retrieve their own pups as opposed to aliens (Beach and Jaynes, 1956a, b; Wallace *et al.*, 1973). Similarly, galagos (*Galago cros-sicaudatus*) (Klopfer, 1970) and guinea pigs (Porter *et al.*, 1973) contact and investigate their own young more than other infants of the same age. In all of these studies, however, visual and auditory cues to recognition were confounded with olfactory cues. It is possible that the licking of the young by the mother after parturition establishes an identification based on the similarity of the odors of the young to those of the mother (see Section X).

Species that have temporary or permanent pair bonds as adults may show preferences for the individual odor of their mates. Epple (1974a) states that stable pair bonds form between male and female marmoset monkeys and that this pair bonding may be enhanced by "partner marking." Epple (1974a) further speculates that, "Pregnant female [marmosets] . . . apparently produce odors that attract males and that might strengthen the bond between mates" (p. 269). There is also some indication that gerbil maternal odors or pregnancy odors may bring male gerbils into a paternal state and inhibit pup killing (Elwood, 1977).

Another role for individual odors in mate recognition has been suggested by research on the "Coolidge effect." The Coolidge effect occurs when a sexually exhausted male presented with a novel female becomes rearoused to copulate, but if presented with the female with which he has recently copulated to exhaustion, he shows no interest. Although rats may initially be attracted to a particular sex partner, this partner may be less attractive when compared to a novel partner after the animal is sexually satiated. Male rats with polygamous sexual experience are more attracted to the odor of novel females, whereas monogamous males show no preferences for the odor of either the old or the new female. Polygamous female rats spend more time investigating the novel male's odor than that of their original sex partner (Carr *et al.*, 1970a). If the male is interrupted in the middle of a sexual encounter, however, he will show a preference for the odor of his original sex partner as opposed to that of a novel female (Krames and Mastromatteo, 1973).

Animals losing a fight might benefit from avoiding the individual that won the fight, but would not need to avoid all other conspecifics. Likewise, territorial animals might learn the individual odors of their neighbors. Peters and Mech (1975) have suggested that wolves may acquire an aversion to territories of conspecifics through agonistic encounters between packs. Gerbils that have been defeated in agonistic encounters avoid the nest odors of the animals that defeated them (Nyby *et al.*, 1970). Defeated male gerbils also scent-mark less in the home cage of animals that have defeated them in fights (Yahr, 1977). Defeated male hamsters, however, do not avoid the odor of males that beat them in a fight (Solomon and Glickman, 1977).

If animals learn to avoid the odors of individuals that have defeated them in fights, the parameters for such learning have not been investigated. These parameters include the amount of aggression or "negative reinforcement" received, the type of odor avoided, and the nature of the avoidance response. A defeated male may avoid scent-marking over the odor of the individual that defeated him, but might still show a great deal of investigation of this odor. Rats can readily learn conditioned responses to odors (Nigrosh *et al.*, 1975), and future studies concerned with "social conditioning" and the responses of defeated rats, mice, gerbils, or hamsters to the odors of dominant animals might use a more conventional avoidance conditioning procedure, pairing an odor of a dominant male with different levels of "punishment" by aggressive responses.

C. CHEMICAL AND MICROBIOLOGICAL EVIDENCE

Among individuals of the same species, age, and sex there is considerable variation in the relative concentrations of the chemical components in scent gland samples. Albone *et al.* (1974) found individual differences in the concentrations of acetic, propionic, isobutyric, *n*-butyric, pentanoic, and isocaproic acids in the anal gland sac secretions of red foxes. Analysis of the fatty acid profiles from both the left and right anal glands of individual foxes using GLC, TLC, and MS showed that the secretions from the two glands showed considerable variability over time so that the intraindividual variations were as great as the interindividual variations (Albone and Perry, 1976). Therefore, these authors could find so simple chemical basis for "chemical fingerprints" of individual foxes in the anal gland secretions. Multivariate chemical profiles, which take more of the variability of the secretions into consideration, may indicate the components of individuality in fox anal gland secretion.

Gorman (1976) found individual differences in the fatty acid profiles from the anal sacs of mongooses (*Herpestes auropunctatus*). The acids found in these secretions were: acetic, propionic, isobutyric, *n*-butyric, isovaleric, and *n*-valeric. Mongooses are able to learn to discriminate between the odors of different mixtures of these acids, indicating that *if* individual odors in mongooses result from the relative concentrations of these fatty acids, then differences in these odors could be used for individual recognition (Gorman, 1976).

Two hypotheses have been put forward to explain the origin of differences in individual odors. The chemical hypothesis proposes that each animal produces a chemically unique glandular secretion upon which the same bacteria act to produce a unique combination of carboxylic acids. The fermentation hypothesis proposes that all individuals produce chemically similar secretions and that individual differences result from to the action of different bacteria inhabiting their glands (Gorman, 1976). The latter mechanism is more likely in the mongoose as 13 species of bacteria have been found in the anal glands of these species (Gorman *et al.*, 1974).

In species with more than one scent gland, there may be individual differences in the secretions from some of the glands but not others. This has been found to be the case among rabbits; the anal and chin glands secrete individually specific substances, but the inguinal glands do not (Goodrich and Mykytowycz, 1972).

Future research on individual odors should concentrate on the ability of conspecifics to make behavioral discriminations using the odors of specific glands. Bioassays concentrating on the components of the secretions from these glands then might be used to investigate the chemical and microbiological contributions to individual odors.

III. Colony Odors

A. Behavioral Responses

Barnett (1967) suggested that "in a settled group [of rats], there is a 'colony odour,' which enables a male to distinguish a member of his own family or group from strange rats" (p. 91). Some experimental results tend to support this suggestion. Male rats prefer the odors of familiar male and female cagemates to odors of unknown rats (Krames and Shaw, 1973).

Colony odors are, however, difficult to define because there is no evidence that they are the result of a glandular secretion. The source of a colony odor may result in part from the food or nest material used by a particular group of animals. The colony odor may depend on the social structure of the group, or on the possibility that each animal recognizes each other member of its group by its individual odor. To be designated a colony odor, it must be shown that colony members react to an odor common to all independent of their reactions to the individual odors of the colony members. Carr *et al.* (1976) found that male rats were able to discriminate between the general body odor of two familiar adult male cagemates, but not between those of two familiar juveniles. This study indicates that rats living in small groups may recognize other *adult* group members by their individual "olfactory signatures" rather than by a group or colony odor.

One source of colony odors among flying phalangers may be the individual odor of the dominant male. Dominant males scent-mark the group's common territory, the nest site, and other colony members with their frontal and sternal glands and with urine. If a female is experimentally smeared with the frontal or sternal gland secretion or urine of an alien male, the infants display defense reactions (Schultz-Westrum, 1965, 1969).

In some species of primates, carnivores, and ungulates the colony is an extended family, consisting of one mated pair and their offspring from a number of matings so that the animals within the group are genetically closely related to one another (Bertram, 1976; Wilson, 1975). In these cases, colony odors may be "kinship" odors as well, and recognition of colony members would be equivalent to recognition of close relatives. Porter *et al.*, (1978) have found that 1-month-old spiny mice (*Acomys cahirinus*) prefer to huddle with their siblings rather than alien conspecifics. Zinc sulfate anosmia elimiantes this preference for siblings, suggesting that these mice use odors to identify their littermates.

Adult mammals may mark their young and thus be able to identify them as family members. Lactating female gerbils (Wallace, *et al.*, 1973; Yahr, 1976) and rabbits (Mykytowycz, 1968) scent-mark their pups. Other colony members in addition to the mother also mark the young. In tree shrews, both parents mark the young with their chin glands (Martin, 1968). In carnivorous species such as mongooses, this marking may serve to assure that the young are treated as colony members and not eaten as prey (Rasa, 1973). Female rabbits do not attack their own young and are only slightly aggressive toward juveniles belonging to their own colony (Mykytowycz and Dudzinski, 1972). However, they are aggressive toward alien juveniles and may attack their own young if they have been smeared with urine from unfamiliar rabbits (Mykytowycz, 1968).

Colony odors in humans may reflect ethnic or racial differences in odor. Based on numerous anecdotes, Ellis (1905, pp. 59–61) suggested that different races have different odors. For example, the Nicobarese Indians of Africa were able to distinguish members of six different tribes by their odors. Since human odors arise from bacterial action on sweat it is possible that group odors in humans may arise from a common microflora (Albone *et al.*, 1977).

B. EFFECTS OF EXPERIENCE

The development of colony odor preference has been tested in only one case. Leon (1975) examined the preference for "colony mother odors" of young rats whose mother was on a sucrose diet (thus producing no maternal odor) and who were raised in a colony room with other litters whose mother did produce the maternal odor. The 16-day-old pups showed a preference for the odor of the colony mothers over the odor of nonlactating females. These results suggest that the maternal odor might be colony or clan specific and the development of colony odor preferences may depend on the maternal odors that the young encounter (Leon, 1975). (See Section X.)

C.   CHEMICAL AND MICROBIOLOGICAL EVIDENCE

Stoddart *et al.* (1975) found that the flank secretions of the water vole (*Arvicola terrestris*) differed in chemical pattern between animals from different families (that is, different litters) and different geographical areas. Twenty different chemicals were identified by their GLC peaks, and each peak height was measured in order to establish that the flank gland secretions from 8–12-week-old males and females from four different laboratory-reared litters differed. The analysis of peak heights from 16 different geographical areas also showed significant differences, but these are not easily interpreted because the samples differed with respect to the sex, age, and reproductive state of the odor donors.

While this analysis gives some indication that colony-specific odors may be an extension of litter-specific patterns of chemicals in the glandular secretion, there is no behavioral evidence to support the chemical analysis. Such evidence might show that animals are more attracted to the actual components of gland secretions from their own littermates as opposed to odors from aliens, or that less aggression is shown to aliens that have been rubbed with the active component of a familiar littermate's secretions.

A bacteriological theory for the production of colony-specific odors postulates that a group of animals share a common microbiological population in the skin glands. Through contact these bacteria cross-infect all members of the colony and act on their chemical secretions to produce a characteristic group odor (Albone *et al.*, 1974; Albone and Perry, 1976). Such bacterial transfer could occur through scent marking, common food sources, and communal nesting. One would expect members of the same family to have a homogeneous bacterial population and members of different families to have somewhat different bacterial populations. This extension of Gorman's (1976) fermentation hypothesis has not yet been tested.

IV.   SPECIES-TYPICAL ODORS

A.   BEHAVIORAL RESPONSES

In most species of mammals, individuals have little difficulty learning to discriminate the odors of conspecifics from those of individuals from unrelated species. Bowers and Alexander (1967) found that male mice (*Mus*) could learn to discriminate between the body odors of *Mus* and *Peromyscus maniculatus* males. Male gerbils learned to discriminate between the body odors of male gerbils and male rats, mice, hamsters, and collared lemmings (Dagg and Windsor, 1971). Male marmosets (*Callithrix jacchus*) investigate and scent-mark over the odors of strange marmosets more than the odors of mice, but female marmosets do not (Epple, 1970). Male guinea pigs (*Cavia porcellus*) investigate urine odors from conspecifics more than odors from human urine and urine from a related caviomorph rodent, *Galea musteloides* (Beauchamp, 1973).

Male bank voles (*Clethrionymys* spp.) are attracted to the odors of their own subspecies over those of other vole species (Godfrey, 1958; Rauschert, 1963). Male and female *Peromyscus maniculatus* both show heterosexual preferences for the odors of their own species when tested against *Peromyscus polionotus* individuals, but neither male nor female *P. polionotus* show homospecific odor preferences (Moore, 1965). In this experiment the estrus condition of the female was not controlled: in some tests both female odor donors were in estrus; in other cases only one female was in estrus. *Peromyscus polionotus* males were generally attracted to the odor of the estrous female, regardless of her species.

Lemur fulvus males show no preferences among the odors of *L. fulvus* subspecies, but are able to discriminate *L. fulvus* odors from the odors of other lemur species (*L. catta, Hapelemur griseus, Propithecus verreauxi*; Harrington, 1979; Fig. 2). The odors consisted of general body odors, urine, and feces and were taken from nonbreeding individuals.

Müller-Schwarze and Müller-Schwarze (1975) showed that black-tailed deer (*Odocoileus hemionus columbianus*) sniff and investigate the tarsal gland of their

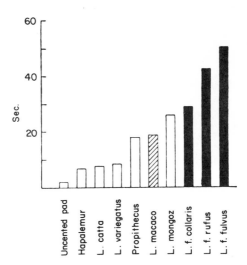

FIG. 2. Mean time investigating the odors of various lemur species by two *L. fulvus* males. Black bars indicate *L. fulvus* subspecies. (From Harrington, 1979.)

own subspecies more than that of the mule deer (*O. h. hemionus*), a closely related subspecies. Similarly, mule deer sniff and investigate the tarsal gland from their own subspecies more than that of black-tailed deer, suggesting that they can use the odor of this gland to discriminate between the two species.

Responses to species-typical odors may be determined by olfactory experience during infancy and adolescence. This is discussed in Section X on maternal odors.

B. CHEMICAL AND MICROBIOLOGICAL EVIDENCE

Chemical and bioassay methods for determining the active components of the ventral gland of the gerbil were reported by Thiessen *et al.* (1974). First, sebum was fractionated by TLC into ten discrete bands, only one of which was odorous. Next gas chromatography followed by mass spectrometry were used to identify the active component in this band as methyl phenylacetate. The ten fractions from the TLC then were divided into four fractions: ''whole'' (all ten scrapings), ''critical'' (three scrapings, including the active component), ''remainder'' (the seven remaining scrapings), and ''control'' (ether and silica gel). Gerbils were first trained in a Skinner box to suppress bar pressing to whole fractions. They then were tested with the other three fractions and showed suppression only when presented with the critical fraction. In a preference test, whole-gland extracts and phenylacetic acid were investigated more than control and remainder fractions.

Using GLC and MS, the principal component of the tarsal gland secretion of the male black-tailed deer was identified as *cis*-4-hydroxydodec-6-enoic acid lactone (Brownlee *et al.*, 1969). This and other related lactones then were sprayed over the tarsal tufts of a male deer, and the tarsal gland sniffing and licking responses of male and female deer recorded. The frequency of responses to the natural acid lactone isolated from the tarsal gland and to a synthetic product were almost the same as those to the entire male scent, whereas six other lactones were investigated no more intensively than the petroleum ether control (Müller-Schwarze, 1969, 1971). Using the same techniques the two active components of the subauricular gland of the male pronghorn deer (*Antilocapra americana*) were identified as 2-methylbutyric acid and isovaleric acid (Müller-Schwarze *et al.*, 1974).

Stoddart (1977) found that there were only slight chemical differences between the flank gland secretions of closely related *Arvicola* subspecies and the caudal glands of closely related *Apodemus* subspecies. There were large differences in chemical components between *Arvicola* and *Apodemus*, however. It seems to be difficult to isolate the specific chemical differences in secretions from very closely related subspecies, and much easier to isolate chemical differences between more unrelated species.

Studies by Albone and his co-workers have identified bacteria in the anal glands of foxes, lions, and dogs that produce the volatile fatty acids responsible for the species-typical odors (Albone, 1977; Albone et al., 1974; Albone and Perry, 1976). Comparative studies of the anal glands of cats, lions, tigers, polecats, minks, and skunks show species differences in their chemical and bacteriological products (Albone et al., 1977; Albone and Gronneberg, 1977; Albone, et al., 1976). Preliminary behavioral studies indicate that the action of bacteria makes fox odors more attractive to other foxes (Albone et al., 1977; Macdonald et al., 1978).

V. AGE-SPECIFIC ODORS

A. BEHAVIORAL RESPONSES

Males of many mammalian species treat juveniles differently from adults. If an alien male rat is introduced into a small colony of rats or a cage containing only one other male, it is often injured or killed, but younger rats are not attacked (Barnett, 1967, p. 83). Adult male mice attack other adult males but not juveniles of either sex (Mackintosh, 1970). Dixon and Mackintosh (1976) found that territorial male mice reduced their attacks toward adult males swabbed with the urine of adult or juvenile females but not toward adult males swabbed with water or urine from juvenile males. Failure to find that the urine of juvenile males reduced attacks on adult males may have occurred because the odor of the intruding adult was not masked by the odor of the juvenile male. The behavior of the introduced males may also have increased the probability of their being attacked, irrespective of their odor. The use of castrated males as test subjects may eliminate these problems (Mugford and Nowell, 1971a, b).

There are two categories of evidence to suggest that the odor of juveniles may differ from that of adults: (1) scent gland secretion in many species does not occur until puberty, and (2) sex-specific physiological primer effects are not stimulated by odors from juveniles. In rabbits the weights of the odor-producing inguinal and anal glands are very low until the age of 65–70 days, when they increase rapidly, reaching maximum weights at about 475 days of age (Mykytowycz, 1966a,b). Hesterman and Mykytowycz (1968; cited by Mykytowycz, 1970) found that, when human subjects were asked to rate from low to high the odor intensity of anal gland secretions from rabbits, ratings were directly related to the age of the rabbits from which the secretions were taken.

Exposing gonadotrophin-injected juvenile female mice to the odors of adult males increases both the probability that ovulation will occur and the number of ova released when compared with gonadotrophin-injected females that are not exposed to males. The odor of an immature male (under 30 days old) does not produce these increases in ovulation and number of ova released (Zarrow et al., 1970).

Age-specific odors may fall into general classes such as infant (preweanling), juvenile (prepubertal), adult, and old age. Differences in odor among such age classes may result from changes in gonadal hormone secretions. Little research has been done on the changes in odor that occur at different ages, or on the role of experience in the differential responses of adults to the odors from animals of different ages.

B. CHEMICAL AND MICROBIOLOGICAL EVIDENCE

There is little research on the chemical changes that occur in glandular secretions of juveniles at puberty. Stoddart (1973) found that the caudal gland secretion of juvenile male *Apodemus* was similar to that of adult females and juvenile females. He suggests that the secretions of juvenile males may show a transformation from female-like (compare Fig. 3d and 3f) to malelike as adulthood approaches (compare Fig. 3a and 3d). Developmental changes may involve the production of new chemical compounds as well as changes in the relative proportions of other chemicals.

VI. SEX-SPECIFIC ODORS

A. BEHAVIORAL RESPONSES

Male and female C57B1 mice can learn to discriminate between the general body odors of male and female mice of the same strain in less than half the trials required for learning a discrimination between two individuals of the same sex (Bowers and Alexander, 1967; Fig. 4). Gerbils can discriminate between male and female gerbils by their body odor and by their urine odor (Dagg and Windsor, 1971). Pigs have been trained to discriminate between urine odors of boars (intact males) and sows (intact females) and, between the urine odors of hogs (castrated males) and gilts (immature females) (Meese *et al.*, 1975).

In preference tests, male guinea pigs are attracted more to female than to male urine odor (Beauchamp, 1973). Both male and female saddle-backed tamarins prefer to investigate and scent-mark over odors from animals of the opposite sex (Epple, 1974a, b, c), but male lemurs (*L. fulvus*) prefer to investigate the odors of male conspecifics over those of females (Harrington, 1974). Both sexes of rats prefer to investigate urine odors from animals of the opposite sex (Brown, 1977).

Territorial male mice attack other adult males but not adult females (Mackintosh, 1970). Adult male mice rubbed with female urine are attacked less and mounted more than males rubbed with water (Dixon and Mackintosh, 1971; Connor, 1972). Conversely, adult males attack females that are swabbed with male urine (Connor, 1972). Urine from castrated males does not elicit aggressive

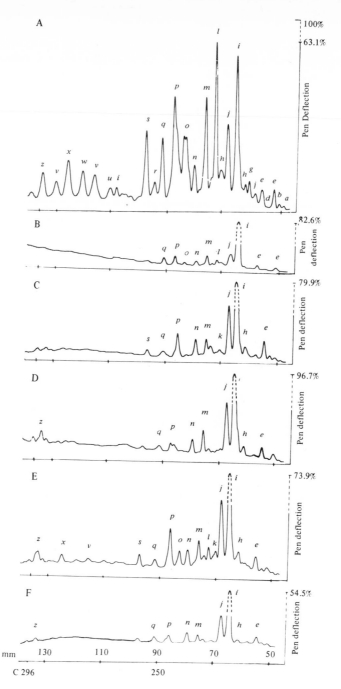

FIG. 3. Chromatograms of caudal glands of *Apodemus flavicollis*. Letters a to z indicate 26 principal components or complexes of components. (A) Adult male caudal gland;

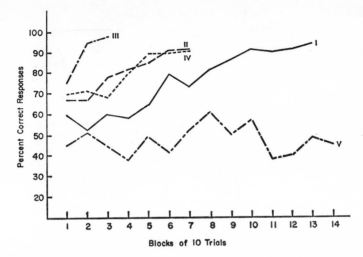

Fig. 4. Acquisition curves for three male and three female *Mus* in the learning of olfactory discrimination between oil of cinnamon and oil of juniper (I); male *Mus* and male *Peromyscus* (II); male and female *Mus* (III); two individual male *Mus* (IV); with no odors (V). (From Bowers and Alexander, 1967.)

responses, nor does urine from spayed females, but urine from testosterone-injected spayed females does elicit attack from males (Mugford and Nowell, 1971a, b). There are indications that the preputial gland secretion rather than urine is the source of the sex-specific odor in mice (Mugford and Nowell, 1971c), but this result has not been supported in all cases (Lee, 1976).

Some authors (Archer, 1968; Mugford and Nowell, 1971a, b) suggest that female odors have aggression-inhibiting properties, whereas male odors have aggression-inducing properties. A communication model would state that the olfactory cues are used as information to distinguish the sex of a conspecific (Bronson, 1976); olfactory signals may be used to gain information that the approaching animal is (1) a conspecific, (2) of the same sex, (3) an adult, and (4) strange. The ensuing aggressive behavior is not "released" by the odor; rather the odor may provide information that leads to an aggressive response (Vandenburgh, 1975).

Ellis (1905, p. 63) noted that human odors may be regarded as secondary sexual characteristics, and this has been borne out by Russell (1976) and Hold and Schleidt (1977), who found that human subjects could perform sex discrimi-

---

(B) adult male nonglandular control sample; (C) pregnant adult female caudal gland; (D) juvenile male caudal gland; (E) juvenile male caudal gland; (F) adult female caudal gland. (From Stoddart, 1973.)

nations based on armpit odors. Discrimination of sex by axillary odors may, however, result from differences in the intensity of these odors rather than actual chemical differences. Doty *et al.* (1978) found that humans were not able to discriminate between males and females by their armpit odors. Subjects tended to identify stronger odors as coming from males, regardless of the sex of the donor. Wallace (1977) has shown that humans can discriminate between males and females by the odor of their hands. Odor intensity was not recorded in this study.

B. PHYSIOLOGICAL RESPONSES

Olfactory stimuli have been shown to produce endocrinological changes in mice in a number of situations and have been inferred in other cases. In all cases these endocrine changes appear to be caused by sex-specific odors (see Table II). Such effects are generally referred to as *primer effects* and often attributed to ''primer'' pheromones. Primer pheromones may affect conspecifics of the same or the opposite sex, although the heterosexual responses have been emphasized more than the homosexual responses.

1. Acceleration of Puberty in Juvenile Males

Prepubertal male mice housed with intact adult females show heavier testis and accessory sex glands than males housed with ovariectomized females (Fox, 1968) or in unisexual groups of juvenile males (Vandenbergh, 1971). Male mice which cohabit with adult females from 20 to 36 days of age show accelerated sexual maturation, but males exposed only to female urine do not (Maruniak *et al.*, 1978). If there is a female odor that accelerates puberty in juvenile males it has not yet been demonstrated.

TABLE II
KNOWN AND SUSPECTED PRIMER PHEROMONE EFFECTS IN MICE AND OTHER MAMMALS

| | Odor donor | |
Odor receiver	Adult female	Adult male
Juvenile male	Accelerate puberty (suspected)	Delay puberty (suspected)
Juvenile female	Delay puberty	Accelerate puberty
Adult female	Cause prolonged anestrus (Whitten effect) or pseudopregnancy (Lee–Boot effect)	Induce estrus and ovulation (Whitten effect)
Adult male	Induce reproductive behavior	Inhibit reproductive behavior (suspected)
Pregnant female	Terminate pregnancy and induce estrus	

2. Delay of Puberty in Juvenile Males

Juvenile male voles (*Microtus arvalis*) housed near crowded cages of adults show inhibited testicular development after 40 days of exposure (Lecyk, 1967). Juvenile male *P. leucopus* show reduced testicular development when exposed continuously to the odors from grouped adult *P. leucopus* of both sexes (Rogers and Beauchamp, 1976a). Prepubertal male mice housed with adult males have significantly lower testis and seminal vesicle weights than males raised alone or with females (Vandenbergh, 1971). Whether this delay of development results from a specific odor has not yet been determined.

3. Delay of Puberty in Juvenile Females

Olfactory cues from adult male and female mice living in dense populations inhibit puberty in juvenile *P. leucopus* females (Rogers and Beauchamp, 1976a). Odors from *Mus* had little effect on the development of *P. leucopus* females; thus, the effect appears to be species specific. Grouped female mice (*Mus*) reach puberty later than isolated females (Vandenbergh, 1973), and bedding soiled by grouped females will inhibit puberty in juvenile females (Drickamer, 1974). The source of the odor that inhibits puberty in juvenile females appears to be the urine of both adult and juvenile females (McIntosh and Drickamer, 1977).

4. Acceleration of Puberty in Juvenile Females

Female mice (*Mus*) that are exposed to either adult males, their soiled bedding, or urine show first estrus and enlarged uteri as many as 20 days before females exposed to adult females, immature males, or castrated males (Bronson and Desjardins, 1974; Drickamer and Murphy, 1978; Lombardi et al., 1976; Vandenbergh, 1967, 1969; Vandenbergh, et al., 1975) (Fig. 5). The presence of an adult male or his urine alters the level of gonadotrophic hormones in juvenile females. Exposing prepubertal female mice to male urine for 48 hr produces an elevated level of luteinizing hormone within 30 min, a gradual increase in prolactin levels, and a decrease in follicle-stimulating hormone (Bronson and Maruniak, 1976).

Acceleration of estrus has also been found in female rats housed with males (Vandenbergh, 1976). There is no direct evidence that this acceleration will occur with male rat odors alone, but Sato et al. (1974) have found that olfactory bulbectomy delays sexual maturation in rats.

5. Inhibition of Estrus in Adult Females

Female mice (*Mus*) housed in groups of 30 or in cages soiled by grouped females show prolonged diestrus (Champlin, 1971; Whitten, 1959), while females housed in groups of 4–6 show an increased frequency of pseudopregnancy (Lee and Boot, 1955, 1956). This suppression of estrus seems to result

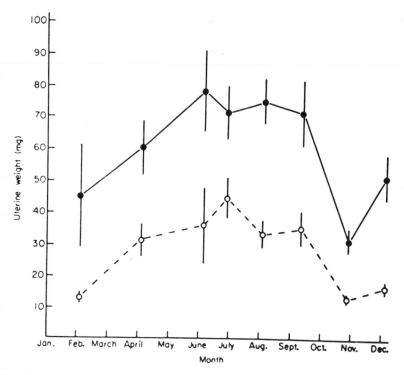

Fig. 5. Seasonal changes in uterine weight of different groups of juvenile mice exposed to male urine (●) or water (o). (From Vandenbergh et al., 1975.)

from an inhibition of gonadotrophic hormone secretion (Whitten and Champlin, 1973). Prolonged anestrus does not occur in grouped females that have had their olfactory bulbs removed (Whitten, 1959).

In groups of female rats the estrus cycles become synchronized, and the odors of grouped females are sufficient to induce this synchrony (McClintock, 1974). All female groups in humans also show synchronous ovulatory cycles. McClintock (1971) found that the menstrual cycles of college girls were synchronized with those of their roommates and closest friends. The role of olfactory stimuli in menstrual synchrony in humans has not yet been determined.

6. Induction of Estrus in Adult Females

Female mice that are anestrous as a result of group housing begin to mate within three days of exposure to adult males, their urine, or soiled bedding (Bronson, 1971; Bronson and Whitten, 1968; Whitten, 1966, 1969; Whitten and Bronson, 1970). Estrus induction by exposure to male urine does not occur in anosmic females (Whitten, 1956; Zarrow et al., 1970).

Johns et al. (1978) have shown that exposure to male urine for as little as 30 min will induce ovulation in females that are acyclic and in constant estrus after being kept in constant light. Lesions of the vomeronasal organ prevent urine-induced ovulation from occurring. This demonstrates that the vomeronasal system rather than the primary olfactory system is involved in the stimulation of gonadotrophin release.

Five-day cyclic female Wistar rats have their cycle reduced to four days following exposure to male urine (Aron and Chateau, 1971). The effect of male urine is to accelerate follicular growth and reduce the duration of diestrus (Chateau *et al.*, 1976). The presence of males also reduces the length of the ovulatory cycle in humans. Female college students who see males three or more times a week show shorter menstrual cycles than females who see males fewer than three times per week (McClintock, 1971). Whether this results from olfactory stimuli is unknown.

Adult males stimulate estrus synchrony in seasonally breeding animals such as domestic sheep (Schinkel, 1954), goats (Shelton, 1960), and pigs (Signoret, 1970a). The male triggers ovulation at the end of the seasonal anestrous period, and this effect does not occur in anosmic females (Morgan *et al.*, 1972).

Among voles (*Microtus*), which are induced ovulators, the presence of a male or his soiled bedding will stimulate ovulation (Milligan, 1974, 1975; Richmond and Conaway, 1969).

7. Induction of Reproductive Behavior in Adult Males

Male rats cohabiting with females have heavier testis and accessory reproductive gland weights and higher testosterone levels than males living in isolation or in unisexual groups (Drori and Folman, 1964; Pranzarone, 1969; Purvis and Haynes, 1972). It is not established that olfactory stimuli alone produce this effect (Folmon and Drori, 1966). Male mice show increased plasma testosterone levels after 30–60 min of cohabitation with a novel female (Macrides *et al.*, 1975), and this effect occurs even when the male is separated from the female by a double wire-mesh screen (Batty, 1978). In the case of male hamsters, exposure to female vaginal secretions alone causes an increase in testosterone secretion after 30 min (Macrides *et al.*, 1974). Male hamsters with high testosterone levels investigate female vaginal odors more than males with low testosterone levels (Macrides *et al.*, 1974). The odor of the female thus appears to activate male sexual arousal through an elevation of testosterone levels.

Female odors may stimulate sexual activity in seasonally breeding males. Hamsters show testicular atrophy and lack of sexual behavior when exposed to short day lengths, and the presence of an estrous female or her soiled bedding accelerates testicular growth when day length is increased (Vandenbergh, 1977).

8. Inhibition of Reproductive Behavior in Adult Males

Schultz-Westrum (1969) suggested that the odor of a dominant male flying phalanger (*Petaurus breviceps*) causes a stress reaction in subordinates that inhibits their sociosexual behavior (scent-marking, patrolling a territory, mating, and aggressiveness). All of these activities increase in subordinates when the dominant male is removed from the colony.

Stress reduces testosterone levels in rhesus monkeys (Mason, 1968) and mice (Bronson, 1976; Macrides *et al.*, 1975). Male mice and rhesus monkeys defeated in aggressive encounters have heavier adrenal glands and higher ACTH and corticosterone levels than dominant males (Archer, 1970; Bronson, and Eleftheriou, 1964; Rose *et al.*, 1971). Male mice exposed to the odors of grouped males also show increased adrenal weights (Archer, 1969). Baldwin (1968) has suggested that olfactory cues from dominant squirrel monkeys (*Saimiri sciureus*) may inhibit gonadal hormone secretions in subordinates, but there is as yet no direct evidence that odors alone can inhibit testosterone secretion and reproductive behavior in other males.

9. Pregnancy Block

A female mouse that has mated within 4 days will have her pregnancy interrupted if exposed to a novel intact male, his soiled bedding, or his urine (Bronson, 1968, 1971; Parkes and Bruce, 1961, 1962; Whitten and Bronson, 1970). This effect does not occur in anosmic females (Bruce and Parrott, 1960), and it appears to be caused by an olfactory induced stimulation of gonadotrophin release, which blocks implantation and induces estrus (Chapman *et al.*, 1970). Pregnancy block may be a side effect of the male-induced estrus (Chapman *et al.*, 1970).

Male-induced pregnancy block also occurs in deermice (Bronson and Eleftheriou, 1963) and two species of voles (Milligan, 1976; Stehn and Richmond, 1975).

C. EFFECTS OF EXPERIENCE AND HORMONES

Sexual experience is not necessary for animals of either sex to discriminate betwen the odors of males and females. Sexually naive male and female rats (Brown, 1977) and male hamsters (Landauer, *et al.*, 1978) are attracted more to the odors of heterosexual conspecifics.

The hormone levels of male and female mammals, however, do determine their responses to odors of the opposite sex. Gonadectomy reduces the time that male and female rats spend investigating the odors of opposite sex conspecifics (Brown, 1977). Male hamsters with low testosterone levels spend less time investigating female odors than males with high testosterone levels (Macrides *et al.*, 1974).

All of the nine known or suspected primer effects may be produced through modification of gonadotrophic hormone secretion in the receiving animal (Bronson, 1976; Rogers and Beauchamp, 1976b). Increased luteinizing hormone (LH) levels produce estrus in females and increase testosterone levels in males, whereas decreased LH levels produce anestrus in females and low testosterone levels in males. Such changes in LH secretion could be mediated through the release of LH releasing hormone from the hypothalamus (Bronson, 1976).

The ability of odors to produce pregnancy block and other primer effects may also depend on experiential factors. If a female is bred with a male of strain A, the odor of a male of strain B is far more likely to cause pregnancy block than the odor of a second male of strain A. This is true even when the female is of strain B and the stud male is of strain A (Dominic, 1966; Parkes and Bruce, 1961, 1962). If the stud male is removed within 3 hr of mating, the pregnancy block is less likely to occur than if the male remains with the female for 24 hr (Lott and Hopwood, 1972). The female thus appears to "learn" some characteristics of the odor of the stud male so that she can identify the "strange" male odor (Bronson, 1976). In sheep, a novel male will accelerate the seasonal estrus period, whereas a familiar male that has been with the flock has no effect (Hulet, 1966). The role of experience in other primer effects is unknown.

D. CHEMICAL AND MICROBIOLOGICAL EVIDENCE

Analyses by GLC have shown sex differences in the chemical secretions from the preputial glands of rats and mice (Gawienowski, 1977; Sansone-Bazzano *et al.*, 1972), the tarsal glands of black-tailed deer (Brownlee *et al.*, 1969), the flank glands of *Arvicola* (Stoddart *et al.*, 1975), the caudal glands of *Apodemus* (Stoddart, 1973, 1974; see Fig. 3), the ventral gland of *Gerbillus compestris* (Stoddart, 1977), and the anal and chin glands of rabbits (Goodrich and Mykytowycz, 1972).

Identifiable chemical differences between sexes are likely to be found in many scent gland secretions, but seldom has a behavioral bioassay been performed to determine whether animals can use the sex-specific components for sexual identification. With reference to Fig. 6, for example, one might test whether male and female deer react differently to fractions containing peaks 66 and 26. Similarly, one might test whether *Apodemus* able to discriminate between males and females solely on the basis of chemicals from peaks t to z, which occur in the male caudal gland but not the female (see Fig. 3).

Urine is difficult to analyze with chromatographic and spectrometric methods because it contains so many specific components. Nevertheless, Beauchamp and Beruter (1973) have fractionated the urine of male and female guinea pigs and shown that specific fractions are attractive to conspecifics of the opposite sex.

Most of the evidence suggests that the primer pheromones of mice are also carried in the urine. The source of the pheromone responsible for the endocrinological changes underlying the acceleration of puberty in female mice is thought to be an androgen-dependent protein in the males' urine (Colby and

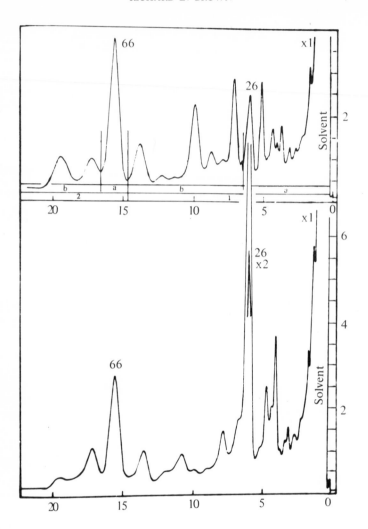

FIG. 6. Gas chromatograms of male (top) and female (bottom) black-tail deer (*Odocoileus hemionus columbianus*) tarsal gland distillate. Abscissa: retention time in minutes. (From Brownlee *et al.*, 1969.)

Vandenbergh, 1974; Vandenbergh *et al.*, 1975). Further analysis indicates that the active compound in male mouse urine for estrus acceleration contains at least six compounds, none of which is a steroid metabolite. It is possible that the active component is a hormone-dependent protein produced in the kidney or liver (Vandenbergh *et al.*, 1976). Urinary protein levels are higher in male than in female mice and are reduced by castration (Marchlewska-Koj, 1977).

It is possible that all of the primer effects of male mouse urine result from these urinary proteins. Marchlewska-Koj (1977) has shown that the same proteins in male urine produce pregnancy block, estrus induction (Whitten effect), and acceleration of puberty in female mice. The preputial glands also contain a pheromone that blocks pregnancy (Marchlewska-Koj, 1977) and induces estrus in grouped females (Bronson, 1971, 1976). Possibly the same combination of androgen-dependent proteins occurs in both the urine and preputial gland secretion.

VII. Rut Odors

A. Behavioral Responses

Rut odors are those given off by sexually active animals and distinguish them from animals that are sexually inactive. Female mammals that are sexually receptive only during their estrus periods produce special odors during this time. In many species the male also has a specific rut period during which his secretions have special odors. In species that breed year round, adult males are distinguished from juvenile or castrated males by their odor.

1. Males

Darwin (1887) noted that scent glands in mammals might be used for defence or protection (in skunks and shrews), but he was more interested in their increased activity during the rutting season. He noted that scent glands in antelopes and deer were larger in males than females, did not start to produce secretions until the animals were over a year old, and failed to emit a secretion at all if the animal was castrated while still young. Darwin concluded that "there can be no doubt that they [scent glands] stand in close relation with the reproductive functions" (p. 529). He further stated that "when only the male emits a strong odor during the breeding season, it probably serves to excite or allure the female" (p. 530). Finally, Darwin (1887) pointed out that the function of these odors can be understood through evolution as follows:

> The odor emitted must be of considerable importance to the male, inasmuch as large and complex glands, furnished with muscles for everting the sack, and for closing or opening the orifice, have in some cases been developed. The development of these organs is intelligible through sexual selection, if the most odoriferous males are the most successful in winning the females, and in leaving offspring to inherit their gradually-perfected glands and odors (p. 530).

A number of other observers have made similar observations. During musth (the rut period), male Asiatic elephants (*Elaphus maximus maximus*) secrete a viscous dark-colored fluid from the temporal gland that is "of considerable interest to the female'' (Eisenberg *et al.*, 1971, p. 200). Quay (1953) found that male kangaroo rats (*Dipodomys merriami*) show most secretory activity from their dorsal skin gland during the breeding season. Other species of kangaroo rats, however, showed no such seasonal fluctuation in dorsal gland activity. Red deer stags (*Cervus elaphus*) develop a musky smell in their urine during the rutting season (Lincoln, 1971). Testosterone was undetectable in the testes of deer stags from April to June, was low in August, and then reached a maximum during rut in September–October (Lincoln *et al.*, 1970). This suggests that the rut odor in the urine of deer stags results from this increase in testosterone level.

Evidence that females prefer the odors of sexually active males to those that are not sexually active (that is, castrated) has come from species such as rats and mice in which males breed year round. Both intact and ovariectomized female rats can be trained to discriminate between the body odors of intact and castrated male rats (Carr and Caul, 1962). Females cannot discriminate between intact male rat odors and odors from castrated males given testosterone injections, indicating that the differences in odor is hormone controlled. The preferences of females for the odors of sexually active males depends on the experience and hormonal state of the female (see below).

The source of the odor of sexually active male rats and mice may be the urine or a secretion from the preputial glands. Female mice (*Mus*) show a preference for the urine odor of intact males as opposed to urine odor of castrated males (Scott and Pfaff, 1970). Bronson and Caroom (1971) suggest that the odor of sexually active mice is secreted by the preputial glands. In a four-choice preference test, female mice were most attracted to the odor from male preputial gland secretions when it was compared with no odor, male urine, and female urine. It is therefore possible that the urine of male mice is attractive to females because it contains preputial gland secretions. The preputial gland of the male rat also produces secretions attractive to females (Orsulak and Gawienowski, 1972), and females can use these secretions to discriminate sexually active (intact) from sexually inactive (castrated) males (Gawienowski *et al.*, 1975).

A number of studies (see Sink, 1967) indicate that the preputial glands of adult male pigs produce a powerful odor attractive to estrous females. Signoret (1970b) gave female pigs a 5-min test of their preference for either males or females in a T maze. Estrous gilts spent more time near the males, but anestrous females showed no preferences. In another test, the boar's preputial fluid was not preferred over a no-odor control. The preputial odor of the boar apparently does not act as an attractant for females, but does induce the ''standing reaction'' that the sow shows prior to copulation (Signoret, 1976).

2. Females

A great deal of observational and experimental evidence suggests that estrous (receptive) females smell differently from diestrous (nonreceptive) females and that the former smell is attractive to males. In many species, estrous females have a swollen vulva or urinate more frequently than nonestrous females. Vaginal secretions and urine are thus suspected to be the primary sources of estrous female odor. There is no evidence that estrus gives rise to increased secretions from specialized skin glands. The increase in estrogen level that occurs during estrus, in fact, may cause a decrease in skin gland secretions, but the evidence on this is contradictory (see the discussion of Strauss and Ebling, 1970).

Male hamsters scent-mark over the odors of the diestrous female hamsters more than they do over the odors of estrous females (Johnston, 1975a). The flank glands of female hamsters secrete less during estrus (see discussion by Johnson, 1973, p. 525). Male hamsters investigate the vaginal secretions of estrous females more than their urine odors, which are not investigated any more intensively than water (Johnston, 1974). Male hamsters appear to be equally attracted to the vaginal odors of estrous and diestrous females (Darby, et al., 1975; Landauer et al., 1978). The attraction of male hamsters to female vaginal secretions thus seems independent of any particular hormonal state of the female.

In dogs, estrous female odors are secreted mainly in the urine, but also occur in vaginal secretions. Adult male dogs investigate the urine of estrous females more than that of nonestrous females (Beach and Gilmore, 1949) and prefer the vaginal odors of estrous females to those of ovariectomized females (Beach and Merari, 1970). In order to determine the relative attractiveness of estrous female odors from different sources, Doty and Dunbar (1974) presented male beagles with anal sac secretions, vaginal secretions, and urine from females in different hormonal states. The males were able to discriminate between estrous and diestrous females using both vaginal secretions and urine, but spent more time investigating urine than anal sac or vaginal secretions (see Fig. 7).

Male rats are attracted to female preputial gland secretions and can use the preputial odor to discriminate estrous from anestrous females (Orsulak and Gawienowski, 1972; Gawienowski et al., 1976). In order to control for such secretions in the urine, Lydell and Doty (1972) presented male rats with female urine that was externally voided and urine that was obtained directly from the bladder. Male rats preferred the urine odors from estrous females over those from diestrous females in both cases. Thus, the bladder urine itself contains attractive components. It is possible, however, that preputial gland and vaginal secretions enhance the differences in odor between estrous and anestrous female rats.

Male mice are able to discriminate between estrous and nonestrous females by their vaginal odors but not their urine odors (Hayashi and Kimura, 1974). Gerbils appear to be able to use urine odors to discriminate between estrous and nonestrous females (Pettijohn, 1974). It is very likely that rams use olfactory cues to find estrous ewes because olfactory bulbectomized rams mount ovariectomized ewes as often as estrous ewes but intact males mostly mount estrous ewes (Lindsay, 1965).

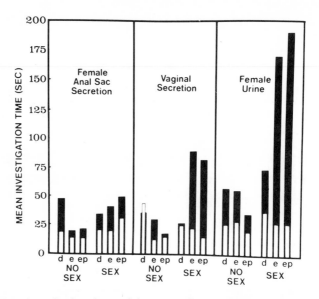

FIG. 7. Mean investigation times of three sexually experienced (sex) and three sexually inexperienced (no sex) male beagles for odors of females. Black bars represent average time spent investigating the odor port containing the female odor during six 2½-min trials. White bars represent average time spent investigating the juxtaposed distilled water stimulus. Total possible = 900 sec; d, no hormone treatment given female donors; e, prior estrogen injection given to female donors; ep, estrogen and progesterone injections given to female donors. (From Doty and Dunbar, 1974.)

McArthur *et al.* (1972) studied two female bonnet monkeys (*Macaca radiata*) for two 28-day menstrual cycles and found that, as ovulation approached, the sexual skin became redder, there was an increase in urinary estrogen level, and an increase in the amount of vaginal secretion. New World primates do not show conspicuous vaginal swellings or reddening of the vulva when in estrus, but observational evidence (summarized by Epple, 1974a, 1976) suggests that they produce special odors in the urine and vaginal secretions at this time.

When in estrus, female rhesus monkeys (*Macaca mulatta*) have swollen vulvas and produce a strong-smelling vaginal discharge (Michael and Keverne, 1968). Male monkeys will work at bar pressing to gain access to an estrous female, or to an ovariectomized female covered with vaginal secretions from estrous females, but will not bar press for access to ovariectomized females that have been rubbed with saline (Michael and Keverne, 1968, 1970). Males also copulate more often with ovariectomized females that have had vaginal secretions applied to them than they will with saline- or ether-rubbed females (Keverne and Michael, 1971; Michael and Keverne, 1970).

Goldfoot *et al.* (1976), however, found that the vaginal odors of estrous female rhesus monkeys had no effect on the sexual behavior of males when applied to ovariectomized females. These authors suggest that the responses of males toward odorized females depend more on experiential factors than on odors. There are large individual differences in the responses of male rhesus monkeys to the odors of estrous females (Keverne, 1976). Some males are responsive to the endocrine and odorous changes in the female and modify their sexual behavior accordingly, while other males are less sensitive to the female and copulate at a high rate, irrespective of her hormonal state (Keverne, 1976). Some males fail to copulate at all.

In humans, Ellis (1905, p. 63) notes that both men and women often give off strong odors during sexual excitement. The odors produced by men during this sexual excitement are said to resemble rancid butter by some and chloroform by others, whereas the odor of sexual excitement in women was said to be of an acid or hircine (goatlike) character. Female vaginal odors change over the menstrual cycle in both intensity and pleasantness. Vaginal odors are judged to be the least intense and least unpleasant during the middle (ovulatory) phase of the menstrual cycle and more intense and unpleasant during the menstrual and premenstrual phases (Doty *et al.*, 1975). With reference to this study, Globus and Cohen (1975) have suggested that the responses to vaginal odors may depend on the state of sexual excitement of the observer, but this has not yet been tested. Female urine odors also change in intensity and pleasantness over the menstrual cycle (Doty, 1977). They are judged most intense during the preovulatory phase and least unpleasant during the early luteal phase. Urine odors were judged to be more unpleasant than vaginal odors. Human breath odors may also change over the menstrual cycle, but no researchers have investigated this odor source yet.

B. Effects of Experience and Hormones

Responses of test animals to rut odors depend on variables such as sexual experience, hormonal state, and age. Responses to odors may also depend on the odor environment in which the animals are housed (owing to possible physiological primer effects).

1. Sexual Experience

The role of sexual experience in determining the preferences of males for the odors of estrous and nonestrous females is unclear. Some studies have found sexual experience necessary for male rats, mice, and dogs to show a preference for estrous female odors (Carr *et al.*, 1965; Doty and Dunbar, 1974; Hayashi and Kimura, 1974; Lydell and Doty, 1972; Stern, 1970). In other studies, both experienced and sexually naive males show preferences for estrous female odors. This has been found in male mice (Rose and Drickamer, 1975), hamsters

(Johnston, 1975b; Macrides *et al.*, 1977), and rats (Landauer *et al.*, 1977). In still other cases, sexually experienced male rats showed no preferences for the odors of estrous or diestrous females (Brown, 1977; Pranzarone, 1969).

The responses of female mice to the preputial gland odor of male mice is enhanced by sexual experience (Bronson and Caroom, 1971; Caroom and Bronson, 1971). Sexual experience does not seem to be necessary for female rats or mice to show preferences for the urine odors of sexually active (intact) males over those of castrated males (Brown, 1977; Carr *et al.*, 1965; Hayashi and Kimura, 1973). Carr *et al.*, (1965) state that the "difference between male and female rats concerning the role played by learning (i.e. previous sexual experience) in determining their reactions to sex odors is in accord with the generally accepted view that the male pattern is more labile than the female pattern" (p. 376). The nature of this learning in males has not been studied systematically. In some studies, males have experienced sexually receptive females, but were never paired with nonreceptive females (Stern, 1970; Lydell and Doty, 1972). In other cases, males were paired on some days with estrous females and on other days with nonestrous females (Carr *et al.*, 1965), and may have learned through social reinforcement to discriminate between these two odors. If the male is exposed to sexually receptive females for long periods of time, he may learn that nonreceptive females will become receptive, and he will be attracted to the odors of both estrous and nonestrous females (Brown, 1977). Such also appears to be the case with male hamsters (Landauer *et al.*, 1978).

2. Hormonal State

Castrated male rats, whether sexually naive or sexually experienced, show no preference for odors of estrous females over those of diestrous females (Brown, 1977; Carr *et al.*, 1965; Carr *et al.*, 1966; Pfaff and Pfaffman, 1969; Stern, 1970). Castration eliminates the preference of male mice and hamsters for the odors of estrous females (Gregory *et al.*, 1975; Rose and Drickamer, 1975). Hormone therapy with testosterone (or estrogen) produces preferences for female odor over no odor in both sexually experienced and naive castrated male rats (Brown, 1978a) and preferences for estrous female odors over those of nonestrous females in sexually experienced, castrated male rats (Stern, 1970). Testosterone injections also facilitate the attraction of male dogs and male hamsters for the odors of estrous females (Anisko, 1977; Gregory *et al.*, 1975).

Intact female rats, whether sexually experienced or naive, prefer the odors of intact males to those of castrated males (Brown, 1977; Carr *et al.*, 1975). Ovariectomized females show no preference (Brown, 1977). The hormonal control of female odor preferences is unclear. In some cases, hormone-injected ovariectomized females show preferences for odors of intact males (Carr *et al.*, 1965); in other cases hormone therapy does not facilitate such preferences (Brown, 1978a). There is an interaction between sexual experience and hormonal state that affects odor preferences in female rats. Sexually experienced estrous females are most likely to approach odors from intact males, whereas sexually naive diestrous females are least likely to do so (Carr *et al.*, 1965). The responses of ovariectomized females are not equivalent to those of diestrous intact females: ovariectomized females spend very little time investigating any conspecific odors (Brown, 1977, 1978a).

3. Age of Test Subjects

Prepubertal male rats show no preferences for estrous over diestrous female odors (LeMagnen, 1952). Given testosterone injections, they do show a preference for these odors (Carr *et al.*, 1970c). Juvenile female rats made sexually receptive through hormone injections show preferences for odors of intact males over castrates, whereas nonreceptive juvenile females do not (Carr *et al.*, 1970c). Thus, since age is related to hormone secretion, prepubertal animals may lack the "motivation" to seek out the odors of animals of the opposite sex.

4. Housing Effects

Because primer pheromones have been shown to alter the hormone levels of adult and immature animals of both sexes (see Section VI), and because hormonal state affects sex and rut odor preferences, it is possible that differential housing conditions might alter odor preferences. The majority of odor preference experiments provide information on the cagemates of test animals but neglect to give information on whether males and females were housed in the same or in separate animal rooms. A number of studies state explicitly the housing room conditions. Scott and Pfaff (1970) housed female mice "in a separate room several doors away from the male and castrate male mice" (p. 407). Krames and Shaw (1973) kept "alien" and "own-group" test male rats in different housing rooms "to prevent the possibility of the subjects having any prior olfactory experience with [the alien odor]" (p. 445). Caroom and Bronson (1971) housed female test mice in rooms without any males. Stern (1970) housed test male rats in the same housing room as females and stated that "no attempt was made to isolate the males from female odors" (p. 519).

In a study directed at this problem, Pranzarone (1969) housed male rats in isolation, in all-male groups, and in mixed-sex groups. He found preferences for the odor of estrous female rats over diestrous females in the isolated and segregated males, but not in males housed with females. Pranzarone suggests that the cohabiting males had habituated to female odors. As suggested above, it is more likely that these males "learned" that diestrous females became estrous and so were highly attracted to all female odors.

Brown (1977) housed half of his male rats in like-sex groups in rooms holding only males and half in rooms holding both males and females. Housing room differences had no effect on the odor preferences of sexually experienced or sexually naive males. Similarly, there were no differences in the odor prefer-ences of female rats housed in mixed-sex or like-sex rooms. It is possible, however, that during some period in the animal's life, housing conditions will affect hormone levels and thus affect behavior.

Female mice born and housed for the first 28 days of life in a room housing both males and females show a preference for the odor of intact over castrated males, whereas females born and housed in rooms containing no males show a preference for the odors of castrated males. The latter preference is reversed after sexual experience (Hayashi and Kimura, 1978). Exposing females to male odors betwen 28 and 60 days of age does not produce a preference for intact over castrate male odors (Hayashi and Kimura, 1978). Early olfactory exposure may serve to "imprint" animals to particular odors (see Section X on maternal odors).

C. CHEMICAL AND MICROBIOLOGICAL EVIDENCE

The identification of rut odors has concentrated mainly on the estrous odors of females. The vaginal secretion of the hamster contains a combination of com-pounds, two of which are attractive to male hamsters. These compounds, di-methyl disulfide and dimethyl trisulfide, are not as attractive, however, as the whole gland secretion and other compounds may contribute to the attractiveness of hamster vaginal odors (Singer et al., 1976). In behavioral tests, dimethyl disulfide smeared on anesthetized males does not elicit sexual advances from male hamsters, but whole vaginal extracts do (Macrides et al., 1977). Dimethyl disulfide, therefore, has only some of the attractant properties of estrous female vaginal secretion. Whether dimethyl disulfide is the product of bacterial action has not been investigated.

The active component of the female rhesus monkey vaginal odor, called "copulin," consists of fatty acids that have been identified as acetic, propionic, isobutyric, n-butyric, and isovaleric (Curtis et al., 1971). A synthetic sample of these five volatile aliphatic acids applied to the sexual skin of ovariectomized females causes males to show a significant increase in the number of mounts and ejaculations. These acids are produced by unidentified microorganisms in the vaginal fluid (Michael and Bonsall, 1977). This synthetic mixture may not be as effective at eliciting male copulation as normal vaginal secretions, and the addi-tion of extra "enhancing" chemicals increases its attractiveness to males. Keverne (1976) found that males copulated with "copulin"-rubbed females in 37% of his tests, whereas copulation occurred in 65% of the tests when copulin plus enhancing chemicals were used. Using copulin alone, however, Michael et al. (1977) found that 75% of the males copulated. Thus, in spite of individual differences in male responses to female odors and specific preferences for particular sex partners, it appears that the synthetic mixture of aliphatic acids can be used by male rhesus monkeys to discriminate sexually active from sexually inactive females.

Human vaginal secretions also contain volatile fatty acids identified as acetic, propionic, butanoic, methybutanoic, and methylpentanoic (Michael *et al.*, 1974). There is a significant increase in volatile fatty acid secretion during the late follicular phase of the menstrual cycle (Days 10–15) and a decrease during the luteal phase. There are large individual differences in the amounts of volatile fatty acids secreted: some women are "nonproducers," whereas others produce high amounts and are designated as "producers" (Michael *et al.*, 1974). Many of the volatile fatty acids found in human vaginal secretions appear to be produced by bacteria of which the most prominent genus is *Lactobacillus* (Morris and Morris, 1967). Subjects who produce high levels of fatty acids may have different bacterial populations than nonproducers (Preti and Huggins, 1975). Changes in hormone levels during the ovulatory cycle alter the amount of carbohydrates in the vaginal secretions of both monkeys (Bo, 1970) and humans (Gregoire *et al.*, 1971). Since the vaginal bacteria produce the volatile acids from these carbohydrates, the vaginal odor is altered when carbohydrate levels change.

VIII. Odors and Social Status

A. Behavioral Responses

In species with dominance hierarchies, it is possible that the dominance status of individuals can be identified by their odors. Odor differences between dominant and subordinate animals could be the result of scent gland size, amount of secretion, or frequency of scent marking. Dominant male guinea pigs produce more perineal gland secretion than second- or third-ranked males and also scent-mark more than these subordinates (Beauchamp, 1974). Dominant rabbits of both sexes have heavier anal, inguinal, and submandibular glands than subordinates of the same body size (Mykytowycz and Dudzinski, 1966). Dominant male mice (*Mus*) possess larger preputial glands than subordinates (Bronson and Marsden, 1973).

Saddle-backed tamarins of both sexes investigate the scent marks of dominant males more than those of subordinates (Epple, 1973, 1974a). Because dominant males mark more than subordinates, however, preferences may be affected by the amount of odor deposited rather than a qualitative difference between the odors of dominant and subordinate males. Male rats are able to discriminate between the body odor of dominant and subordinate males and appear to prefer the odor of the subordinate (Krames *et al.*, 1969). Mice can discriminate between the body odor of dominant and subordinate males, and between the odors of subordinate males and isolates, but not between the odors of dominant males and isolates (Carr *et al.*, 1970b). This suggests that the odor of the subordinate mouse changes, possibly as a result of an increase in adrenal corticosteroid secretion. Socially subordinate male mice have heavier adrenal glands and lower testosterone levels than dominant males (Archer, 1970; Jones and Nowell, 1974; Lombardi and Vandenbergh, 1977). This change in hormone secretion may alter the body odor of the subordinate male.

Acceleration of estrus does not occur in juvenile female mice that are exposed to the urine of subordinate males. The uterine weights of these females are significantly lower than those of females exposed to the urine of dominant males; in fact, they do not differ from the uterine weights of females exposed to water (Lombardi and Vandenbergh, 1977).

The results of experiments by Jones and Nowell seem to suggest that it is the odor of the dominant male rather than of the subordinate that changes. In these studies male mice were attracted to the urine odor of isolated male mice (Jones and Nowell, 1973a), but avoided the urine odor of dominant males (Jones and Nowell, 1973b, 1974). Estrous females were attracted more to the odor of urine from dominant males than to that from subordinate males. Little work has been done to examine the chemical or microbiological differences between the odorous secretions of dominant and subordinate animals using GLC or MS.

B.   Effects of Experience and Hormones

In a social hierarchy, dominant males do the majority of scent marking and males that lose an aggressive encounter show a reduction in scent marking (Beauchamp, 1974; Bronson, 1976; Epple, 1974; Maruniak *et al.*, 1974; Nyby *et al.*, 1970). This may be the result of lowered testosterone levels.

Ralls (1971) has argued that the primary motivation for scent marking in mammals is aggression. Such aggression may arise when a territory resident or a dominant animal (not necessarily a territory resident) is confronted with an alien conspecific of the same sex. If animals learn to avoid the odors of individuals that are dominant to them (as a result of defeats in agonistic confrontations or otherwise), the use of such odor displays could reduce the probability of an actual fight occurring. When fights do occur, the odor of a dominant animal (or a territory resident) might intimidate subordinates (or intruders) so that the intensity of the fight and the number of injuries to each animal are reduced (Epple, 1974a). Since dominance hierarchies may change, the social status function of a particular odor may depend on the outcome of recent aggressive encounters and individual recognition (Section II).

IX. Odors of Stress

A. Fear and Alarm Odors

When a male black-tailed deer (*Odocoileus hemionus columbianus*) is chased by a dog or a human, it releases an odor from the metatarsal gland on the outside of its hind leg. Black-tailed deer fawns show some avoidance of objects covered with metatarsal gland secretion (Müller-Schwarze, 1971). The eastern woodchuck (*Marmota monax*) extrudes its anal glands when alarmed, and the odor from this gland inhibits activity in other woodchucks (Haslitt, 1973). In both of these examples, alarm or fear odors (Angstgeruch) are released directly into the air, but there may be other methods of odor dispersal. For example, Steiniger (1950) reported that brown rats (*Epemys norvegicus*) defecate and urinate on poisonous baits, and these are subsequently avoided by all other rats.

Mice that have received an electric shock or are stressed with an injection of NaCl produce an odor that is avoided by other mice (Carr *et al.*, 1971; Müller-Velten, 1966; Rottman and Snowdon, 1972). Rats can be trained to discriminate between the odors of shocked and unshocked rats (Valenta and Rigby, 1968), suggesting that rats produce a different odor when stressed. The findings of King *et al.*, (1975) indicate, however, that the odors of stressed rats are not alarm odors in that they do not induce avoidance responses in other rats.

There is convincing evidence for an actual alarm substance in rats in their response to blood and muscle. Rats show more emotional reactions and avoidance responses when they encounter blood and muscle homogenates from dead rats than when presented with no tissue (Stevens and Saplikoski, 1973). The nature of the alarm substance in the blood and muscle of rats has yet to be investigated. Fright reactions do not occur to samples of rat brain tissue nor to samples of guinea pig blood and muscle tissue (Stevens and Gerzog-Thomas, 1977). The alarm substance does not appear to be an odor released from the intact animal; it may be similar to the alarm odor (Schreckstoff) released from the cut skin of fish (Pfeiffer, 1963).

B. Effects of Experience

King *et al.* (1975) found that rats which had experienced shock themselves avoided the odors of stressed (shocked) rats, but rats which had not been shocked showed no such avoidance. Similarly, Mackay-Sim (1978) found that the avoidance responses of rats to the body odors and blood of stressed rats depended on whether the subjects themselves had experienced stress.

The aversive responses to odors of stress may be determined by experience or conditioning. Adult rats learn to avoid odors associated with illness-inducing stimuli (Domjan, 1973), and male hamsters that are poisoned after exposure to the vaginal secretions of adult females are hesitant to approach this secretion and are hesitant to copulate with estrous females (Johnston and Zahorik, 1975; Johnston *et al.*, 1978; Zahorik and Johnston, 1976).

There has been little research on the stimuli eliciting fear or alarm odors and the responses of animals to these odors in naturalistic situations. Anecdotal reports suggest that prosimians and other primates produce fear odors (Fossey, 1978; Manly, 1974; White, 1978), but no experimental evidence is available. Similarly, humans may produce fear odors, but no research has been oriented toward identifying them. No research seems to have been done on the chemical and microbiological bases of alarm odors of any species of mammal.

C. FRUSTRATION ODORS

A number of studies suggest that rats produce an odor when they are frustrated (see review by Schultz and Tapp, 1973). Rats reinforced with food for running down a straight alley produce an "odor of frustration" when the food is not present. Rats are able to discriminate between the odors of frustrated and nonfrustrated rats, but not between pairs of nonfrustrated rats (Morrison and Ludvigson, 1970). The odor of the frustrated rat is sufficiently discriminable to serve as a stimulus for an arbitrary response. Rats jump a hurdle to escape from a start box faster in the presence of the frustration odor than in the presence of no odor or the odor of a nonfrustrated rat. This occurs whether or not the test rat has experienced frustration itself (Collerain, 1978; Collerain and Ludvigson, 1977). The frustration odor may be an odor of general emotionality, and thus similar to stress or alarm odors. The source of the frustration odor has not been identified, nor has a frustration odor been found in any mammal other than the rat.

X. MATERNAL ODORS

A. BEHAVIORAL RESPONSES

The infant mammal shows a progression of odor-directed responses toward its mother from birth to weaning. Odors are used for nipple attachment in newborn animals, for sibling and home orientation in mobile infants, for discriminating lactating from nonlactating females, and for discriminating the infant's own mother from other lactating females (Leon, 1978a; Rosenblatt, 1976, 1978). The maternal odor may also aid the mother in discriminating between her own and alien young.

Altricial mammals appear to use olfactory cues to locate the nipple prior to attachment. These olfactory cues arise from the mother's saliva and amniotic fluids and are spread over the nipples when the mother grooms. Newborn rats fail to attach to nipples that have been washed or covered with foreign odors (Teicher and Blass, 1977). Infant rats only a few hours old are able to discriminate between the saliva of their own mother (or any lactating female) and that of a virgin female, and also between the amniotic fluid and urine of their mother. Mother's urine does not stimulate nipple attachment (Teicher and Blass, 1977). Bilateral olfactory bulbectomy and intranasal zinc sulfate anosmia both produce deficiencies in nipple attachment and suckling behavior in rats as young as 2 days of age (Singh and Tobach, 1975; Singh et al., 1976). Olfactory cues are also important for nipple attachment and suckling in infant rabbits, kittens, and puppies (Rosenblatt, 1976, 1978).

Huddling by altricial infants serves as a means of thermoregulation and is disrupted by zinc sulfate anosmia and olfactory bulbectomy (Alberts, 1978, Singh and Tobach, 1975). Huddling has been found to be under tactile, thermal, and olfactory control in rats and cats, with temperature being the most important in the first few days of life, and olfaction showing increased importance between 5 and 10 days of age (Alberts, 1978; Freeman and Rosenblatt, 1978a). At this second stage of development, the young altricial mammal begins to use olfactory stimuli to discriminate between familiar and unfamiliar objects and to initiate movements toward specific objects before the eyes open (Rosenblatt, 1976; Tobach, 1977).

Before 6–8 days of age locomotor responses do not appear to be reliable indices of olfactory discrimination ability in rats (Leon and Moltz, 1971; Tobach, 1977). Studies recording movement, ultrasonic distress vocalizations, and heart rate have found that rats between 5 and 12 days of age can discriminate between maternal odors and those of nonlactating females or nonrat control odors (clean air or sawdust) (Compton et al., 1977; Oswalt and Meier, 1974; Schapiro and Salas, 1970). While the infant rats can discriminate between familiar and unfamiliar odors at 1 day of age and between lactating and nonlactating female odors at 5 days of age, the age at which they can discriminate between their own mother's odor and that of another lactating female is not known. Rosenblatt (1978) suggests that the maternal odor has two components: a general lactating female odor, and an individual odor specific to each female (compare individual odors). Kittens between the ages of 8 and 12 days of age are able to discriminate between their own home cage odor and the odor of another litter of the same age. These kittens vocalize more in an unfamiliar cage and do not settle in the home area. Fostering 8-day-old kittens onto a strange mother for 48 hr confuses their home orientation so that they fail to orient to their new "home" odor or their original home odor (Freeman and Rosenblatt, 1978b). Similarly, dogs between 8 and 21 days of age show increased vocalizations in unfamiliar environments and may recognize olfactory cues associated with the home (Scott et al., 1974).

Juvenile rats, mice, hamsters, rabbits, and cats are able to discriminate between the odors of lactating and nonlactating females (Breen and Leshner, 1977; Devor and Schneider, 1974; Gregory and Pfaff, 1971; Leon and Moltz, 1971, 1972; Mykytowycz and Ward, 1971; Rosenblatt *et al.*, 1969). Spiny mice (*Acomys cahirinus*) at 1 day of age prefer their own home cage bedding and that of other litters the same age over clear bedding and bedding from the cages of nulliparous females (Porter and Ruttle, 1975). *Acomys* females produce a maternal odor up to 30 days postpartum as long as nursing pups are still present (Porter and Doane, 1976).

In the rat, the maternal odor originates in a light gold-colored secretion produced in the caecum, excreted through the anus, and called caecotrophe (Leon, 1974). The production of caecotrophe is dependent on prolactin secretion. Injection of the prolactin-inhibiting drug, ergocornine, prevents caecotrophe production, whereas adrenalectomy and ovariectomy have no effect (Leon and Moltz, 1973).

Caecotrophe not only acts to attract pups to the mother but may also be used to attract pups to their littermates and nest. Sixteen-day-old rat pups prefer the odor of their littermates to an empty goal box but not to the odor of unfamiliar 16-day-old rat pups (Leon, 1974). Rat pups are not attracted to 16-day-old pups from sucrose-fed mothers who produce no caecotrophe odor (Leon, 1974). Caecotrophe may be deposited on the pups and nest by the mother, or the pups, by eating their mother's feces, may produce their own caecotrophe odor (Leon, 1974). Infant rats begin to produce their own caecotrophe at about 26 days of age, and this is attractive to 19-day-old rat pups (Leon and Behse, 1977).

Galef and Heiber (1976) have replicated some of Leon's experiments and found that 20-day-old Long–Evans rat pups are attracted to the odors of both nulliparous and lactating females when each is paired with no odor. Brown (1978b) found, however, that, when fresh feces from lactating and nonlactating females were collected using the method of Galef and Heiber (1976) and controlled for weight, 16–20-day-old rats show a preference for lactating over nonlactating female odors.

Nulliparous female rats produce a maternal odor when housed continuously with young foster pups (a procedure called concaveation) but males do not, even though they act "maternally" (Leidahl and Moltz, 1975). The maternal odor of concaveated females is dependent on prolactin secretion, and 16-day-old rats cannot distinguish between the odors of concaveated females and those of normal lactating females (Leidahl and Moltz, 1977). Prolactin is thought to act on the liver to alter the bile acids in the lactating female. Male rats that have had bile from lactating females injected into their caecum produce a "maternal" odor as long as the injections are continued (Moltz and Leidahl, 1977). Leon (1978b), however, suggests that bile does not influence the synthesis of the maternal odor but increases the emission of this odor by acting as a laxative (see review by Leon, 1978a).

The particular caecotrophe odor that a given mother emits is dependent on the food she eats (Leon, 1975). Different diets induce the synthesis of caecotrophes with different odors. Thus, each mother may produce a different caecotrophe odor (unless they all eat exactly the same food, which may occur in the laboratory, but is unlikely in the wild). The maternal odors of gerbils are also modified by the food eaten (Skeen and Thiessen, 1977).

Eight- to 24-week-old squirrel monkeys (*Saimiri sciurius*) are able to discriminate their own mother from other lactating females by her odor alone (Kaplan *et al.*, 1977). Human infants are able to discriminate between the odor of their mother's breast pad and a clean breast pad when 2–7 days of age (Macfarlane, 1975). By 6 weeks of age human infants are able to discriminate between their own mother's breast odor and that of another lactating mother (Russell, 1976).

In addition to the infant's ability to use olfactory cues for nipple, home, and mother recognition, the mother uses odors to recognize her own offspring. This occurs in many species of ungulates including caribou, elk, goats, sheep, and deer (see reviews by Lent, 1974, and Grau, 1976). In many of these species mothers drive away unfamiliar infants that approach them. Licking the young after birth may serve to "imprint" the mother to the young and thus form a "maternal bond." Cross fostering of infant sheep and goats to new mothers is successful if the infant's odor is disguised with the odor of the mother or her milk (Hersher *et al.*, 1963b; Lent, 1974). If young domestic goats are removed from their mother directly after parturition and kept separated for 1 hr, the mother pushes them away when they attempt to suckle. If the mother is allowed to lick and nuzzle her kid for as little as 2–5 min before it is removed, separation for up to 3 hr does not affect maternal behavior (Klopfer *et al.*, 1964; Klopfer and Gamble, 1966).

Female mouse-eared bats (*Myotis myotis*) may recognize their young by smell (Kolb, 1977).

No chemical or microbiological analysis of maternal odors seems to have been done.

B. EFFECTS OF EXPERIENCE: OLFACTORY IMPRINTING

Mammalian infants are able to discriminate between the odors of lactating and nonlactating females before they can discriminate the odor of their own mother from that of other lactating females. The development of species and colony odor recognition may occur through a process of olfactory imprinting whereby the infant first develops a preference for the maternal odor, and this preference is then generalized to other conspecifics. Such a process may be similar to the imprinting of young birds to the auditory and visual characteristics of their species. Stimuli having these characteristics elicit approach by infants, filial responses during adolescence, and sexual responses in adults (Bateson, 1966; Hinde, 1970; Lorenz, 1937; Sluckin, 1972). Research on olfactory imprinting has investigated the effects of early olfactory experience on adult odor preferences and whether or not there are "critical periods" for the imprinting to occur. Because social experiences during adolescence may modify the effects of early imprinting (Hinde, 1970, pp. 524–526), the housing conditions of "imprinted" animals during adolescence are particularly important in evaluating the effects of early odor experience on adult odor preferences.

In mammals, olfactory imprinting studies have been conducted with natural odors and with synthetic odors applied to litters during infancy. When synthetic odors have been used there is the possibility that natural animal odors are confounded with the synthetic odor. This is especially likely to occur if the synthetic odor is presented when the animals are very young and they are then housed with unscented conspecifics for some period of time before being tested. Mainardi *et et al.* (1965), for example, scented litters of mice with 'Violetta di Parma' perfume when they were between 5 and 20 days of age and then reared these mice in groups of like-sex animals until 90 days of age. They were then tested for a choice between unscented and scented conspecifics of the opposite sex. Female odor donors were brought into estrus with hormone injections so that they produced natural estrous odors as well as perfume odors. In this case, neither the males raised with perfumed parents nor the unscented controls showed a preference for females having the odor with which they were reared. Likewise, females reared with perfume showed no preferences for perfumed males over natural males, but control females did show a preference for natural males over perfumed males. The confusion of which odors are imprinted, natural, or perfume has also occurred when rats were reared with red rose cologne or methyl salicylate (Marr and Gardner, 1965).

The results of studies using natural animal odors demonstrate more clearly that olfactory imprinting occurs. A black-tailed deer fawn that was reared and bottle-fed by a mother surrogate having the odor of pronghorn deer showed preferences for the odor of pronghorn deer as well as for live pronghorn over its own species (Müller-Schwarze and Müller-Schwarze, 1971). Cross-fostering experiments have shown that adult animals prefer the odors encountered in their infancy. Quadango and Banks (1970) cross-fostered 2-day-old pygmy mice (*Baiomys taylori ater*) and laboratory mice (*Mus*) and then housed them individually after weaning. When tested at 75 days of age with the odors from animals of the opposite sex, cross-fostered male and female *Mus* both preferred *Biaomys* odors, while control *Mus* of both sexes preferred *Mus* odors. Cross-fostered female *Baiomys* preferred *Mus* odors, whereas cross-fostered male *Baiomys*, like the control *Baiomys*, preferred *Baiomys* odors. Cross-fostered infant *Peromyscus leucopus* and *Onchomys torridus* showed a preference for the odors of the species with which they were reared (McCarty and Southwick, 1977). Cross-fostered spiny mice (*Acomys*) show preferences for their *Mus* families after only 48 hr of exposure (Porter *et al.*, 1977).

Investigations of the olfactory "critical period" have been conducted only with synthetic odors. Rats reared with ethyl benzoate or acetophenone odors 3–10 days of age or 11–18 days of age show preferences for these odors when tested at 45–60 days of age, while rats exposed to these odors between 20 and 29 days of age show no preference for them (Marr and Lilliston, 1969). It is possible that preferences emerged in this study because the test animals were housed individually after weaning.

The length of time between the odor exposure and testing as well as the housing methods may affect the demonstration of imprinting. Guinea pigs reared with either ethyl benzoate or acetophenone odors for the first 15 or 22 days of life show preferences for their rearing odor 1 day after the end of odor exposure (Carter and Marr, 1970). Guinea pigs reared with synthetic odors for shorter periods, from 1 to 3, 4 to 6, or 7 to 9 days of age preferred normal guinea pig odors when tested at 23 days of age (Carter and Marr, 1970). In this study, the "imprinted" animals were housed in groups and thus exposed to natural guinea pig odors between the last odor exposure and the test period. Lack of evidence for olfactory imprinting in another guinea pig experiment (Carter, 1972) may also have been the result of experiences during adolescence.

Experiments with the precocial spiny mouse (*Acomys*) suggest that brief olfactory experiences in the first few days of life are retained, at least for a few days, in spite of exposure to conflicting natural odors. One-day-old spiny mice exposed to cumin or cinnamon odors for 1 hr show preferences for that odor 24 hr later (Porter and Etscorn, 1974). *Acomys* exposed to cinnamon odors at 2 or 3 days of age prefer this odor when 6 days old, but show no preference if exposed to cinnamon at 4 or 5 days of age (Porter and Etscorn, 1976). These results suggest that *Acomys* pups are the most responsive to odors that they experience during the first three days of life, even though the interval in which they could experience interfering natural odors was much longer than that for animals which had odor experiences at 4 and 5 days of age.

One function of early olfactory experience may be to imprint the infant with the odor of its colony or group. In species in which the colony members are all closely related, the colony odor will be a "kinship odor." The adult animal thus may use his or her olfactory knowledge to prevent inbreeding as well as to ensure conspecific breeding. A female thus should choose to mate with a male of the same species but not from the same colony or family as herself. This prediction has been tested by Gilder and Slater (1978), who found that female mice from two different strains preferred the odors from unrelated males of the same strain to the odors of their brothers or males of a different strain.

Studies in which olfactory imprinting has been found have either tested the imprinted animals a short time after the odor experience or housed the animals individually until testing in adulthood. Cohabitation with conspecifics between olfactory experience and odor testing interferes with the imprinting process. Olfactory imprinting with a perfume odor enhances the sexual arousal value of that odor in adult male mice that have mated with females having the same perfume odor. Males experiencing a perfume odor in infancy show no arousal when presented with that odor if they have had sexual experience with unperfumed females (Nyby *et al.*, 1978).

Infant rats can be "imprinted" to odors without reinforcement (Leon, Galef, and Behse, 1977) and can learn to avoid odors that have been associated with illness (Rudy and Cheatle, 1977). Leon (1978a) suggests that simple familiarity with an odor will enhance its attractiveness and novel odors are aversive. The infant rat thus appears to be capable of more than one kind of learning and may develop odor preferences or aversions through different types of experience.

XI. Summary

This review has integrated research on three questions related to olfactory communication in mammals:

1. What information is contained in odorous secretions?
2. What are the chemical and microbiological bases of this information?
3. What determines the reactions of conspecifics to this odorous information?

I have categorized nine different types of information contained in mammalian social odors. These are the species, age, sex, colony membership, individuality, social status, reproductive state, maternal state, and stress odors. These types of information may not be independent: social status, colony, and individuality may be identified on the basis of individual differences in other odors.

The information available in chemical secretions has been difficult to identify. Social odors are modified by diet and hormone levels and by bacterial action. While chemical bases have been found for species, sex, age, rut, maternal, and individual odors, the exact identity of the odor source is known in only a few cases, and even the "active components" seldom elicit the same responses as whole gland secretions. There are no chemicals that have been identified as the bases of colony, social status, or stress odors.

When the chemicals and bacteria responsible for producing the social odors have been identified, the responses of test animals show large individual differences. Responses to olfactory stimuli depend on hormonal and experiential factors. The relationships between olfactory primer effects, hormone levels, early olfactory "imprinting," and social learning during adolescence and adulthood have only begun to be investigated.

Theoretical models in the study of population regulation, sexual selection, kinship recognition, altruism, parental care, and territoriality infer that animals recognize particular individuals and specific relationships, and such recognition may depend to some extent on the information contained in olfactory signals.

Acknowledgments

The preparation of this paper was undertaken while I was a Postdoctoral Research Fellow in the Animal Behaviour Research Group at the University of Oxford. I would like to thank David McFarland for research facilities at Oxford; Bob Rodger, John Fentress, Graham Goddard, and Marian Dawkins for comments on earlier drafts of the manuscript; and Robert A. Hinde and Jay S. Rosenblatt for editorial assistance. I am grateful to the National Research Council of Canada and to the Department of Psychology, Dalhousie University for their support of this work.

References

Alberts, J. R. 1974. Producing and interpreting experimental olfactory deficits. *Physiol. Behav.* **12,** 657–670.

Alberts, J. R. 1978. Huddling by rat pups: Multisensory control of contact behavior. *J. Comp. Physiol. Psychol.* **92,** 220–230.

Albone, E. 1977. Ecology of mammals—a new focus for chemical research. *Chem. Brit.* **13,** 92–99.

Albone, E. S., and Grönneberg, T. O. 1977. Lipids of the anal sac secretions of the red fox, *Vulpes vulpes,* and of the lion, *Panthera leo. J. Lipid Res.* **18,** 474–479.

Albone, E. S., and Perry, G. C. 1976. Anal sac secretion of the red fox, *Vulpes vulpes;* volatile fatty acids and diamines: Implications for a fermentation hypothesis of chemical recognition. *J. Chem. Ecol.* **2,** 101–111.

Albone, E. S., Eglinton, G., Walker, J. N., and Ware, G. C. 1974. The anal sac secretion of the red fox (*Vulpes vulpes);* its chemistry and microbiology. A comparison with the anal sac secretion of the lion (*Panthera leo). Life Sci.* **14,** 387–400.

Albone, E. S., Robins, S. P., and Patel, D. 1976. 5-aminovaleric acid, a major free amino acid component of the anal sac secretion of the red fox, *Vulpes vulpes. Comp. Biochem. Physiol.* **55B,** 483–486.

Albone, E. S., Gosden, P. E., and Ware, G. C. 1977. Bacteria as a source of chemical signals in mammals. *In* "Chemical Signals in Vertebrates" (D. Müller-Schwarze and M. M. Mozell, eds.), pp. 35–43. Plenum, New York.

Albone, E. S., Gosden, P. E., Ware, G. C., Macdonald, D. W., and Hough, N. G. 1978. Bacterial action and chemical signalling in the red fox (*Vulpes vulpes)* and other mammals. *In* "Flavor Chemistry and Animal Foods (R. W. Bullard, ed.), pp. 78–91. American Chemical Society Symposium Series, No. 67.

Altmann, D. 1969. "Harnen und Koten bei Säugetieren." Ziemsen Verlag, Wittenberg Lutherstadt.

Anisko, J. J. 1977. Hormonal substrate of estrous odor preference in beagles. *Physiol. Behav.* **18,** 13–17.

Archer, J. 1968. The effects of strange male odours on aggressive behaviour in male mice. *J. Mammal.* **49,** 572–575.

Archer, J. 1969. Adrenocortical responses to olfactory social stimuli in male mice. *J. Mammal.* **50,** 839–841.

Archer, J. 1970. Effects of aggressive behavior on the adrenal cortex in male laboratory mice. *J. Mammal.* **51,** 327–332.

Aron, C., and Chateau, D. 1971. Presumed involvement of pheromones in mating behavior in rat. *Horm. Behav.* **2,** 315.

Baldwin. B. A., and Meese, G. B. 1977. The ability of sheep to distinguish between conspecifics by means of olfaction. *Physiol. Behav.* **19,** 803–808.

Baldwin, J. D. 1968. The social behavior of adult squirrel monkeys (*Saimiri sciureus*) in a seminatural environment. *Folia Primatol.* **9,** 281–314.

Barnett, S. A. 1967. "A Study in Behaviour." Methuen, London.

Bateson, P. P. G. 1966. Characteristics and context of imprinting. *Biol. Rev.* **41,** 177–220.

Batty, J. 1978. Acute changes in plasma testosterone levels and their relation to measures of sexual behavior in the male house mouse (*Mus musculus*). *Anim. Behav.* **26,** 349–357.

Beach, F. A., and Gilmore, J. 1949. Response of male dogs to urine from females in heat. *J. Mammal.* **30,** 391–392.

Beach, F. A., and Jaynes, J. 1956a. Studies of maternal retrieving in rats. I. Recognition of young. *J. Mammal.* **37,** 177–180.

Beach, F. A., and Jaynes, J. 1956b. Studies of maternal retrieving in rats. III. Sensory cues involved in the lactating female's response to her young. *Behaviour* **10,** 104–125.

Beach, F. A., and Merari, A. 1970. Coital behavior in dogs. V. Effects of estrogen and progesterone on mating and other social behavior in the bitch. *J. Comp. Physiol. Psychol. Monogr. 70* (1, pt. 2).

Beauchamp, G. K. 1973. Attraction of male guinea pigs to conspecific urine. *Physiol. Behav.* **10,** 589–594.

Beauchamp, G. K. 1974. The perineal scent gland and social dominance in the male guinea pig. *Physiol. Behav.* **13,** 669–673.

Beauchamp, G. K., and Berüter, J. 1973. Source and stability of attractive components in guinea pig (*Cavia porcellus*) urine. *Behav. Biol.* **9,** 43–47.

Beauchamp, G. K., Doty, R. L., Moulton, D. B., and Mugford, R. A. 1976. The pheromone concept in mammalian chemical communication: A critique. *In* "Mammalian Olfaction, Reproductive Processes and Behavior" (R. L. Doty, ed.), pp. 144–160. Academic Press, New York.

Bertram, B. C. R. 1976. Kin selection in lions and in evolution. *In* "Growing Points in Ethology" (P. P. G. Bateson and R. A. Hinde, eds.), pp. 239–280. Cambridge Univ. Press, London and New York.

Bethe, A. 1932. Vernachlässigte hormone. *Naturwissenschaften* **20,** 177–181.

Bo, W. J. 1970. The effect of progesterone and progesterone-estrogen on the glycogen deposition in the vagina of the squirrel monkey. *Am. J. Obstet. Gynecol.* **107,** 524–530.

Bowers, J. M., and Alexander, B. K. 1967. Mice: Individual recognition by olfactory cues. *Science* **158,** 1208–1210.

Breen, N. F., and Leshner, A. I. 1977. Maternal pheromone: A demonstration of its existence in the mouse (*Mus musculus*). *Physiol. Behav.* **18,** 527–529.

Bronson, F. H. 1968. Pheromonal influences on mammalian reproduction. *In* "Perspectives in Reproduction and Sexual Behavior" (M. Diamond, ed.), pp. 341–361. Indiana Univ. Press, Bloomington.

Bronson, F. H. 1971. Rodent pheromones. *Biol. Reprod.* **4,** 344–357.

Bronson, F. H. 1976. Urine marking in mice: Causes and effects. *In* "Mammalian Olfaction, Reproductive Processes and Behavior" (R. L. Doty, ed.), pp. 119–143. Academic Press, New York.

Bronson, F. H., and Caroom, D. 1971. Preputial gland of the male mouse: Attractant function. *J. Reprod. Fertil.* **25**, 279–282.

Bronson, F. H., and Desjardins, C. 1974. Relationships between scent marking by male mice and the pheromone-induced secretion of the gonadotropic and ovarian hormones that accompany puberty in female mice. *In* "Reproductive Behavior" (A. Montagna and W. A. Sadler, eds.), pp. 157–178. Plenum, New York.

Bronson, F. H., and Eleftheriou, B. E. 1963. Influence of strange males on implantation in the deermouse. *Gen. Comp. Endocrinol.* **3**, 515–518.

Bronson, F. A., and Eleftheriou, B. E. 1964. Chronic physiological effects of fighting in mice. *Gen. Comp. Endocrinol.* **4**, 9–14.

Bronson, F. H., and Marsden, H. M. 1973. The preputial gland as an indicator of social dominance in male mice. *Behav. Biol.* **9**, 625–628.

Bronson, F. H., and Maruniak, J. A. 1976. Differential effects of male stimuli on follicle-stimulating hormone, luteinizing hormone, and prolactin secretion in prepubertal female mice. *Endocrinology* **98**, 1101–1108.

Bronson, F. H., and Whitten, W. K. 1968. Oestrus-accelerating pheromone of mice: Assay, androgen-dependency, and presence in bladder urine. *J. Reprod. Fertil.* **15**, 131–134.

Brown, R. E. 1975. Object-directed urine-marking by male rats (*Rattus norvegicus*). *Behav. Biol.* **15**, 251–254.

Brown, R. E. 1977. Odor preference and urine-marking scales in male and female rats: Effects of gonadectomy and sexual experience on responses to conspecific odors. *J. Comp. Physiol. Psychol.* **91**, 1190–1206.

Brown, R. E. 1978a. Hormonal control of odor preferences and urine marking in male and female rats. *Physiol. Behav.* **20**, 21–24.

Brown, R. E. 1978b. Rearing environment and the generalization of odor preferences in infant rats. Paper presented at the third ECRO Congress, Pavia, Italy.

Brown, W. L., Jr. 1968. An hypothesis concerning the function of the metapleural glands in ants. *Am. Nat.* **102**, 188–191.

Brownlee, R. G., Silverstein, R. M., Müller-Schwarze, D., and Singer, A. G. 1969. Isolation, identification and function of the chief component of the male tarsal scent in black-tailed deer. *Nature (London)* **221**, 284–285.

Bruce, H. M., and Parrott, D. M. V. 1960. Role of olfactory sense in pregnancy block by strange males. *Science* **131**, 1526.

Cain, D. P. 1974. The role of the olfactory bulb in limbic mechanisms. *Psychol. Bull.* **81**, 654–671.

Caroom, D., and Bronson, F. H. 1971. Responsiveness of female mice to preputial attractant: Effects of sexual experience and ovarian hormones. *Physiol. Behav.* **7**, 659–662.

Carr, W. J., and Caul, W. F. 1962. The effect of castration in the rat upon discrimination of sex odors. *Anim. Behav.* **10**, 20–27.

Carr, W. J., Loeb, L. S., and Dissinger, M. L. 1965. Responses of rats to sex odors. *J. Comp. Physiol. Psychol.* **59**, 370–377.

Carr, W. J., Loeb, L. S., and Wylie, N. R. 1966. Responses to feminine odors in normal and castrated male rats. *J. Comp. Physiol. Psychol.* **62**, 336–338.

Carr, W. J., Krames, L., and Costanzo, D. J. 1970a. Previous sexual experience and olfactory preference for novel versus original sex partners in rats. *J. Comp. Physiol. Psychol.* **71**, 216–222.

Carr, W. J., Martorano, R. D., and Krames, L. 1970b. Responses of mice to odors associated with stress. *J. Comp. Physiol. Psychol.* **71**, 223–228.

Carr, W. J., Wylie, N. R., and Loeb, L. S. 1970c. Responses of adult and immature rats to sex odors. *J. Comp. Physiol. Psychol.* **72**, 51–59.

Carr, W. J., Roth, P., and Amore, M. 1971. Responses of male mice to odors from stressed vs. nonstressed males and females. *Psychonom. Sci.* **25**, 275-276.

Carr, W. J., Yee, L., Gable, D., and Marasco, E. 1976. Olfactory recognition of conspecifics by domestic Norway rats. *J. Comp. Physiol. Psychol.* **90**, 821-828.

Carter, C. S. 1972. Effects of olfactory experience on the behaviour of the guinea pig (*Cavia porcellus*). *Anim. Behav.* **20**, 54-60.

Carter, C. S., and Marr, J. N. 1970. Olfactory imprinting and age variables in the guinea pig (*Cavia porcellus*). *Anim. Behav.* **18**, 238-244.

Champlin, A. K. 1971. Suppression of oestrus in grouped mice: The effects of various densities and the possible nature of the stimulus. *J. Reprod. Fertil.* **27**, 233-241.

Chapman, V. M., Desjardins, C., and Whitten, W. K. 1970. Pregnancy block in mice: Changes in pituitary LH, LTH, and plasma progesterone levels. *J. Reprod. Fertil.* **21**, 333-337.

Chateau, D., Roos, J., Plas-Roser, S., Roos, S., and Aron, C. 1976. Hormonal mechanisms involved in the control of oestrus cycle duration by the odour of urine in the rat. *Acta Endocrinol.* **82**, 426-435.

Claesson, A., and Silverstein, R. M. 1977. Chemical methodology in the study of mammalian communications. *In* "Chemical Signals in Vertebrates" (D. Müller-Schwarze and M. M. Mozell, eds.), pp. 71-94. Plenum, New York.

Colby, D. R., and Vandenbergh, J. G. 1974. Regulatory effects of urinary pheromones on puberty in the mouse. *Biol. Reprod.* **11**, 268-279.

Collerain, I. 1978. Frustration odor of rats receiving small numbers of prior rewarded running trials. *J. Exp. Psychol. (Anim. Behav. Proc.)* **4**, 120-130.

Collerain, I. J., and Ludvigson, H. W. 1977. Hurdle-jump responding in the rat as a function of conspecific odor of reward and nonreward. *Anim. Learn. Behav.* **5**, 177-183.

Compton, R. P., Koch, M. D., and Arnold W. J. 1977. Effect of maternal odor on the cardiac rate of maternally separated infant rats. *Physiol. Behav.* **18**, 769-773.

Connor, J. 1972. Olfactory control of aggressive and sexual behavior in the mouse (*Mus musculus*). *Psychonom. Sci.* **27**, 1-3.

Curtis, R. F., Ballantine, J. A., Keverne, E. B., Bonsall, R. W., and Michael, R. P. 1971. Identification of primate sex pheromones and the properties of synthetic attractants. *Nature (London)* **232**, 396-398.

Dagg, A. I., and Windsor, D. E. 1971. Olfactory discrimination limits in gerbils. *Can. J. Zool.* **49**, 283-285.

Darby, E. M., Devor, M., and Chorover, S. L. 1975. A presumptive sex pheromone in the hamster: Some behavioral effects. *J. Comp. Physiol. Psychol.* **88**, 496-502.

Darwin, C. 1887. "The Descent of Man" (2nd ed). Murray, London.

Devor, M., and Schneider, G. E. 1974. Attraction to home-cage odor in hamster pups: Specificity and changes with age. *Behav. Biol.* **10**, 211-221.

Dixon, A. K., and Mackintosh, J. H. 1971. Effects of female urine upon the social behavior of adult male mice. *Anim. Behav.* **19**, 138-140.

Dixon, A. K., and Mackintosh, J. H. 1976. Olfactory mechanisms affording protection from attack to juvenile mice (*Mus musculus*). *Z. Tierpsychol.* **41**, 225-234.

Dominic, D. J. 1966. Observations on the reproductive pheromones in mice. II. Neuroendocrine mechanisms involved in the olfactory block to pregnancy. *J. Reprod. Fertil.* **11**, 415-421.

Domjan, M. 1973. Role of ingestion in odor-toxicosis learning in the rat. *J. Comp. Physiol. Psychol.* **84**, 507-521.

Doty, R. L. 1975. Determination of odor preferences in rodents: A methodological review. *In* "Methods in Olfactory Research" (D. G. Moulton, J. W. Johnson, and A. Turk, eds.), pp. 395-406. Academic Press, London.

Doty, R. L. 1977. A review of recent psychophysical studies examining the possibility of chemical communication of sex and reproductive state in humans. *In* "Chemical Signals in Vertebrates" (D. Müller-Schwarze and M. M. Mozell, eds.), pp. 273-286. Plenum, New York.

Doty, R. L., and Dunbar, C. 1974. Attraction of beagles to conspecific urine, vaginal and anal sac secretion odors. *Physiol. Behav.* **12**, 825-833.

Doty, R. L., Ford, M., Preti, G., and Huggins, G. R. 1975. Changes in the intensity and pleasantness of human vaginal odors during the menstrual cycle. *Science* **190**, 1316-1318.

Doty, R. L., Orndorff, M. M., Leyden, J., and Kligman, A. 1978. Communication of gender from human axillary odors: Relationship to perceived intensity and hedonicity. *Behav. Biol.* **23**, 373-380.

Drickamer, L. C. 1974. Sexual maturation of female house mice: Social inhibition. *Dev. Psychobiol.* **7**, 257-265.

Drickamer, L. C., and Murphy, R. X., Jr. 1978. Female mouse maturation: Effects of excreted and bladder urine from juvenile and adult males. *Dev. Psychobiol.* **11**, 63-72.

Drori, D., and Folman, Y. 1964. Effects of cohabitation on the reproductive system, kidneys and body composition of male rats. *J. Reprod. Fertil.* **8**, 351-359.

Eisenberg, J. F., and Kleiman, D. G. 1972. Olfactory communication in mammals. *Annu. Rev. Ecol. Syst.* **3**, 1-32.

Eisenberg, J. F., McKay, G. M., and Jainudeen, M. R. 1971. Reproductive behavior of the asiatic elephant (*Elaphus maximus maximus*). *Behaviour* **38**, 193-225.

Ellis, H. H. 1905 (1920). "Studies in the Psychology of Sex," Vol. IV. "Sexual Selection in Man." Davis, Philadelphia, Pennsylvania.

Elwood, R. W. 1977. Changes in the responses of male and female gerbils (*Meriones unguiculatus*) towards test pups during the pregnancy of the female. *Anim. Behav.* **25**, 46-51.

Epple, G. 1970. Quantitative studies on scent marking in the marmoset (*Callithrix jacchus*). *Folia Primatol.* **13**, 48-62.

Epple, G. 1973. The role of pheromones in the social communication of marmoset monkeys (*Callithicidae*). *J. Reprod. Fertil. Suppl.* **19**, 447-454.

Epple, G. 1974a. Olfactory communication in South American primates. *Ann. N.Y. Acad. Sci.* **224**, 261-278.

Epple, G. 1974b. Primate pheromones. *In* "Phermones" (M. C. Birch, ed.) pp. 366-385. North Holland Publ., Amsterdam.

Epple, G. 1974c. Pheromones in primate reproduction and social behavior. *In* "Reproductive Behavior" (W. Montagna and W. A. Sadler, eds.), pp. 131-155. Plenum, New York.

Epple, G. 1976. Chemical communication and reproductive processes in nonhuman primates. *In* "Mammalian Olfaction, Reproductive Processes and Behavior" (R. L. Doty, ed.), pp. 257-282. Academic Press, New York.

Fiedler, W. 1964. Die Haut der Säugetiere als Ausdrucksorgan. *Stud. Gen.* **17**, 362-390.

Folman, Y., and Drori, D. 1966. Effects of social isolation and of female odours on the reproductive system, kidneys and adrenals of unmated male rats. *J. Reprod. Fertil.* **11**, 43-50.

Fossey, D. 1978. His name was digit. *Int. Primate Protect. League Newslett.* **5**(2), 2-7.

Fox, K. A. 1968. Effects of prepubertal habitation conditions on the reproductive physiology of the male house mouse. *J. Reprod. Fertil.* **17**, 75-85.

Freeman, N. C. G., and Rosenblatt, J. S. 1978a. The interrelationship between thermal and olfactory stimulation in the development of home orientation in newborn kittens. *Dev. Psychobiol.* **11**, 437-457.

Freeman, N. C. G., and Rosenblatt, J. S. 1978b. Specificity of litter odors in the control of home orientation among kittens. *Dev. Psychobiol.* **11**, 459-468.

Galef, B. G., Jr., and Heiber, L. 1976. Role of residual olfactory cues in the determination of feeding site selection and exploration patterns of domestic rats. *J. Comp. Physiol. Psychol.* **90**, 727-739.

Gawienowski, A. M. 1977. Chemical attractants of the rat preputial gland. *In* "Chemical Signals in Vertebrates" (D. Müller-Schwarze and M. M. Mozell, eds.), pp. 45–60. Plenum, New York.

Gawienowski, A. M., Orsulak, P. J., Stacewicz-Sapuntzakis, M., and Joseph, B. 1975. Presence of sex pheromone in preputial gland of male rats. *J. Endocrinol.* **67**, 283–288.

Gawienowski, A. M., Orsulak, P. J., Stacewicz-Sapuntzakis, M., and Pratt, J. J., Jr. 1976. Attractant effect of female preputial gland extracts on the male rat. *Psychoneuroendocrinology* **1**, 411–418.

Gilder, P. M., and Slater, P. J. B. 1978. Interest of mice in conspecific male odors is influenced by degree of kinship. *Nature (London)* **274**, 364–365.

Glickman, S. E. 1973. Responses and reinforcement. *In* "Constraints on Learning" (R. A. Hinde and J. Stevenson-Hinde, eds.), pp. 207–241. Academic Press, New York.

Globus, G. G., and Cohen, H. B. 1975. Human vaginal odors. *Science* **192**, 96.

Godfrey, J. 1958. The origin of sexual isolation between bank voles. *Proc. R. Phys. Soc. Edinburgh* **27**, 47–55.

Goldfoot, D. A., Dravetz, M. A., Goy, R. W., and Freeman, S. K. 1976. Lack of effect of vaginal lavages and aliphatic acids on ejaculatory responses in rhesus monkeys: behavioral and chemical analyses. *Horm. Behav.* **7**, 1–27.

Goodrich, B. S., and Mykytowycz, R. 1972. Individual and sex differences in the chemical composition of pheromone-like substances from the skin glands of the rabbit, *Oryctolagus cuniculus*. *J. Mammal.* **53**, 540–548.

Gorman, M. L. 1976. A mechanism for individual recognition by odour in *Herpestes auropunctatus (Carnivora: viverridae)*. *Anim. Behav.* **24**, 141–145.

Gorman, M. L., Nedwell, D. B., and Smith, R. M. 1974. An analysis of the contents of the anal scent pockets of *Herpestes auropunctatus (Carnivora: viverridae)*. *J. Zool.* **172**, 389–399.

Grau, G. A. 1976. Olfaction and reproduction in ungulates. *In* "Mammalian Olfaction, Reproductive Processes and Behavior" (R. L. Doty, ed.), pp. 219–242. Academic Press, New York.

Gregoire, A. T., Kandil, O., and Ledger, W. J. 1971. The glycogen content of human vaginal epithelial tissue. *Fertil. Steril.* **22**, 64–68.

Gregory, E. H., and Pfaff, D. W. 1971. Development of olfactory-guided behavior in infant rats. *Physiol. Behav.* **6**, 573–576.

Gregory, E., Engel, K., and Pfaff, D. 1975. Male hamster preference for odors of female hamster vaginal discharges: Studies of experiential and hormonal determinants. *J. Comp. Physiol. Psychol.* **89**, 442–446.

Halpin, Z. T. 1974. Individual differences in the biological odors of the mongolian gerbil (*Meriones unguiculatus*). *Behav. Biol.* **11**, 253–259.

Halpin, Z. T. 1976. The role of individual recognition by odors in the social interactions of the mongolian gerbil (*Meriones unguiculatus*). *Behaviour* **58**, 117–130.

Harrington, J. 1974. Olfactory communication in *Lemur fulvus*. *In* "Prosimian Biology" (R. D. Martin, G. A. Doyle, and A. C. Walker, eds.), pp. 331–346. Duckworth, London.

Harrington, J. E. 1976a. Discrimination between individuals by scent in *Lemur fulvus*. *Anim. Behav.* **24**, 207–212.

Harrington, J. E. 1976b. Recognition of territorial boundaries by olfactory cues in mice (*Mus musculus*). *Z. Tierpsychol.* **41**, 295–306.

Harrington, J. 1979. Responses of *Lemur fulvus* to scents of different subspecies of *L. fulvus* and to scents of different species of lemuriformes. *Z. Tierpsychol.* **49**, 1–9.

Haslitt, G. W. 1973. The significance of anal scent marking in the Eastern woodchuck. *Bull. Ecol. Soc. Amer.* **54**, 43–44. (Abstr.)

Hayashi, S., and Kimura, T. 1973. Respones of normal and neonatally estrogenized female mice to the odor of males. *Sci. Pap. Coll. Gen. Educ. Univ. Tokyo* **23**, 39–43.

Hayashi, S., and Kimura, T. 1974. Sex-attractant emitted by female mice. *Physiol. Behav.* **13**, 563–567.

Hayashi, S., and Kimura, T. 1978. Effects of exposure to males on sexual preference in female mice. *Anim. Behav.* **26**, 290–295.

Hersher, L., Richmond, J. B., and Moore, A. V. 1963a. Maternal behaviour in sheep and goats. *In* "Maternal Behaviour in Mammals" (H. L. Rheingold, ed.), pp. 203–222. Wiley, New York.

Hersher, L., Richmond, J. B., and Moore, A. V. 1963b. Modifiability of the critical period for the development of maternal behavior in sheep and goats. *Behaviour* **20**, 311–320.

Hesterman, E. R., and Mykytowycz, R. 1968. Some observations on the intensities of odors of anal gland secretions from the rabbit *Oryctolagus cuniculus. CSIRO. Wildl. Res.* **13**, 71–81.

Hinde, R. A. 1970. "Animal Behaviour," 2nd ed. McGraw-Hill, New York.

Hold, B., and Schleidt, M. 1977. The importance of human odor in non-verbal communication. *Z. Tierpsychol.* **43**, 225–238.

Hulet, C. V. 1966. Behavioral, social and psychological factors affecting mating time and breeding efficiency in sheep. *J. Anim. Sci. Suppl.* **25**, 5–20.

Irwin, F. W. 1958. An analysis of the concepts of discrimination and preference. *Am. J. Psychol.* **71**, 152–163.

Johns, M. A., Feder, H. H., Komisaruk, B. R., and Mayer, A. D. 1978. Urine-induced reflex ovulation in anovulatory rats may be a vomeronasal effect. *Nature (London)* **272**, 446–448.

Johnson, R. P. 1973. Scent marking in mammals. *Anim. Behav.* **21**, 521–535.

Johnston, R. E. 1974. Sexual attraction function of golden hamster vaginal secretion. *Behav. Biol.* **12**, 111–117.

Johnston, R. E. 1975a. Scent marking by male golden hamster (*Mesocricetus auratus*). I. Effects of odors and social encounters. *Z. Tierpsychol.* **37**, 75–98.

Johnston, R. E. 1975b. Sexual excitation function of hamster vaginal secretion. *Anim. Learn. Behav.* **3**, 161–166.

Johnston, R. E., and Zahorik, D. M. 1975. Taste aversions to sexual attractants. *Science* **189**, 893–894.

Johnston, R. E., Zahorik, D. M., Immler, K., and Zakon, H. 1978. Alterations of male sexual behavior by learned aversions to hamster vaginal secretion. *J. Comp. Physiol. Psychol.* **92**, 85–93.

Jones, R. B., and Nowell, N. W. 1973a. The effect of urine on the investigatory behaviour of male albino mice. *Physiol. Behav.* **11**, 35–38.

Jones. R. B., and Nowell, N. W. 1973b. Aversive and aggression-promoting properties of urine from dominant and subordinate male mice. *Anim. Learn. Behav.* **1**, 207–210.

Jones, R. B., and Nowell, N. W. 1974. A comparison of the aversive and female attractant properties of urine from dominant and subordinate male mice. *Anim Learn. Behav.* **2**, 141–144.

Kalmus, H. 1955. The discrimination by the nose of the dog of individual human odors and in particular of the odors of twins. *Br. J. Anim. Behav.* **3**, 25–31.

Kaplan, J. N., Cubicciotti, D., and Redican, W. K. 1977. Olfactory discrimination of squirrel monkey mothers by their infants *Dev. Psychobiol.* **10**, 447–453.

Karlson, P., and Lüscher, N. 1959. 'Pheromones': A new term for a class of biologically active substances. *Nature (London)* **183**, 55–56.

Keverne, E. B. 1976. Sexual receptivity and attractiveness in the female rhesus monkey. *In* "Advances in the Study of Behavior" (J. S. Rosenblatt, R. A. Hinde, E. Shaw, and C. Beer, eds.), Vol. 7, pp. 155–200. Academic Press, New York.

Keverne, E. B., and Michael, R. P. 1971. Sex-attractant properties of ether extracts of vaginal secretions from rhesus monkeys. *J. Endocrinol.* **51**, 313–322.

King, M. G., Pfister, H. P., and Di Giusto, E. L. 1975. Differential preference for and activation by the odoriferous compartment of a shuttlebox in fear-conditioned and naive rats. *Behav. Biol.* **13,** 175–181.

Kirschenblatt, J. 1962. Terminology of some biologically active substances and validity of the term "pheromones". *Nature (London)* **195,** 916–917.

Klopfer, P. H. 1970. Discrimination of young in galagos. *Folia Primatol.* **13,** 137–143.

Klopfer, P. H., Adams, D. K., and Klopfer, M. S. 1964. Maternal 'imprinting' in goats. *Proc. Natl. Acad. Sci. U.S.A.* **52,** 911–914.

Klopfer, P. H., and Gamble, J. 1966. Maternal 'imprinting' in goats: The role of chemical senses. *Z. Tierpsychol.* **23,** 588–592.

Kolb, A. 1977. Wie erkennen sich Mutter und Junges des Mausohrs, *Myotis myotis,* bei der Rückkehr vom Jagdflug wieder? *Z. Tierpsychol.* **44,** 423–431.

Krames, L., and Shaw, B. 1973. Role of previous experience in the male rat's reaction to odors from group and alien conspecifics. *J. Comp. Physiol. Psychol.* **82,** 444–448.

Krames, L., Carr, W. J., and Bergman, B. 1969. A pheromone associated with social dominance among male rats. *Psychonom. Sci.* **6,** 11–12.

Krames, L., and Mastromatteo, L. A. 1973. Role of olfactory stimuli during copulation in male and female rats. *J. Comp. Physiol. Psychol.* **85,** 528–535.

Landauer, M. R., Wiese, R. E., and Carr, W. J. 1977. Responses of sexually experienced and naive male rats to cues from receptive vs. nonreceptive females. *Anim. Learn. Behav.* **5,** 398–402.

Landauer, M. R., Banks, E. M., and Carter, C. S. 1978. Sexual and olfactory preferences of naive and experienced male hamsters. *Anim. Behav.* **26,** 611–621.

Lecyk, M. 1967. The influence of crowded population stimuli on the reproduction of the common vole. *Acta Theriol.* **12,** 177–179.

Lederer, E. 1949. Chemistry and biochemistry of some mammalian secretions and excretions. *J. Chem. Soc.* 2115–2125.

Lederer, E. 1950. "Odeurs et Parfumes des Animaux." Springer-Verlag, Berlin and New York.

Lee, C. T. 1976. Agonistic behavior, sexual attraction, and olfaction in mice. *In* "Mammalian Olfaction, Reproductive Processes, and Behavior" (R. L. Doty, ed.), pp. 161–180. Academic Press, New York.

Lee, S., van der, and Boot, L. M. 1955. Spontaneous pseudopregnancy in mice. *Acta Physiol. Pharmacol. Neerl.* **4,** 442–443.

Lee, S. van der, and Boot, L. M. 1956. Spontaneous pseudopregnancy in mice II. *Acta Physiol. Pharmacol. Neerl.* **5,** 213–214.

Leidahl, L. C., and Moltz, H. 1975. Emission of the maternal pheromone in the nulliparous female and failure of emission in the adult male. *Physiol Behav.* **14,** 421–424.

Leidahl, L. C., and Moltz, H. 1977. Emission of the maternal pheromone in nulliparous and lactating females. *Physiol. Behav.* **18,** 399–402.

LeMagnen, J. 1952. Les phenomenes olfacto-sexuels chez le rat blanc. *Arch. Sci. Physiol.* **6,** 295–331.

Lent, P. C. 1974. Mother-infant relationships in ungulates. *In* "The Behaviour of Ungulates and its Relation to Management" (V. Geist and F. Walther, eds.), Vol. 1, pp. 14–55. IUCN New Ser., No. 24, Morges, Switzerland.

Leon, M. 1974. Maternal pheromone. *Physiol. Behav.* **10,** 441–453.

Leon, M. 1975. Dietary control of maternal pheronome in the lactating rat. *Physiol. Behav.* **14,** 311–319.

Leon, M. 1978a. Filial responsiveness to olfactory cues in the laboratory rat. *In* "Advances in the Study of Behavior" (J. S. Rosenblatt, R. A. Hinde, C. Beer, and M.-C. Busnel, eds.), Vol. 8, pp. 117–153. Academic Press, New York.

Leon, M. 1978b. Emission of maternal pheromone. *Science* **201,** 938–939.

Leon, M. and Behse, J. H. 1977. Dissolution of the pheromonal bond: Waning of approach response by weanling rats. *Physiol. Behav.* **18,** 393–397.

Leon, M. and Moltz, H. 1971. Maternal pheromone: Discrimination by preweanling albino rats. *Physiol. Behav.* **7**, 265–267.

Leon, M., and Moltz, H. 1972. The development of the pheromonal bond in the albino rat. *Physiol. Behav.* **8**, 683–686.

Leon, M., and Moltz, H. 1973. Endocrine control of the maternal pheromone in the postpartum female rat. *Physiol. Behav.* **10**, 65–67.

Leon, M., Galef, B. G., Jr., and Behse, J. H. 1977. Establishment of pheromonal bonds and diet choice in young rats by odor pre-exposure. *Physiol. Behav.* **18**, 387–391.

Lincoln, G. A. 1971. The seasonal reproductive changes in the red deer stag (*Cervus elaphus*). *J. Zool.* **163**, 105–123.

Lincoln, G. A., Youngson, R. W., and Short, R. V. 1970. The social and sexual behavior of the red deer stag. *J. Reprod. Fertil. Suppl.* **11**, 71–103.

Lindsay, D. R. 1965. The importance of olfactory stimuli in the mating behavior of the ram. *Anim. Behav.* **13**, 75–78.

Lombardi, J. R., and Vandenbergh, J. G. 1977. Pheromonally induced sexual maturation in females: Regulation by the social environment of the male. *Science* **196**, 545–546.

Lombardi, J. R., Vandenbergh, J. G. and Whitsett, J. M. 1976. Androgen control of the sexual maturation pheromone in house mouse urine. *Biol. Reprod.* **15**, 179–186.

Lorenz, K. 1937. The companion in the bird's world. *Auk* **54**, 245–273.

Lott, D. F., and Hopwood, J. H. 1972. Olfactory pregnancy-block in mice (*Mus musculus*): An unusual response acquisition paradigm. *Anim. Behav.* **20**, 263–267.

Lydell, K., and Doty, R. L. 1972. Male rat odor preferences for female urine as a function of sexual experience, urine age, and urine source. *Horm. Behav.* **3**, 205–212.

McArthur, J. W., Ovadia, J., Smith, B. W., and Bashir-Farahmand, J. 1972. The menstrual cycle of the bonnet monkey (*Macaca radiata*). Changes in cervical mucous secretion, vaginal cytology, sex skin and urinary estrogen secretion. *Folia Primatol.* **17**, 107–121.

McCartney, W. 1968. "Olfaction and Odours." Springer-Verlag, Berlin and New York.

McCarty, R., and Southwick, C. H. 1977. Cross-species fostering: Effects on the olfactory preferences of *Onychomys torridus* and *Peromyscus leucopus*. *Behav. Biol.* **19**, 255–260.

McClintock, M. K. 1971. Menstrual synchrony and suppression. *Nature (London)* **229**, 244–245.

McClintock, M. K. 1974. Sociobiology of reproduction in the Norway rat (*Rattus norvegicus*). Estrus synchrony and the role of the female rat in copulatory behavior. Unpublished Ph.D. dissertation, University of Pennsylvania.

McIntosh, T. K., and Drickamer, L. C. 1977. Excreted urine, bladder urine, and the delay of sexual maturation in female house mice. *Anim. Behav.* **25**, 999–1004.

Macdonald, D. W., Hough, N. G., Perry, G. R., and Blizzard, R. 1978. The anal sac of the red fox (*Vulpes vulpes*)—a multidisciplinary investigation. Unpublished manuscript.

Macfarlane, A. 1975. Olfaction in the development of social preferences in the human neonate. *In* "Parent-Infant Interaction" (Ciba Foundation Symposium 33, pp. 103–117). Elsevier, Amsterdam.

Mackay-Sim. A. 1978. Stress-related odors in rats. Unpublished Ph.D. dissertation, Macquarie University, Australia.

Mackintosh, J. H. 1970. Territory formation by laboratory mice. *Anim. Behav.* **18**, 177–183.

Macrides, F., Bartke, A., Fernandez, F., and D'Angelo, W. 1974. Effects of exposure to vaginal odor and receptive females on plasma testosterone in the male hamster. *Neuroendocrinology* **15**, 355–364.

Macrides, F., Bartke, A., and Dalterio, S. 1975. Strange females increase plasma testosterone levels in mice. *Science* **189**, 1104–1106.

Macrides, F., Johnson, P. A., and Schneider, S. P. 1977. Responses of the male golden hamster to vaginal secretion and dimethyl disulfide: Attraction versus sexual behavior. *Behav. Biol.* **20**, 377–386.

Mainardi, D., Marsan, M., and Pasquali, A. 1965. Causation of sexual preferences in the house mouse: The behaviour of mice reared by parents whose odour was artificially altered. *Soc. Ital. Sci. Nat. Museo Civico Storia Mat. Milano* **104**, 825–838.

Manley, G. H. 1974. Functions of the external genital glands of *Perodicticus* and *Arctocebus*. *In* "Prosimian Biology" (R. D. Martin, G. A. Doyle, and A. C. Walkers, eds.), pp. 313–329. Duckworth, London.

Marchlewska-Koj, A. 1977. Pregnancy block elicited by urinary proteins in male mice. *Biol. Reprod.* **17**, 729–732.

Marr, J. N., and Gardner, L. E. 1965. Early olfactory experience and later social behavior in the rat: Preference, sexual responsiveness, and care of young. *J. Genet. Psychol.* **107**, 167–174.

Marr, J. N., and Lilliston, L. G. 1969. Social attachment in rats by odor and age. *Behaviour* **33**, 277–282.

Martin, R. D. 1968. Reproduction and ontogeny in tree shrews (*Tupaia belangeri*) with reference to their general behaviour and taxonomic relationships. *Z. Tierpsychol.* **25**, 409–495, 505–532.

Maruniak, J. A., Coquelin, A., and Bronson, F. H. 1978. The release of LH in male mice in response to female urinary odors: Characteristics of the response in young males. *Biol. Reprod.* **18**, 251–255.

Maruniak, J. J., Owen, K., Bronson, F. H., and Desjardins, C. 1974. Urinary marking in male house mice: Responses to novel environmental and social stimuli. *Physiol. Behav.* **12**, 1035–1039.

Mason, J. W. 1968. Organization of the multiple endocrine responses to avoidance in the monkey. *Psychosom. Med.* **30**, 774–790.

Meese, G. B., Connor, D. J., and Baldwin, B. A. 1975. Ability of the pig to distinguish between conspecific urine samples using olfaction. *Physiol. Behav.* **15**, 121–125.

Mertl, A. L. 1975. Discrimination by scent in a primate. *Behav. Biol.* **14**, 505–509.

Michael, R. P., and Bonsall, R. W. 1977. Chemical signals and primate behavior. *In* "Chemical Signals in Vertebrates" (D. Müller-Schwarze and M. M. Mozell, eds.), pp. 251–271. Plenum, New York.

Michael, R. P., and Keverne, E. B. 1968. Pheromones in the communication of sexual status in primates. *Nature (London)* **218**, 746–749.

Michael, R. P., and Keverne, E. B. 1970. Primate sex pheromones of vaginal origin. *Nature (London)* **225**, 84–85.

Michael, R. P., Bonsall, R. W., and Warner, P. 1974. Human vaginal secretions: Volatile fatty acid content. *Science* **186**, 1217–1219.

Michael, R. P., Zumpe, D., Richter, M., and Bonsall, R. W. 1977. Behavioral effects of a synthetic mixture of aliphatic acids in rhesus monkeys (*Macaca mulatta*). *Horm. Behav.* **9**, 296–308.

Milligan, S. R. 1974. Social environment and ovulation in the vole (*Microtus agrestis*). *J. Reprod. Fertil.* **41**, 35–47.

Milligan, S. R. 1975. Further observations on the influence of the social environment on ovulaation in the vole, *Microtus agrestis*. *J. Reprod. Fertil.* **44**, 543–544.

Milligan, S. R. 1976. Pregnancy blocking in the vole, *Microtus agrestis*. I. Effect of the social environment. *J. Reprod. Fertil.* **46**, 91–95.

Moltz, H., and Leidahl, L. C. 1977. Bile, prolactin and the maternal pheromone. *Science* **196**, 81–83.

Moore, R. E. 1965. Olfactory discrimination as an isolating mechanism between *Peromyscus maniculatus* and *Peromyscus polionotus*. *Am. Mdl. Nat.* **73**, 85–100.

Morgan, P. D., Arnold, G. W., and Lindsay, D. R. 1972. A note on the mating behaviour of ewes with various senses impaired. *J. Reprod. Fertil.* **30**, 151–152.

Morris, C. A., and Morris, D. F. 1967. 'Normal' vaginal microbiology of women of childbearing age in relation to the use of oral contraceptives and vaginal tampons. *J. Clin. Pathol.* **20**, 636–640.

Morrison, R. R., and Ludvigson, H. W. 1970. Discrimination by rats of conspecific odors of reward and nonreward. *Science* **167**, 904–905.

Mugford, R. A., and Nowell, M. W. 1971a. Endocrine control over production and activity of the anti-aggression pheromone from female mice. *J. Endocrinol.* **49**, 225–232.

Mugford, R. A., and Nowell, M. W. 1971b. The relationship between endocrine status of female opponents and aggressive behaviour of male mice. *Anim. Behav.* **19**, 153–155.

Mugford, R. A., and Nowell, M. W. 1971c. The preputial glands as a source of aggression-promoting odors in mice. *Physiol. Behav.* **6**, 247–249.

Müller-Schwarze, D. 1969. Complexity and relative specificity in a mammalian pheromone. *Nature (London)* **223**, 525–526.

Müller-Schwarze, D. 1971. Pheromones in black tailed deer (*Odocoileus hemionus columbianus*). *Anim. Behav.* **19**, 141–152.

Müller-Schwarze, D. 1977. Complex mammalian behavior and the pheromone bioassay in the field. *In* "Chemical Signals in Vertebrates" (D. Müller-Schwarze and M. M. Mozell, eds.), pp. 413–434. Plenum, New York.

Müller-Schwarze, D., and Müller-Schwarze, C. 1971. Olfactory imprinting in a precocial mammal. *Nature (London)* **229**, 55–56.

Müller-Schwarze, D., and Müller-Schwarze, C. 1975. Subspecies specificity of response to a mammalian social odor. *J. Chem. Ecol.* **1**, 125–131.

Müller-Schwarze, D., Müller-Schwarze, C., Singer, A. G., and Silverstein, R. M. 1974. Mammalian pheromone: Identification of active component in the subauricular scent of the male pronghorn. *Science* **183**, 860–862.

Müller-Velten, H. 1966. Über den Angstgeruch bei der Hausmaus (*Mus musculus*). *Z. Vergl. Physiol.* **52**, 401–429.

Mykytowycz, R. 1966a. Observations on odoriferous and other glands in the Australian wild rabbit *Oryctolagus cuniculus* and the hare *Lepus europaeus*. I. The anal gland. *CSIRO Wildl. Res.* **11**, 11–29.

Mykytowycz, R. 1966b. Observations on odoriferous and other glands in the Australian wild rabbit *Oryctolagus cuniculus*, and the hare *Lepus europaeus*. II. The inguinal glands. *CSIRO Wildl. Res.* **11**, 49–64.

Mykytowycz, R. 1968. Territorial marking by rabbits. *Sci. Am.* **218**, 116–126.

Mykytowycz, R. 1970. The role of skin glands in mammal communication. *In* "Advances in Chemoreception" (J. W. Johnston, D. G. Mouton, and A. Turk, eds.), pp. 327–360. Appleton, New York.

Mykytowycz, R., and Dudzinski, M. L. 1966. A study of the weight of odoriferous and other glands in relation to social status and degree of sexual activity in the wild rabbit, *Oryctolagus cuniculus*. *CSIRO Wildl. Res.* **11**, 31–47.

Mykytowycz, R., and Dudzinski, M. L. 1972. Aggressive and protective behaviour of adult rabbits *Oryctolagus cuniculus* towards juveniles. *Behaviour* **43**, 97–120.

Mykytowycz, R. and Ward, M. M. 1971. Some reactions of nestlings of the wild rabbit, *Oryctolagus cuniculus*, when exposed to natural rabbit odors. *Forma Functio* **4**, 137–148.

Mykytowycz, R., Hesterman, E. R., Gambale, S., and Dudzinski, M. L. 1976. A comparison of the effectiveness of the odors of rabbits, *Oryctolagus cuniculus*, in enhancing territorial confidence. *J. Chem. Ecol.* **2**, 13–24.

Nigrosh, B. J., Slotnick, B. M., and Nevin, J. A. 1975. Olfactory discrimination, reversal learning, and stimulus control in rats. *J. Comp. Physiol. Psychol.* **89**, 285–294.

Nyby, J., Thiessen, D. D., and Wallace, P. 1970. Social inhibition of territorial marking in the Mongolian gerbil (*Meriones unguiculatus*). *Psychonom. Sci.* **21**, 310–312.

Nyby, J., Whitney, G., Schmitz, S., and Dizinno, G. 1978. Postpubertal experience establishes signal value of mammalian sex odor. *Behav. Biol.* **22**, 545–552.

O'Connell, R. J. 1977. From insect to mammal: Complications of the bioassay. *In* "Chemical Signals in Vertebrates" (D. Müller-Schwarze and M. M. Mozell, eds.), pp. 377–390. Plenum, New York.

Orsulak, P. J., and Gawienowski, A. M. 1972. Olfactory preferences for the rat preputial gland. *Biol. Reprod.* **6**, 219–223.

Ortmann, R. 1960. Die Analregion der Säugetiere. *Handb. Zool.* **8**(26), 1–68.

Oswalt, G. L., and Meier, G. W. 1975. Olfactory, thermal, and tactual influences on infantile ultrasonic vocalizations in rats. *Dev. Psychobiol.* **8**, 129–135.

Parkes, A. S., and Bruce, H. M. 1961. Olfactory stimuli in mammalian reproduction. *Science* **134**, 1049–1054.

Parkes, A. S., and Bruce, H. M. 1962. Pregnancy-block in female mice placed in boxes soiled by males. *J. Reprod. Fertil.* **4**, 303–308.

Peters, R. P., and Mech, L. D. 1975. Scent-marking in wolves: A field study. *Am. Sci.* **63**, 628–637.

Pettijohn, T. F. 1974. Attractiveness of conspecific urine to the male mongolian gerbil. Paper presented at Animal Behavior Society, Champaign, Illinois.

Pfaff, D. W., and Pfaffman, C. 1969. Behavioral and electrophysiological responses of male rats to female rat urine odors. *In* "Olfaction and Taste: III" (C. Pfaffman, ed.), pp. 260–267. Rockefeller Univ. Press, New York.

Pfeiffer, W. 1963. Alarm substances. *Experientia* **19**, 113–123.

Pocock, R. I. 1910. On the specialized cutaneous glands of ruminants *Proc. Zool. Soc. London* 840–982.

Pocock, R. I. 1916. Scent glands in mammals. *Proc. Zool. Soc. London* 742–755.

Porter, R. H., and Doane, H. M. 1976. Maternal pheromone in the spiny mouse (*Acomys cahirinus*). *Physiol. Behav.* **16**, 75–78.

Porter, R. H., and Etscorn, F. 1974. Olfactory imprinting resulting from brief exposure in *Acomys cahirinus*. *Nature (London)* **250**, 732–733.

Porter, R. H., and Etscorn, F. 1976. A sensitive period for the development of olfactory preference in *Acomys cahirinus*. *Physiol. Behav.* **17**, 127–130.

Porter, R. H., and Ruttle, K. 1975. The responses of one-day old *Acomys cahirinus* pups to naturally occurring chemical stimuli. *Z. Tierpsychol.* **38**, 154–162.

Porter, R. H., Fullerton, C., and Berryman, J. C. 1973. Guinea-pig maternal-young attachment behaviour. *Z. Tierpsychol.* **32**, 489–495.

Porter, R. H., Deni, R., and Doane, H. M. 1977. Responses of *Acomys cahirinus* pups to chemical cues produced by a foster species. *Behav. Biol.* **20**, 244–251.

Porter, R. H., Wyrick, M., and Parkey, J. 1978. Sibling recognition in spiny mice (*Acomys cahirinus*). *Behav. Ecol. Sociobiol.* **3**, 61–68.

Pranzarone, G. F. 1969. The effects of differential exposure of male rats to estrous and diestrous female odors upon approach behavior and physiological development. Unpublished Master's thesis, George Peabody College for Teachers, Nashville, Tennessee.

Preti, G., and Huggins, G. R. 1975. Cyclical changes in volatile acidic metabolites of human vaginal secretions and their relation to ovulation. *J. Chem. Ecol.* **1**, 361–376.

Pribram, K. H., and Kruger, L. 1954. Functions of the "olfactory brain." *Ann. N.Y. Acad. Sci.* **58**, 109–138.

Purvis, K., and Haynes, N. B. 1972. The effect of female rat proximity on the reproductive system of male rats. *Physiol. Behav.* **9**, 401–407.

Quadagno, D. M., and Banks, E. M. 1970. The effect of reciprocal cross fostering on the behavior of two species of rodents, *Mus musculus* and *Baiomys taylori ater*. *Anim. Behav.* **18**, 379–390.

Quay, W. B. 1953. Seasonal and sexual differences in the dorsal skin gland of the kangaroo rat (*Dipodomys*). *J. Mammal.* **34**, 1–14.

Ralls, K. 1971. Mammalian scent marking. *Science* **171**, 443–449.

Rasa, O. A. E. 1973. Marking behaviour and its social significance in the African Dwarf Mongoose (*Helogale undulata rufula*). *Z. Tierpsychol.* **32**, 293–318.

Rauschert, K. 1963. Sexuelle Affinität zwishen Arten und Unterarten von Rotelmäusen (*Clethrionomys*). *Biol. Zentralbl.* **82**, 653–664.

Richmond, M., and Conaway, C. H. 1969. Induced ovulation and oestrus in *Microtus ochrogaster*. *J. Reprod. Fertil. Suppl.* **6**, 357–376.

Rogers, J. G., Jr., and Beauchamp, G. K. 1976a. Influence of stimuli from populations of *Peromyscus leucopus* on maturation of young. *J. Mammal.* **57**, 320–330.

Rogers, J. G., Jr., and Beauchamp, G. K. 1976b. Some ecological implications of primer chemical stimuli in rodents. *In* "Mammalian Olfaction, Reproductive Processes and Behavior" (R. L. Doty, ed.), pp. 181–195. Academic Press, New York.

Rose, E., and Drickamer, L. C. 1975. Castration, sexual experience, and female urine odor preferences in adult BDF_1 male mice. *Bull. Psychonom. Soc.* **5**, 84–86.

Rose, R. M., Holaday, J. W., and Bernstein, I. S. 1971. Plasma testosterone, dominance rank and aggressive behaviour in male rhesus monkeys. *Nature (London)* **231**, 366–368.

Rosenblatt, J. S. 1976. Stages in the early behavioral development of altricial young of selected species of non-primate mammals. *In* "Growing Points in Ethology" (P. P. G. Bateson and R. A. Hinde, eds.), pp. 345–383. Cambridge Univ. Press, London and New York.

Rosenblatt, J. S. 1978. The sensorimotor and motivational bases of early behavioral development of selected altricial mammals. *In* "Ontogeny of Learning and Memory" (N. Spear and B. Campbell, eds.) (in press).

Rosenblatt, J. S., Turkewitz, G., and Schneirla, T. C. 1969. Development of home orientation in newly born kittens. *Trans. N.Y. Acad. Sci.* **31**, 231–250.

Rottman, S. J., and Snowdon, C. T. 1972. Demonstration and analysis of an alarm pheromone in mice. *J. Comp. Physiol. Psychol.* **81**, 483–490.

Rudy, J. W., and Cheatle, M. D. 1977. Odor-aversion learning in neonatal rats. *Science* **198**, 845–846.

Russell, M. J. 1976. Human olfactory communication. *Nature (London)* **260**, 520–522.

Sansone-Bazzano, G., Bazzano, G., Reisner, R. M., and Hamilton, J. G. 1972. The hormonal induction of alkyl glyceral, wax and alkyl acetate synthesis in the preputial gland of the mouse. *Biochim. Biophys. Acta* **260**, 35–40.

Sato, N., Haller, E. W., Powell, R. D., and Henkin, R. I. 1974. Sexual maturation in bulbectomized female rats. *J. Reprod. Fertil.* **36**, 301–309.

Schaffer, J. 1940. "Die Hautdrüsenorgane der Säugetiere." Urban Schwarzenberg, Berlin.

Schaller, G. 1967. "The Deer and the Tiger." Univ. of Chicago Press, Chicago.

Schapiro, S., and Salas, M. 1970. Behavioral response of infant rats to maternal odor. *Physiol. Behav.* **5**, 815–817.

Schinkel, P. G. 1954. The effect of the ram on the incidence of estrous in ewes. *Aust. Vet. J.* **30**, 189–195.

Scholoeth, R. 1956. Zur Psychologie der Begegnung zwischen Tieren. *Behaviour* **10**, 1–79.

Schultz, E. F., and Tapp, J. T. 1973. Olfactory control of behavior in rodents. *Psychol. Bull.* **79**, 21–44.

Schultze-Westrum, T. G. 1965. Innerartliche Verständigung durch Düfte beim Gleitbeutler *Petaurus breviceps papuanus* Thomas (*Marsupialia, Phalangeridae*). *Z. Vergl. Physiol.* **50**, 151–220.

Schultze-Westrum, T. G. 1969. Social communication by chemical signals in flying phalangers (*Petaurus breviceps papuanus*). *In* "Olfaction and Taste. III" (C. Pfaffman, ed.), pp. 268-277. Rockefeller Univ. Press, New York.

Scott, J. P., Stewart, J. M., and DeGhett, V. J. 1974. Critical periods in the organization of systems. *Dev. Psychobiol.* **7**, 489-513.

Scott, J. W., and Pfaff, D. W. 1970. Behavioral and electrophysiological responses of female mice to male urine odors. *Physiol. Behav.* **5**, 407-411.

Shelton, M. 1960. Influence of the presence of a male goat on the initiation of estrous cycling and ovulation in Angora goats. *J. Anim. Sci.* **19**, 368-375.

Signoret, J. P. 1970a. Reproductive behavior of pigs. *J. Reprod. Fertil. Suppl.* **11**, 105-117.

Signoret, J. P. 1970b. Sexual behavior patterns in female domestic pigs (*Sus scrofa*) reared in isolation from males. *Anim. Behav.* **18**, 165-168.

Signoret, J. P. 1976. Chemical communication and reproduction in domestic mammals. *In* "Mammalian Olfaction, Reproductive Processes, and Behavior" (R. L. Doty, ed.), pp. 243-256. Academic Press, New York.

Singer, A. G., Agosta, W. C., O'Connell, R. J., Pfaffman, C., Bowen, D. V., and Field, F. H. 1976. Dimethyl disulfide: An attractant pheromone in hamster vaginal secretion. *Science* **191**, 948-950.

Singh, P. J., and Tobach, E. 1975. Olfactory bulbectomy and nursing behavior in rat pups (Wistar DAB). *Dev. Psychobiol.* **8**, 151-164.

Singh;, P. J., Tucker, A. M., and Hofer, M. A. 1976. Effects of nasal $ZnSO_4$ irrigation and olfactory bulbectomy on rat pups. *Physiol. Behav.* **17**, 373-382.

Sink, J. D. 1967. Theoretical aspects of sex odor in swine. *J. Theoret. Biol.* **17**, 174-180.

Skeen, J. T., and Thiessen, D. D. 1977. Scent of gerbil cuisine. *Physiol. Behav.* **19**, 11-14.

Sluckin, W. 1972 (1964). "Imprinting and Early Learning." Methuen, London.

Solomon, J. A., and Glickman, S. E. 1977. Attraction of male golden hamsters (*Mesocricetus auratus*) to the odors of male conspecifics. *Behav. Biol.* **20**, 367-376.

Stehn, R. A., and Richmond, M. E. 1975. Male-induced pregnancy termination in the prairie vole, *Microtus orchogaster. Science* **187**, 1211-1213.

Steiniger, F. 1950. Über Duftmarkierung bei der Wanderratte. *Z. Angewa. Zool.* **38**, 357-361.

Stern, J. J. 1970. Responses of male rats to sex odors. *Physiol. Behav.* **5**, 519-524.

Stevens, D. A. 1975. Laboratory methods for obtaining olfactory discrimination in rodents. *In* "Methods in Olfactory Research" (D. G. Moulton, A. Turk, and J. W. Johnston, Jr., eds.), pp. 375-394. Academic Press, New York.

Stevens, D. A., and Gerzog-Thomas, D. A. 1977. Fright reactions in rats to conspecific tissue. *Physiol. Behav.* **18**, 47-51.

Stevens, D. A., and Saplikoski, M. J. 1973. Rats' reactions to conspecific muscle and blood: Evidence for an alarm substance. *Behav. Biol.* **8**, 75-82.

Stoddart, D. M. 1973. Preliminary characterisation of the caudal organ secretion of *Apodemus flavicollis. Nature (London)* **246**, 501-503.

Stoddart, D. M. 1974. The role of odor in the social biology of small mammals. *In* "Pheromones" (M. C. Birch, ed.), pp. 297-315. North Holland Publ. Amsterdam.

Stoddart, D. M. 1977. Two hypotheses supporting the social function of odorous secretions of some old world rodents. *In* "Chemical Signals in Vertebrates" (D. Müller-Schwarze and M. M. Mozell, eds.), pp. 333-356. Plenum, New York.

Stoddart, D. M., Aplin, R. T., and Wood, M. J. 1975. Evidence for social difference in the flank organ secretion of *Arvicola terrestris (Rodentia: Microtinae). J. Zool. London* **177**, 529-540.

Strauss, J. S., and Ebling, F. J. 1970. Control and function of skin glands in mammals. *In* "Hormones and the Environment" (G. K. Benson and J. G. Phillips, eds.), pp. 341-371. Cambridge Univ. Press, London and New York.

Teicher, M. H., and Blass, E. M. 1977. First suckling response of the newborn albino rat: The roles of olfaction and amniotic fluid. *Science* **198**, 635–636.

Thiessen, D. D. 1977. Methodology and strategies in the laboratory. *In* "Chemical Signals in Vertebrates" (D. Müller-Schwarze and M. M. Mozell, eds.), pp. 391–412. Plenum, New York.

Thiessen, D. D., and Rice, M. 1976. Mammalian scent gland marking and social behavior. *Psychol. Bull.* **83**, 505–539.

Thiessen, D. D., Regnier, F. E., Rice, M., Goodwin, M., Isaaks, N., and Lawson, N. 1974. Identification of a ventral scent marking pheromone in the male mongolian gerbil (*Meriones unguiculatus*). *Science* **184**, 83–85.

Tobach, E. 1977. Developmental aspects of chemoreception in the Wistar (DAB) rat: Tonic processes. *Ann. N.Y. Acad. Sci.* **290**, 226–269.

Valenta, J. G., and Rigby, M. K. 1968. Discrimination of the odor of stressed rats. *Science* **161**, 599–601.

Vandenbergh, J. G. 1967. Effect of the presence of a male on the sexual maturation of female mice. *Endocrinology* **81**, 345–349.

Vandenbergh, J. G. 1969. Male odor accelerates female sexual maturation in mice. *Endocrinology* **84**, 658–660.

Vandenbergh, J. G. 1971. The influence of the social environment on sexual maturation in male mice. *J. Reprod. Fertil.* **24**, 383–390.

Vandenbergh, J. G. 1973. Acceleration and inhibition of puberty in female mice by pheromones. *J. Reprod. Fertil. Suppl.* **19**, 411–419.

Vandenbergh, J. G. 1975. Hormones, pheromones and behavior. *In* "Hormonal Correlates of Behavior. Vol. 2: An Organismic View" (B. E. Eleftheriou and R. L. Sprott, eds.), pp. 551–584. Plenum, New York.

Vandenbergh, J. G. 1976. Acceleration of sexual maturation in female rats by male stimulation. *J. Reprod. Fertil.* **46**, 451–453.

Vandenbergh, J. G. 1977. Reproductive coordination in the golden hamster: Female influences on the male. *Horm. Behav.* **9**, 264–275.

Vandenbergh, J. G., Whitsett, J. M., and Lombardi, J. R. 1975. Partial isolation of a pheromone accelerating puberty in female mice. *J. Reprod. Fertil.* **43**, 515–523.

Vandenbergh, J. G., Finlayson, J. S., Dobrogosz, W. J., Dills, S. S., and Kost, T. A. 1976. Chromotographic separation of puberty accelerating pheromone from male mouse urine. *Biol. Reprod.* **15**, 260–265.

Wallace, P. 1977. Individual discrimination of humans by odor. *Physiol. Behav.* **19**, 577–579.

Wallace, P., Owen, K., and Thiessen, D. D. 1973. The control and function of maternal scent marking in the mongolian gerbil. *Physiol. Behav.* **10**, 463–466.

White, L. E. 1978. Personal communication.

Whitten, W. K. 1956. The effect of removal of the olfactory bulbs on the gonads of mice. *J. Endocrinol.* **14**, 160–163.

Whitten, W. K. 1959. Occurrence of anoestrus in mice in caged groups. *J. Endocrinol.* **18**, 102–107.

Whitten, W. K. 1966. Pheromones and mammalian reproduction. *In* "Advances in Reproductive Physiology I" (A. McLaren, ed.), pp. 155–178. Academic Press, New York.

Whitten, W. K. 1969. Mammalian pheromones. *In* "Olfaction and Taste III" (C. Pfaffman, ed.), pp. 252–257. Rockefeller Univ. Press, New York.

Whitten, W. K., and Bronson, F. H. 1970. Role of pheromones in mammalian reproduction. *In* "Advances in Chemoreception - Vol. I" (J. W. Johnston, Jr., D. G. Moulton, and A. Turk, eds.), pp. 309–326. Appleton, New York.

Whitten, W. K., and Champlin, A. K. 1973. The role of olfaction in mammalian reproduction. *In* "Handbook of Physiology" (R. O. Greep, ed.), Vol. II, Pt. 1, pp. 109–123. American Physiological Society, Washington, D.C.

Wickler, W. 1967. Socio-sexual signals and their intraspecific imitation among primates. *In* "Primate Ethology" (D. Morris, ed.), pp. 69–147. Weidenfeld & Nicolson, London.

Wilson, E. O. 1975. "Sociobiology." Belknap Press of Harvard Univ. Press, Cambridge, Massachusetts.

Yahr, P. 1976. The role of aromatization in androgen stimulation of gerbil scent marking. *Horm. Behav.* **7**, 259–265.

Yahr, P. 1977. Social subordination and scent-marking in male Mongolian gerbils (*Meriones unguiculatus*). *Anim. Behav.* **25**, 292–297.

Zahorik, D. M., and Johnston, R. E. 1976. Taste aversions to food flavors and vaginal secretion in golden hamsters. *J. Comp. Physiol. Psychol.* **90**, 57–66.

Zarrow, M. X., Estes, S. A., Denenberg, V. H., and Clark, J. H. 1970. Pheromonal facilitation of ovulation in the immature mouse. *J. Reprod. Fertil.* **23**, 357–360.

The Development
of Friendly Approach Behavior in the Cat:
A Study of Kitten–Mother Relations
and the Cognitive Development of the Kitten
from Birth to Eight Weeks*

MILDRED MOELK

BROCKPORT, NEW YORK

*The study of animal behavior has always depended upon individuals who, because of their deep interest and long experience with a particular animal, have provided the kind of accurate observations upon which further study can be undertaken. From these keen observers of animal behavior we have learned that to be able to make and record such observations is itself a creative act, no less creative than the experiments that they inspire in others. They have taught us, too, that sensitivity to an animal, openness to fresh perception, and the self-disciplined objectivity required to make such observations are possessed not only by trained scientists but also by those who have cultivated these qualities in themselves and possess them in their lives as a whole. Mildred Moelk is, we believe, among the latter, and the Editors of *Advances in the Study of Behavior* are pleased to publish, this, her second article on vocalization in kittens in 35 years as an important contribution to the study of behavioral development in the kitten and kitten–mother relationship in the cat.

Copyright © 1979 by Academic Press, Inc.
All rights of reproduction in any form reserved.
ISBN 0-12-004510-9

I. Introductory Frames of Reference

In the author's first report (Moelk, 1944) housecat vocalization was analyzed in terms of intensity, and the sounds uttered by the cat were divided into three main types: (1) sounds made through the nose with the mouth closed (purring and mhrn murmurs); (2) sounds made as the mouth opens and then closes, producing a pattern of vowels (basically a:ou); and (3) sounds made with the mouth held tensely open (strained intensity cries such as growling and hissing). The basic adult vocalization is a vowel pattern opening from a murmur: mhrn-a:ou. All housecat vocalization can be described as a modification of this basic pattern with both halves subject to infinite variation, including one or the other half at zero. (''Mew'' and ''meow'' are tense kitten forms of mhrn-a:ou.) Infinite variation or gradation occurs because, unlike human beings, cats are not held to socially agreed upon specific forms of vocalization (words), and no feline vocalization is a verbal symbol or name for an object or action, but only a reflection of whether the cat is, to varying degrees, (1) in a state of trying to induce someone or something to approach or withdraw (or whether the cat itself is approaching or withdrawing), (2) in a state of certainty and confidence or of uncertainty and bewilderment, or (3) in a state of satisfaction or dissatisfaction.

Kitten vocalization, based on observation of kittens from birth to 11w1d (eleven weeks, one day), was described briefly as a matter of overly intense vowel cries and slowly developing mhrn murmurs. Although by 11w1d kittens showed a considerable amount of mhrn greeting and there seemed no reason why they should not in time develop the same use of mhrns of greeting and request in daily life with human companions as their mother displayed, kitten use of mhrn greeting was much more limited in range than that of the mother, and no kitten had yet begun to use mhrns of request for door opening or food. When the author finally had an opportunity to rear a kitten to adulthood some 30 years later (S15, Rennie), it was discovered that in the 12w0d–19w6d period the kitten makes rapid advances in socialized mhrn use and by 20 weeks has begun to use all the types of mhrns used by the adult S1 (Emy).

In the years between the publication of this analysis of the vocalization of S1 (Emy) and her kittens to 11w1d and the observation of S15's (Rennie's) behavior and vocalization from 5w2d through adulthood, the author made significant progress in the analysis of mhrn vocalization through coming to regard it not just as one type of vocalization but as one of the four ways a cat has of expressing greeting. Purring, rubbing against some vertical post, and rolling on the ground with or without playful clutching or batting can be used interchangeably in many instances with mhrn as greeting. Considering friendly approach behavior in general rather than concentrating on mhrn vocalization alone provided a great increase in the data available for study.

A second significant advance in the analysis of mhrn vocalization that was made before the study of S15's (Rennie's) behavior was begun came through distinguishing between socialized mhrns and unsocialized, undirected by-product of forward motion mhrns. These by-product mhrns form an irreducible base from which socialized mhrn use can develop at any age if the cat meets suitable responses under favorable low-tension circumstances.

II.   THE FOUR PATTERNS OF FRIENDLY APPROACH BEHAVIOR AND THEIR SOCIALIZATION BY CATS AND BY HUMAN COMPANIONS

1.   Purring is a continuous vibration of both the inhaled and exhaled breath. It can be represented by "hrn-rhn-hrn-rhn. . . ."

2.   The independent mhrn murmur is like an expanded single stroke of purring. It is usually uttered in one to four distinctive rolls, most often in only one or two rolls or in pairs of two rolls. These may be complete (mhrn) or incomplete (mhr) rolls: mhrn, mhrnhrn, mhrnhrn mhrnhrn, or mhrn, mhrhrn, mhrhrn mhrhrn, etc. The rolls are usually easy to count, but the simple fact that the cat is uttering a closed-mouth murmur of any length rather than an open-mouth vowel cry is always more important than the specific number or length or combination of mhrn rolls.

3.   Rubbing consists of the cat rubbing its head, shoulder, and body against some vertical post such as a human leg, door post, or furniture. The adult cat usually rubs first one side of its body and then the other in a figure eight pattern. Sometimes the cat goes through these motions without having any vertical support to rub against.

4.   Rolling consists of the cat lying down on its side on a horizontal surface and rolling its body from side to side. In kittenhood rolling is associated with clutching play with a partner as the kitten lies on its back, but in adulthood rolling is often a gesture releasing energy without physical contact with a partner and without any clutching or batting with the paws. What starts as play in kittenhood becomes merely rolling in adulthood.

The cat has a very low flight threshold even in domesticity, and it takes very little in the way of noise or motion to cause a cat to run off. Any encounter, however friendly, is likely to arouse energy for flight in the adult cat, and therefore the cat remaining at the point of encounter in a nonhostile manner often requires the release of aroused energy in some manner other than flight. Rolling, rubbing, mhrning, and purring, singly or in various combinations, serve the purpose of releasing varying degrees of energy in a friendly manner without the cat leaving a point of interest. This is the most impersonal interpretation of feline friendly approach behavior. In more human terms it can be said that the cat has four ways of saying "Hello": purring, rolling (play), rubbing, and mhrning.

In the wild, cats of housecat size are solitary hunters of small prey, and as adults they make friendly contact with their own kind only with the opposite sex during brief periods of courtship and mating, and in the case of the female only, with her own young during the 10–12 months that a litter remains with the mother. Only one litter per year is reared in the wild, and if the preceding year's litter seeks to remain with the mother, she will drive the yearlings away. Human beings, however, offer continuous life-long friendly and nurturing behavior toward their pet cats, and the cats reciprocate by responding with and initiating friendly approach behavior toward their human companions daily. Under human encouragement pet cats are sometimes able to maintain friendly relations with other adult cats in their own households. Very occasionally a cat may develop a friendship with a cat from another household and greet it with mhrns, but the cat addressed is seldom able to respond at the right moment with a mhrn, and so no mhrn greeting exchange habit results.

Purring, rubbing, and rolling are not susceptible to very much variation beyond length and intensity, and therefore are not subject to much expansion of use and meaning. Mhrn murmurs, however, are very sensitive to the response of a partner, and with suitable response, mhrns can develop a wide range of intonation and purposive use. Aside from varying the number of rolls, mhrns can be uttered with (1) a special initial emphasis ('mhrn), which conveys "Come to me" in tones varying from coaxing to command, (2) a level emphasis of many uses, or (3) a sinking final emphasis that indicates approval, confirmation, or acceptance of an offer of service (Moelk, 1944). Add to this the initial mhrns which open into the whole range of infinitely variable vowel pattern cries (mhrn-plus-vowel pattern), and mhrn will be found to provide the cat with an enormously varied range of vocal expression.

In the two situations in which adult cats display friendly approach behavior toward other cats—both males and females toward the opposite sex while courting and mating and females toward the young in their care—friendly response to friendly approach is guaranteed by physiological changes that require one of the pair of cats to offer sexual or nurturing service and the other to need to receive sexual or nurturing service. In both of these situations mhrn use greatly increases.

Because male cats make themselves available for mating much more often and for longer periods of time than the one to two weeks a year a female cat is in estrus, the male cat may very well have almost as many occasions in a year to utter mhrns and 'mhrns in the course of courting as the female cat does in the course of only two weeks of courting plus rearing the subsequent litter of kittens. Mhrns are not "mating" cries but only the same friendly ("Come here" or "I'm here") approach behavior that can be used daily to obtain routine food or door opening from a human companion. The pattern of the vowel cries uttered during courtship is not basically different from everyday mhrn-a:ou cries of demand and complaint, but courtship crying takes on a "liquid" tone, which in its most intense form can change both parts of the mhrn-a:ou cry into a distinctive umlauted ö form (ϕ-ϕ:) and in less tense form can give an underlying liquid quality to the initial or independent mhrn as well. Since the cat's mouth seems to water at such times, the transformation may quite literally be due to rounding the mouth to contain the liquid that arises.

From the viewpoint of the female cat, courting lasts a week or more with the female controlling the immediate presence and behavior of the males with her mhrn, 'mhrn, mhrn (plus vowel pattern), or growl vocalization and her accompanying approach or withdrawal behavior. From the viewpoint of both female and male cats, but especially the male, courting and mating involve a great deal of variable success and failure, resulting in much solitary complaint vocalization as well as in quarreling among rival males. All in all, courtship with its seeking and often falling to find a ready partner brings out both the male and the female cat's entire range of vocalization, not just the special liquid and ö (ϕ) tones peculiar to courting behavior.

The courting situation is one in which a cat wishes to remain in a particular spot even though it is under the pressure of high tension. Consequently, rubbing and rolling behavior frequently occur as a means of releasing some of the pressure. Although some of this rubbing and rolling may have a very frenzied intensity, this technique is no different from the gentler rubbing and rolling with which a cat may respond to a greeting from a friendly human companion.

The housecat mother greets and calls her kittens with pleasant mhrns and mhrns when she rejoins them in the nest, and she directs the behavior of older, runabout kittens with 'mhrns that range in intonation from coaxing to firm command. The mother tends to respond with mrhn to any mhrn or cry uttered by her kittens. Occasional complaint cries and even growls from the mother when kittens misbehave do not have much effect on kitten behavior, but the mother's mhrns and 'mhrns attract kittens to her.

Human mothers also greet and soothe their infants with murmuring vocalization, and humans tend to direct the same sort of mothering tone of voice toward animals as they do toward human infants. Although human beings cannot purr or utter a pleasant murmur on the inhaled breath as cats do, murmured human speech is very similar in sound, intonation, use, and purpose to feline mhrn vocalization. Consequently, human murmuring vocalization is as useful and as successful in interacting with cats as it is in interacting with human infants.

III. REVIEW OF LITERATURE

Leyhausen, who has probably studied housecats and all Felidae more extensively and intensively than any one else, has described the friendly approach behavior displayed by cats toward cats (Leyhausen, 1956a), based on observation of 30 cats kept together in rooms and of numerous free-roaming pets as (1) seeming to pay no attention to one another, (2) sleeping entwined together, (3) grooming each other, (4) rubbing against each other (Köpfchengeben), (5) friendly greeting after long absence, and (6) in young animals, playing together. In very friendly greeting he describes cats as running alongside each other purring and rubbing against each other with tail raised high, turning around and rubbing again so that the anal region is presented. He points out that Felidae can behave in a friendlier manner toward human beings who behave in the correct manner toward them much more easily than they can toward other cats.

In another publication Leyhausen (1956b) states that the friendly behavior that cats display toward human beings derives in part from the sexual sphere and in part from the family sphere, and he lists the cat's friendly behavior as (1) hard rubbing with the head (Köpfchengeben), (2) flank rubbing (Flankenrieben), (3) licking, (4) sniffing, and (5) for females, the entire coquettish courting behavior. Mhrn murmurs get mentioned only in the description of sexual behavior.

Questions of dominance and sexual behavior, that is, political and sexual power, seem to have been the chief focus of early studies of animal behavior, even in the case of cats that live solitary lives in the wild, and female sexual behavior seems to have been regarded as submissive and subordinate. It is not surprising that the sexual behavior of cats should attract human attention first because during this behavior human beings lose control over their pets. However, overemphasis on sexual behavior distorts and limits comprehension of friendly approach behavior in cats. When the development of purring, rubbing, rolling (play), and mhrn murmurs is traced from birth to maturity, it can be seen that the kitten always has some form of these friendly approach patterns in its repertoire of behavior from birth, along with its suckling technique, and that courtship behavior is only a special, temporarily intensified instance of friendly approach behavior. Rather than saying that daily friendly approach behavior toward human companions is derived from the sexual and family spheres of feline behavior, it seems more accurate and useful to say that the friendly approach behavior of courtship and (for the female) of rearing kittens as well as the friendly approach behavior displayed daily toward human companions all result from circumstances which encourage and permit the surfacing of friendly approach patterns of behavior that a cat always has in its repertoire from birth.

More recently Schneirla *et al.* (1963) studied and reported on the behavior of feline mothers and kittens in terms of the mechanics of suckling in relation to deprivation. Although they state that mother and kittens are constantly interacting in a manner that involves subtle visual, tactual, and olfactory clues, they report only on the kitten's suckling and on its returning to the home spot in the nest when removed by human experimenters to some more distant point. What I wish to point out in this article is that it is the vocal and touch behavior that accompanies suckling behavior—the mother's mhrns of greeting and 'mhrns of call, the touch of her tongue, and the kitten's purring and paddling while suckling contentedly—that are the origin of all future friendly approach behavior in the cat.

Almost no attention has been paid to vocalization in reports of laboratory studies of feline behavior. Only intense crying is mentioned except for some references to mhrn murmurs during courting and mating in the narrowest sense. In a summary article on the behavior of housecats Rosenblatt and Schneirla (1962) include a chart entitled "Synchronization of Male and Female Courtship, Copulation, and Postcopulatory Reaction" adapted from Rosenblatt and Aronson (1958), which covers a total time range of 2–14 min. The male is described as uttering a "mating call" more fully described in the text as a "short, chirplike mating call" and the female's vocalization is described as "low throaty vocalization." Both of these vocalizations, in fact, are varying intonations of mhrns and 'mhrns. They also describe the female as rubbing along the floor and rolling. The original article reveals that the female was administered estrogen to induce the estrus state. She therefore was in no condition to control the courtship and mating situation for a week or more as free-roaming female cats do.

The popular books of Joy Adamson (1960, 1961, 1964, 1969) on a pet lioness and a pet female cheetah that were rehabituated to life in the wild, where each bore and reared wild cubs, provide scattered evidence that lions and cheetahs utter murmurs similar to housecat mhrns to their young, to potential mates, and to human companions. They also use play, rubbing, and rolling as greeting. Although lion cubs take more than three years to reach maturity, they can follow the mother about for long distances and start to eat solid food at six weeks of age just as housecat kittens do.

Schaller has studied wild lions in the Serengeti. In saying that the "greeting pattern, including circling, rubbing, raising the tail, swivelling the rump toward the other, and rolling on the back have a striking similarity to the sexual behavior of the female" and that this greeting pattern is a ritualized form of sexual behavior, Schaller (1972, p. 88) is following Leyhausen in deriving friendly approach behavior in lions from female sexual behavior, and is therefore subject to the same criticism that it is more useful to regard sexual behavior as a special form of friendly approach behavior rather than vice versa. The tail up, eyes closed greeting behavior (which in 1944 Moelk called "near-purr" behavior in cats greeted on the street) is displayed by both male and female cats and is simply the completely open, friendly, relaxed opposite of any cat's tightly crouching, tense, hostile posture. The cat may have friendly approach behavior in its repertoire only because it ultimately needs this behavior to make sexual contact in order for the species to survive, but its friendly approach patterns (minus the drive for sexual contact) are always present in the kitten and cat, ready to surface in any friendly approach situation. It is friendly approach situations with respect to fellow cats that are greatly lacking in the daily life of cats, not the ability to display friendly approach behavior. Wild Felidae are apparently always much more tense and on guard, much more ready to utter hostile warning vocalization, than domestic cats are.

With reference to vocalization Schaller (1972) says that "most sounds made by lions belong to three graded systems linked by intermediates not only within the same system but also between them, thereby enabling the animals to communicate subtle changes in the intensity and nature of their emotions" (p. 103). This is indeed the correct approach to feline vocalization, although Schaller does not attempt to interrelate the various murmur sounds and murmur plus vowel pattern sounds he describes. The types of leonine vocalization that he describes which seem to fall within the purring and mhrn patterns both in sound production and use are (1) puffing, which is the equivalent of the tiger's Prusten, (2) purring, (3) humming (when two lions rub cheeks, licking each other or playing gently), a sound of contentment, and (4) grunting or soft roaring. Grunts are used by lionesses to call cubs, and occasionally a cub grunts in return.

Schaller (1972) also had the opportunity to observe other wild Felidae in the Serengeti. He states that small cats near housecat size such as the serval and wild cat proved difficult to observe because of their shyness and nocturnal habits. In another publication Schaller (1973) describes a female cheetah uttering "low chirrs" and "chirps" to her cubs signifying "Come here" and a male cheetah "chirping" during courtship.

Both Schaller (1972) and Adamson (1960) report some purring from lions. Cheetahs apparently purr as easily as housecats (Adamson, 1969).

All in all, there seems every reason to believe that all Felidae have purring, murmuring, rolling, and rubbing patterns in their repertoires of friendly approach behavior. Just how much a particular type of cat or individual cat makes use of each type of friendly approach behavior varies with the genus and the individual animal's circumstances. The different kinds of Felidae seem to differ not in the length of their nest stage (suckling technique) or weaning stage (development of dead prey handling technique) but in the length of time after 8w0d that it takes them to be ready to hunt their type of live prey for themselves in the prey's own territory.

Batting or rolling play and rubbing behavior can be seen rather frequently in television camera studies of many of the Felidae, and all observers are likely to agree on when an animal is engaged in playing or rubbing. Hearing vocalization is more of a problem because the sound that is easily recognized as purring in a housecat may come out of the television set like the ominous rumble of thunder from larger Felidae, and a "moan" or a "soft grunt" from a lion may sound like anything but a pleasant murmur to an unaccustomed ear. Adamson (1960, 1961, 1964), who has had much experience in living with lions, refers to the "mhn, mhn" or "soft grunt" uttered by a lioness to her cubs as a very pleasant sound. In order to interpret any animal vocalization one needs to have enough experience with that type of animal to know what kind of behavior is likely to follow a particular form of vocalization.

In the case of the housecat kitten S15 (Rennie), whose behavior was studied into adulthood by the author, the last type of mhrn use to appear was a 'mhrn of request for food in his twentieth week, in place of loud, prolonged vowel cries of demand. Because there is rarely an opportunity to hear mhrns of request or even greeting from any but one's own pets and because obtaining solid food is a very competitive, high-tension activity for wild Felidae it did not seem possible that direct evidence for mhrn use and especially for a 'mhrn of request would ever come from a television set. However, an *American Sportsman* program aired on Springfield, Massachusetts Channel 40 on February 10, 1974 included a segment dealing with cheetahs in the Serengeti in which this did occur. The purpose of the film was to demonstrate that a female cheetah did not get to eat most of the kills she made because other animals took her captures away from her. A male cheetah was shown taking a gazelle away from a female and then eating it as the female and her two large cubs looked on. One of the cubs was very eager for some of this food. It crouched down on its belly and slowly and cautiously crawled toward the feeding male, repeatedly uttering coaxing 'mhrhrn, 'mhrhrhrn sounds, half a dozen times or more. Given such direct audible and visual evidence from the wild for such a late-developing use of mhrn as a coaxing 'mhrn of request for food, there seems no reason to doubt that all the simpler uses of mhrn would also be found in wild cheetahs if enough observations could be made under enough different circumstances.

The presence of a stranger usually creates a situation too tense for mhrn utterance, but over the course of 35 years and more the author has heard enough mhrns from housecats that were not her own pets and has seen enough rubbing and rolling behavior to be certain that there was nothing unique about the behavior of the cats that were her subjects of study.

It was, however, the books by Spitz (1957, 1965) and Escalona (1968) on the relationship between human mothers and their infants that were most helpful in discovering explanations for the development of friendly approach behavior in cats. Escalona's work will be discussed below in relation to the role of the mother cat and the familiar human companion. Spitz's theory of three organizers of human infant behavior was very helpful in recognizing how kitten behavior is organized. Although it is difficult to pin down exact parallels, there are many haunting similarities between the development and behavior of kittens and that of human infants.

Since the basic purpose of this review of literature is to point out the universality of the use of mhrn murmurs in Felidae, perhaps this is the best point at which to mention that, although the vowel cries of Siamese cats seem rather different from those of other housecats on the surface (the normal Siamese cry is a prolonged kitten-type wail), mhrns uttered by Siamese cats are the same as those uttered by other housecats.

IV. Subjects and Type of Data Collected

There are two sets of data for analysis of kitten development in the 0–7w6d period. One consists of descriptive notes made on various litters of the author's pet cat S1 (Emy), principally her 19th–23rd litters (L19A, L20A–C, L21A–C, L22A–B, L23A–C) with some scattered notes on earlier and later litters. At the time when these notes were made the author was only trying to devise a phonetic method of writing down the vocalization of cats and attempting to discover which types of vocalization were used in which situations.

The second set of data consists of much more detailed and frequent descriptive notes made on a male kitten, S15 (Rennie), from 5w2d to 7w6d, many years later. By this time the pattern of kitten development for the 0–7w6d period was clear from analysis of the notes of S1's (Emy's) kittens, and the author was watching for all four patterns of friendly approach behavior in S15 (Rennie).

S15 (Rennie) was the only male kitten available within the author's circle of acquaintances at the time the acquisition of a house and garden made adopting a kitten possible. Therefore, no personal choice was involved in his selection except that a male kitten was wanted for comparison with the female S1 (Emy). At the time S15 (Rennie) was adopted, the author had no idea whether or not he would become a user of mhrns of greeting and request in daily life as S1 (Emy) had been. Emy had been almost a year old when she joined the author's household as a stray, and she was ten years old before the study of her vocalization was begun, and so it was not known how she had acquired her use of mhrns. S15 (Rennie), in fact, became fully as frequent and nearly as skillful a user of mhrns as S1 (Emy) had been, and there seems to be no reason to doubt that any other kitten adopted in his place would have done the same.

V. ROLE OF THE HUMAN PARTNER AND OF THE NATURAL MOTHER

Escalona's (1968) study of human mothers and infants found that both babies and mothers have many varieties of temperament and that many different styles of mothering and of relationships between different varieties of mothers and different varieties of infants result in the rearing of a normal child. For instance, some human mothers talk to their babies a great deal but touch them only a necessary minimum amount. Other human mothers touch their babies often, but seldom speak to them. There are probably almost as many different styles of mothering pet animals as there are of mothering human infants. However, human beings usually tend to interact with kittens more by touch than by voice because (1) everyone knows how to cause a cat to purr by petting it, (2) everyone is aware of the playfulness of kittens and how to encourage their playful response by touch and motion, and (3) there is little one can do to prevent a cat from rubbing against human legs or other posts. However, very few persons seem ever to have noticed mhrn vocalization and therefore do not respond to it or encourage it. At the age when a kitten is usually transferred to a new home under human care only, it is at a peak of playfulness while mhrn utterance is at a minimum. A habit of petting and play touch contact therefore easily becomes established between the human companion and kitten, and continues. When mhrn murmurs receive no response, the cat must use more intense open-mouth vowel vocalization to attract human attention and get what it wants.

The author has made two attempts to help interested persons encourage the socialized use of mhrns in their young cats. Both attempts resulted at first in considerable success under special effort on the part of the human partner and nearly daily questioning by the author as to whether the cat had uttered any mhrns and how they were responded to. But, when no longer scrutinized regularly, this mhrn use soon disappeared because the human partner reverted to her old habits. A human's style of mothering even pets is apparently so deeply a part of his or her personality that it cannot easily be changed without very persistent and long continued conscious effort. Any human partner who is not unconsciously inclined to address a good deal of quiet speech toward cats, to respond verbally to anything vaguely resembling a greeting on the cat's part, and to make offers of service verbally before going into action is not likely to find himself or herself with a pet that uses mhrns of greeting and request in daily life.

The kitten, like the human infant, develops through interaction with its mother or human mother substitute, and any one who takes on the care of an animal that is not able to maintain itself in the wild is serving as a mother substitute. Although too little attention was paid to the behavior of the maternal partner in the notes on S1's (Emy's) kittens and on S15 (Rennie), the examples cited will, it is hoped, be enough to give some idea of the style of mothering offered by both the natural mother and the author–observer (O) as human companion or mother substitute. Unless silence is specified, it should be taken for granted that S1 (Emy) always uttered mhrns of greeting and 'mhrns of call as she approached her kittens and that the human companion-observer (O) also always murmured quiet greetings when approaching kittens before touching them. When L19A (Zeb) was 5w1d old, it was noted that all of S1's (Emy's) vocalization to her kitten, except for a cry of pain when he bit her too hard in play, was of the mhrnhrn type. Her mhrns were "richer than usual, with more shades of meaning, more expression, more tones, more variety. The mhrnhrn is a very kindly sound." This increased use of mhrns occurred also with her human companions, and only occasionally, when very impatient, did she open her mouth into a vowel cry. (Possibly because of Emy's maternal duties her human companions were quicker than usual to do what she wished.)

The behavior required from the human partner in order to have a cat that uses mhrns of greeting and request to human companions in daily life is (1) much verbal greeting initiated toward the cat, (2) greeting response to any mhrn uttered, (3) readiness to open at least one door and to supply at least one type of food quickly enough for the cat not to become impatient enough to open its mouth in a vowel cry, but not so quickly that the cat does not have to vocalize at all, and (4) a habit of making offers of service verbally ("Do you want . . .?," "Would you like . . .?") and waiting a moment for a reply, which gives the cat a chance to utter a mhrn of pleased anticipation with the effect of "Yes" in acceptance of the offer.

VI. BIRTH TO FOUR WEEKS FOUR DAYS, S1'S (EMY'S) KITTENS: NEST STAGE (SUCKLING TECHNIQUE)

Because all subsequent friendly approach behavior develops out of the earliest interactions between mother and kitten, it is important to take a close look at the kitten's earliest behavior, vocalization, and experiences.

Unless otherwise noted S1's (Emy's) kittens were born and reared in a large wooden box that was about 3 ft in each dimension with half the top and half the front removed for entrance. It stood raised on legs in an unheated cellar with earthen walls and floor, which therefore tended to reflect outdoor cold and dampness. Woolen material such as an old sweater was needed for bedding to supply sufficient warmth. S1 (Emy) was given food and water only in the kitchen, and she went outdoors to relieve herself. A small wooden box in front of the raised nest box served Emy as a step and O as a seat. The kittens were visited by O at least once each morning and once each evening. Each kitten was picked up carefully at least once a day with one warm hand grasping and the other supporting the kitten's body and feet, and O spoke to the kitten soothingly as she picked it up and held it. After kittens were able to run about, they were brought up or allowed to come up into a large kitchen each evening and sometimes at other times of day when the kitchen was free of human activity. During the box stage the other three members of S1's (Emy's) human family did no more than occasionally look at the kittens. Because the kitchen was a large sunny room with an easychair in one corner and a door leading directly to the outdoors on both the yard and the street sides of the house, there was usually someone in the room ready to open doors for S1 (Emy) to go or come from cellar and outdoors as she pleased.

A. PARTURITION

Human companions are present in the domestic kitten's world from birth as noises acceptable to the mother. The presence of a familiar favorite human companion can serve the feline mother in labor as a tension reducer, reducing cries of pain to a very low minimum and inducing much purring in their place even though the human companion has nothing to offer but encouraging and reassuring murmured words. Kittens then hear a murmuring human voice before they hear their own mother's mhrn murmurs.

The kitten cries as soon as its head is cleared of its birth sack by its mother's cleansing tongue. Littermates can vary greatly in the amount of vocalizing they do from birth, some being very quiet and others very noisy. The newborn, blind kitten must drag itself along by its forelegs and move its head back and forth until it finds a milk-giving nipple on its mother's body. The most the mother can do to help is to lie down so as to encircle the kitten. As the kitten pushes forward seeking and failing to find the milk-giving nipple, it usually cries and continues to cry until it locates a nipple. The kitten is strongly geared from birth to move forward to locate food for itself. Frustrated forward movement results in vocalization, and this is the basic source of energy for vocalization throughout a cat's life. A kitten or cat that could always simply move forward and immediately obtain whatever it sought would never have occasion to vocalize. The kitten's complaint–demand vocalizing attracts the mother's attention. She licks the kitten, utters mhrns and 'mhrns to it, and may encircle it in a new and more helpful manner. The newborn kitten's earliest pushing forward to locate the food and comfort it needs establishes the mother, which is identical with that necessary food and comfort, as the focus of the kitten's attention. On the mother's side it seems to be lactation that establishes the mother's bond with her kittens. One young female described by Wilson and Weston (1947) would not stay with her first two litters of kittens. It was not until her third litter that she could nurse her kittens, and she remained with and took care of this litter.

B. FOOD BEHAVIOR (SUCKLING TECHNIQUE OF TOUCH–SEEK AND
     CRY–FIND AND SUCKLE–PADDLE AND PURR)

Once each kitten, clean and dry, has located a milk-producing nipple, all is well, and the mother rests for hours encircling her kittens. When the mother first leaves her sleeping kittens and then returns to them, the kittens experience what will be the basic drama of their lives during the nest period. If the kittens are not already awake, crying, and searching for milk when their mother returns to the nest, they start searching and crying as soon as her tongue touches them. The mother utters mhrns of greeting and 'mhrns of call as she jumps up into the box and as she licks the kittens. Finally all the kittens locate milk and settle down to suckle and sleep with their mother curled around them. The mother's licking tongue enables the kittens to void their waste products and keep the nest dry and clean.

Besides uttering cries of discomfort and displeasure kittens have the ability to purr when they feel especially comfortable. Spontaneous purring while suckling does not become audible to human ears until kittens are three or four days old. The first purring is "one-cylinder," an audible vibration of one stroke of breathing only, not yet a continuous purring of both the inhaled and exhaled breath. By the end of the first week purring while suckling was heard from all S1's (Emy's) litters and kittens.

In between open-mouth vowel crying and relaxed closed-mouth purring there are other sounds. At about the same time that spontaneous purring while suckling is first heard, the tense grunting often uttered along with vowel cries as kittens push forward to reach their mother relaxes into less tense "mhrhrhr . . ." trills. These are not independent mhrns but a more continuous and somewhat cranky sound. More efficiency in suckling after a few days of experience and some anticipation of suckling probably account for the relaxation of tense grunting into more anticipatory mhrhrhr trills.

As the kitten suckles and purrs, it is also likely to be paddling its feet against its mother's body in a gentle manner. Thus, during the nursing–suckling interaction between mother and kitten the mother offers both touch (her licking, cleansing tongue) and sound (mhrns and 'mhrns) while the kitten responds with first the sounds of frustrated searching (crying), then the touch contact of suckling, and finally the sound of purring and the touch contact of its paddling feet against the mother's body as it suckles. The full technique displayed by the kitten in obtaining its food by suckling is touch (of mother's tongue)–seek and cry–find and suckle–paddle and purr.

Although at first seek-and-cry behavior is more prominent than paddle-and-purr, and, although the kitten at first is responding to the touch of its mother's tongue rather than to her mhrn murmurs, the mother's mhrns and the touch of her tongue and the kitten's purring and paddling are all part of the total suckling pattern. This makes it possible in the future for a similar touch or sound to arouse any one of the four friendly approach patterns. All feline friendly approach behavior derives from this initial nursing–suckling touch-and-sound relationship between mother and kitten and kitten and mother.

In their second week kittens are such experts at locating milk on their mother's body that "much purring, little crying" is common, and they can shift almost on the same breath from a piercing cry upon being touched by their mother's tongue to suckling and purring.

Normally kittens do not need or use supplementary food before about 5w0d, and the entire 0–4w4d nest period is one of dependence on suckling only. During the last years of S1's (Emy's) 14½-year life she developed a mammary tumor that tended to make her milk inadequate for more than one kitten in the latter part of the nest period, and so warm milk by the dish was supplied to the multikitten litters under study before 5w0d. Using a suckling technique kittens can manage to suck up some other moist soft food before they can eat dry solid food at about 6w0d, but none of this constitutes true weaning.

The change in food behavior that does take place for all kittens in the 0–4w4d nest stage is that, after kittens are able to walk about easily, they can silently take the initiative in approaching their mother to suckle when she is within reach and willing instead of always having to wait in the nest and cry until their mother comes to them.

C. Motor Development, Use of Space, and Play

The amount of space kittens use in their earliest weeks obviously depends on how much moving about they are capable of doing. Play is also very closely related to motor development because any new motor skill is soon put to use in play. Play is also dependent on the length of time a kitten can stay awake without becoming hungry, whereupon play turns to food seeking. At first newborn kittens are awake only to suckle, and the earliest play is the paddling against the mother's body that often accompanies suckling along with purring. When kittens are awake but not hungry enough to suckle, they lie on their backs and paw at their mother's head and purr as she licks them. Paddling becomes clutching, kicking, and rolling play set off by a touch of the mother's tongue or anything similar—a human companion's finger or a littermate's nose. In time this rolling play, which first appears in moments of comfortable relaxation, will become the cat's basic defense posture, freeing all four feet and the mouth for use against an attacker.

In the brief but daily lengthening moments when kittens are awake but not hungry, the touch–seek and cry–find and suckle–paddle and purr suckling pattern shortens to touch–paddle (play) and purr, or touch–paddle (play), or touch–purr.

In its second week the kitten manages to stand up on all four feet and walk instead of merely dragging itself about by means of its forelegs only. At the beginning of its third week the kitten can sit up alone and soon can walk from the center sleeping spot to the front of the 3-ft square nest box. Before long it can walk all around in the box.

Kittens are born with sealed eyelids. S1's (Emy's) kittens usually began opening their eyes at seven days of age, but eyesight does not really become operative until the third week, when kittens begin looking up in the right direction for their mother's return to the box and begin watching littermates walk about in the box.

Once a kitten can sit up easily in its third week, it begins to lick its own fur and other objects such as bedding or littermates. It also begins to paw at objects such as a protruding corner of bedding. This is the beginning of right-side-up pawing, batting, and eventually pouncing play as motor ability increases. The motions of this type of play will become the motions of serious attack in the adult cat. Thus, both defense and attack are first practiced as pleasurable play and friendly contact behavior in moments of relaxation with mother and littermates.

Although littermates are very convenient as playmates and as warm cushions to sleep on, the singleton kitten develops as satisfactorily as kittens with littermates because it is play with the mother rather than play with littermates that is basic to kitten development. The space in the 3-ft square nest box remained sufficient for kittens in their fourth week because right-side-up play mostly takes the form of the kitten throwing itself up against its seated mother's shoulders and head. Standing up or throwing itself up against the side of the cellar box over which the mother disappeared and reappeared served much the same purpose as jumping up at the mother in that it provided exercise while maintaining the kitten's focus on its mother. As the kitten throws itself up at its mother's shoulder and head or clutches at and chews on her tail and legs, the kitten begins to become familiar with its mother as something more than a bag of warm milk that comes and goes. All right-side-up play is a step in the general direction toward the next stage of obtaining food, that of seizing dead prey brought by the mother to her kittens.

In the 0–4w4d period it is in the cosy hollow of the nest, of the mother's encircling body, or of the familiar human companion's lap that a kitten is most playful. The young kitten is most comfortable and relaxed when in a cosy hollow with much of its body in contact with another body or surface.

In their fourth week all of S1's (Emy's) litters of kittens were allowed to walk on the flat, hard surface of a large kitchen for the first time because (1) the mother asked to have them brought up into the kitchen (L19A at 3w0d), or (2) the human companion took the initiative (L15A–B, L20A–C, and L21A–C at 3w2d, and L23A–C at 3w4d), or (3) in the case of two singleton kittens reared in a small box behind the kitchen coal stove for warmth, because the kitten itself took the initiative in climbing out of a small, low box bed (L25A at 3w4d, L17A at

3w5d). However, fourth-week excursions on-to the kitten-the vast, hard, flat surface of the kitchen floor were not always successful, and kittens usually soon had to be returned to their nest box either because they cried in distress or because their mother asked to have them be returned. It was not until the first half of the fifth week that all litters were comfortably able to enjoy periods of play on the kitchen floor. L22A-B (Grinetta, Juniper) were transferred by their mother from the cellar box to a nest on the floor of a nearly empty coal bin at 2w6d, apparently because O had not changed the bedding in the box, which had become too damp from the air during a period of cold rain. Because these kittens then had plenty of room in which to expand gradually the amount of space that they used, as kittens do in the wild, they were not brought up into the kitchen until 4w5d. O discovered the new nest at 2w6d because L22A (Grinetta) came walking out of the coal bin, crying at the sound of O's voice. Although these kittens had no physical barrier preventing them from leaving the coal bin, it was not until 4w2d that they first began to leave the bin to follow or come to meet their mother and O.

It was possible for nest-stage kittens to use the large, flat space of the kitchen only because various substitute home spots or nests were provided to mediate between their cosy nest box and the broad, hard surface of the floor. The familiar human companion's lap or the cushioned kitchen easy chair in which their mother frequently settled were completely acceptable substitute nests. In more active moments a piece of carpeting on which the mother often sat on the floor, the low-sided market basket in which the kittens were carried upstairs and which was left on the floor, and, of course, their mother and O, wherever they were, all served as home sites to which kittens could return. Play and exploration took place as brief excursions away from and back to the mother or other home site.

At first it takes intense concentration on the kitten's part merely to walk on a flat surface, tail held stiffly up in the air for balance. As motor development progresses and familiarity with the new environment increases, the kitten not only can walk easily but begins to run rather than walk. Before long, in the first half of its fifth week, the kitten at times has energy to spare for playing at running by running about in stiff-legged bucking-bronco poses, and this type of play plus the mhrns the kitten occasionally utters while running about in play indicate that the kitten feels at home in its new larger environment.

With the beginning of spring-jumping from the hind legs in the first half of the fifth week the kitten's repertoire of motions is complete. Further development in both motor and play skills takes the form of more precise muscular control, greater speed, larger size, and new combinations of familiar motions.

It is impossible to overestimate the importance of play in the development of kittens and as a clue to their development. The reason for ending the nest stage at 4w4d and beginning a new stage with 4w5d can be seen most easily as a difference in the manner in which kittens play. Before 4w5d S1's (Emy's) kittens are described as playing one at a time without getting any response from a littermate. If two kittens clutched and kicked at each other playfully, it was because they happened to be close enough to touch each other when both were in a playful mood. It is not until after 4w5d that kittens develop the eye–paw coordination that enables them to bat at small movable objects, including litter-mates' paws, purposefully and accurately. Before 4w5d play is mostly whole body contact play, which originates either from within the kitten or in response to external touch.

D.   FRIENDLY APPROACH BEHAVIOR

1.   Purring

Purring occurs at the contented end of the suckling pattern: touch–seek and cry–find and suckle–paddle and purr. By the end of the first week purring while suckling had been heard from all of S1's (Emy's) kittens. As kittens become able to stay awake for brief periods of time in their second week without seeking food, a touch of the mother's tongue or any similar touch may set off purring. A gentle stroking touch from a human finger can evoke purring at as early as two days of age, a day or two earlier than spontaneous purring while suckling is heard. At 2d as L26A (Jemmy) was lying in O's hand, grunting and nuzzling but only opening his mouth into a vowel cry very occasionally, and O was talking to him and petting him with one finger, Jemmy dropped into purring, perhaps about six rolls. The purring was a relaxation of the grunting into purring.

In form the earliest purring is an audible vibration of one stroke of breathing only, the outstroke, not a continuous vibration of both the inhaled and exhaled breath as in later purring. On 2w1d when O held L23B (Lefric) in her hand, petting him and speaking to him, he breathed audibly with "odd little separate sounds that unified into a purr a couple of times." On 2w3d L21C (Jeda) "purred about four pairs of purrs before she discovered that it had not been her mother who had touched her." On 2w4d, as Lefric searched for milk on O's arm, he kept making "clicking, licking noises with his mouth and noises in his nose that were the repeated beginnings of a purr, the exhaled part. Any one would have turned into full continuous purring if he had found milk." During the third week the kitten's purring changes from the disjointed one-stroke form to continuous vibration of both the inhaled and exhaled breath.

Changes in the use of purring occur because of anticipation of success in finding milk to suckle. In the first week purring occurs only after milk has been flowing down the kitten's throat for some time. With increased anticipation and skill in locating a milk-producing nipple purring comes to occur ever earlier with respect to obtaining food. In their third week kittens can purr as soon as they start to suckle and may be ready to purr before they actually locate a milk-giving nipple, as L23B (Lefric) was at 2w4d as he searched for milk on O's arm. Purring in situations that do not involve food and food seeking shows the same progression due to anticipation. At this point it should be remembered that the mother always uttered mhrns when rejoining and licking her kittens, and O also always murmured to kittens before and while touching them. In time through anticipation a voice greeting from either would be as effective as a touch in arousing kittens from sleep to searching and crying or to rolling and purring.

In the third week there begin to be instances of a kitten's initiating purring without have received a touch. At 2w0d L21B (Gilda) was especially active and noisy, but apparently her vowel cries did not mean much as complaints, for several times she dropped into purring for no apparent reason. At 2w4d L20B (Hilary) purred a second as he sat up all alone. On 2w5d when O held all three L21A–C kittens on her lap, L21B (Gilda) turned over on her back, kicked out her legs, and purred. L21A (Peleg) did the same thing a few minutes later. In its third week the kitten is able to initiate separate bits of purring when it is in a relaxed but alert condition, just as it is able to initiate fuller and more independent mhrns.

Purring was taken for granted and not especially looked for in the later note taking on S1's (Emy's) kittens. It can safely be assumed, however, that kittens continue to purr almost every time they suckle and in quiet moments when they are petted by a human companion. The failure to make notes on purring from 3w0d–4w4d means that the opportunity was lost to notice whether kittens purred in response to voice greeting only from either the mother or human companion before 4w5d and whether kittens began to initiate purring greetings on sight before 4w5d.

2. Play As Greeting

As the kitten grows in size and strength, weak paddling against the mother's body becomes increasingly skillful clutching and rolling play. When touched in greeting the kitten can respond with playful rolling and clutching, and at a later age it can initiate rolling and clutching, thus in effect initiating play greeting.

Right-side-up pawing and body throwing play occurs on the kitten's initiative. Particularly when directed toward the mother this pouncing type of play can be greeting initiative. When on 4w3d L21B (Gilda), while running around in play, came upon her mother sitting on the floor, Gilda jumped and slapped at her, and when her mother gave her a few licks of her tongue, Gilda made little purring sounds. Gilda thus initiated a play greeting, the mother responded with touch greeting, and Gilda responded to this touch greeting with purring greeting.

Since no one but a littermate really enjoys being pounced on, batted at, and bitten, pouncing play does not have much future as greeting no matter how friendly a mood the kitten may be in when offering this type of greeting.

Unlike purring, play serves many other purposes besides greeting, in the earliest weeks principally that of exercising increasing motor skills.

3. Mhrn

The independent mhrn (mhrhrn, etc.) sound has a very controlled, wordlike quality. It has a definite ending, whereas purring, mhrhrhr . . ., trills, grunting, and crying in young kittens usually are continuous sounds that are repeated until stopped by the kitten finding milk, becoming exhausted, falling asleep, etc.

From 2w0d on kittens are found sitting up in their nest and sometimes uttering a mhrn. At this age they have a new ability to sit up and to start and stop walking, a new ability to stop action, which also seems to account for the appearance of independent mhrns and brief runs of nontouch purring.

On 2w2d both L20A (Gerrit) and L20C (Joella) uttered a mhrn roll at separate times, Gerrit a quick round one, Joella a very quiet one as she sat up by herself. When L20A (Gerrit) and L20B (Hilary) woke up on O's lap at the same time a little later on 2w2d, they both uttered "mnrns" several times while nuzzling O's arm, then cried, then fell asleep again. On 2w3d Gerrit uttered a "mhrrn" quietly while sitting still. On 2w4d Hilary purred for a second as he sat up alone and made sounds half way between a grunt and a vowel cry. When O put her hand into the box to pick him up after he had cried, looking up, he came toward her hand uttering "mhrrn" sounds. On 2w4d while the mother was in the box with L20A–C and O approched and spoke, Gerrit popped his head up and cried with vowel cries preceded by mhrns. After O had picked Gerrit up and then returned him to his mother in the box and all three kittens had settled down to suckling, O petted and spoke to the mother to make her purr. She responded first with a "mhrn." Gerrit also at once popped up his head in response to O's voice and came toward O, uttering loud mhrns and vowel cries. (These were not one-breath mhrn-plus-vowel-pattern cries.) When L19A (Zeb) was 2w6d old, as O approached the box and spoke, Zeb responded with "something like mrnrn" and his mother also responded to O with "mhrn." This note was the first note that O made on kitten vocalization because this was the first kitten vocalization that could be recognized as being very similar to the adult vocalization of the mother.

In their fourth week cellar box reared kittens continued to respond with mhrn to the human companion's verbal greeting. When O approached the box and spoke when L19A (Zeb) was 3w3d old, Zeb responded with "mrnrn, mrnrn." When O spoke to L21A–C (Peleg, Gilda, Jeda) at 3w4d, they "greeted O with mhrn sounds." A 3w6d note on L20A–C (Gerrit, Hilary, Joella) reports that, whenever O approached the box and first spoke to these kittens, Gerrit always popped up with a scratchy mhrn sound along with vowel cries.

To anyone who likes cats nothing is more charming or rewarding than to hear a cat, and especially a tiny kitten, respond to one's verbal greeting with a pleasant mhrn murmur. It can be safely assumed that O always rewarded such mhrn greeting responses with whatever the kitten uttering the mhrn most wanted and enjoyed—to be picked up and held for a while, spoken to, petted, played with, etc.

A habit of mhrn greeting response to the familiar human companion's voice seemed very well established in the third and fourth weeks. It was therefore surprising to find that instead of increasing after 3w6d, mhrn response disappeared until some time after 9w0d. It was the effort to solve this mystery that led to the discovery of the importance of mhrn utterance as a clue to the development of kittens (see Section VII,A,6).

The basic line of mhrn use, which kittens and cats never lose, would prove to be the uttering of accidental, unsocialized, by-product mhrns, neither offering nor expecting response, while in forward motion during a state of relative relaxation. At about 4w0d the kitten can run about in play outside its nest sometimes with energy to spare for running in stiff-legged bucking-bronco poses or uttering a mhrn. Both of these behaviors indicate that the kitten feels comfortable and happy in its expanded surroundings.

On 4w1d L23A–C (Nichette, Lefric, Jorian) were immediately playful when brought up into the kitchen in the evening, especially Nichette, who ran around in "comical, unbalanced" (that is, bucking-bronco) poses, once uttering a "mhrhrn" to herself as she did so. On 4w3d L21B (Gilda) was the first to jump out of the carrying basket on to the kitchen floor expertly with one hop, and then after standing still a moment to get her bearings, she uttered a little "mhrhrn" as she started to run in play.

Another situation in which young kittens (and older cats) utter a by-product mhrn is when leaving an object of play interest. At 3w6d as L23C (Jorian) quickly turned away from a protruding corner of bedding that he had been playfully pawing, he uttered a "mhrhrnhrn." In such instances it seems as if withdrawing attention from a play object and turning away from it releases some energy, which escapes through the nose in an accidental mhrn murmur.

In later kittenhood and in adulthood cats frequently utter a by-product mhrn as they jump up on or down from a piece of furniture. The independent mhrn is also closely related to purring. It is like a thickened, independent stroke of purring. If a cat happens to draw a deep breath for any reason while purring, perhaps just because it turns its head, a mrhn sound can result. The author has never spent two quiet hours alone with a strange cat without hearing a by-product mhrn from it as it jumped up on or down from furniture or while purring. This is the irreducible base of mhrn use, which, with suitable response and low tension circumstances, can be socialized into greeting and other uses.

In their 2w0d–4w4d period S1's (Emy's) kittens were heard to utter (1) by-product of motion mhrn, (2) mhrn of response to the human companion's greeting voice, and (3) mhrn of response to the touch of their mother's tongue.

4. Rubbing

At the end of the nest stage in the first half of the fifth week (4w0d–4w4d) when kittens can first run about freely in play outside their nest for some time at a stretch without needing to suckle, they are comfortably separate enough from their mother to be able to initiate a batting play greeting to her and to be able to respond to a greeting touch of her tongue with a mhrn or a bit of purring, and then continue running about in play. It would seem that kittens also then should be separate enough from their mother to be able to take the initiative in rubbing their heads and shoulders against her briefly in the fourth type of friendly approach behavior, rubbing. No instance of rubbing happened to be recorded in the 0–4w4d notes on the behavior of S1's (Emy's) kittens, but rubbing was not being looked for at the time.

Lindemann and Rieck (1953), reporting on a litter of European wildcat kittens reared by a housecat foster mother, cite an instance of rubbing against the foster mother at 4w5d, an age established on the basis of only one kitten in the litter of three having had its eyes open when they were found in the wild compared with the age at which similar wildcat kittens born in a zoo opened their eyes.

In 1976 Helen Budinski Moelk of Rochester, New York was asked to watch for the first rubbing behavior in her pet cat's second litter of kittens, born during the night of August 3/4. She reported (personal communication) that the lead kitten, the "first in nearly everything," rubbed against its mother on September 2 (4w1/2d). The second and third kittens did so on September 3 (4w2/3d), the same day on which the kittens did "the most moving around." The earliest instance of rubbing on the part of the fourth and fifth kittens of this litter was not noticed.

Since L22A–B (Grinetta, Juniper), the litter reared from 2w6d on in a ground-level nest in a nearly empty coal bin where there was no physical barrier to their leaving this roomlike bin, whose doorway was in direct line with the cellar stairs, did not begin to walk out of the bin to follow or meet their mother and O until 4w2d of age, it would seem that rubbing against the mother occurs when kittens are first able to make relatively long excursions away from the nest at 4w2d–4w4d.

In rubbing briefly against its mother a kitten is initiating a momentary restoration of soothing body contact with the mother at a time when the kitten needs a little comforting but is not interested in actually suckling. What precedes rubbing in the kitten's life cycle is the almost continuous body contact with mother, nest, and littermates, which the nest-stage kitten requires to be comfortable. Although independent exploratory and play excursions become longer each day, in the first half of its fifth week a kitten still spends most of its time sleeping, suckling, or playing in close body contact with its mother and/or littermates in a cosy hollow.

5. Summary

At the end of the nest stage, in the first half of its fifth week, the kitten has four types of friendly approach behavior in its repertoire that can be used as greeting response or greeting initiative: purring, rolling or batting play, rubbing, and mhrn murmur. Each of these four types can be used (1) as response to the mother's touch, (2) as response to the familiar human companion's touch, (3) as response to the mother's voice, (4) as response to the familiar human companion's voice, (5) as greeting initiative at the sight of the mother, and (6) as greeting initiative at the sight of the familiar human companion. This makes a total of 24 different kinds of greeting or greeting situations. Even considered as 12 kinds of greeting to one mother figure, the young kitten has a remarkably large and flexible repertoire of greeting behavior.

In their 0–4w4d nest stage S1's (Emy's) kittens were actually noted as exhibiting purring, play, and mhrn responses to their mother's touch, and purring and play responses to O's touch. They exhibited mhrn response to O's voice. They initiated play toward both their mother and O. This makes a total of only 8 types out of a possible 24, but there is some evidence from other sources for the appearance of some of the other types before 4w4d, and it seems very likely that, if an observer systematically looked for all 24 types and provided some favorable circumstances (such as a more open nest so that the mother could be seen and heard by kittens before being felt), kittens would be found to exhibit all 24 types of greeting behavior by 4w4d.

E. QUICK/NOISY–SLOW/QUIET DIFFERENCES IN TEMPERAMENT AND BEHAVIOR

Littermates can differ markedly in their temperaments and in the quantity of their vocalizing from birth. During the nest stage when kittens are doing everything for the first time and a day or two can make a big change in behavior, differences along a quick/noisy–slow/quiet scale may affect performance considerably at any one time. When two kittens or cats are being compared, some apparently very opposite behavior can be accounted for if it can be established that one is a quick/noisy and the other a slow/quiet type.

No two kittens are exactly alike, and each of S1's (Emy's) multikitten litters included a variety of types. The effects of individual variation always must be watched for, but since there is no difference in the basic pattern of the development of friendly approach behavior, the types of individual variation will only be summarized here:

1. From birth on some kittens vocalize very frequently and loudly, while others are very silent although healthy and active.

2. The quiet type kitten tends to be the largest of the litter, and it often opens its sealed eyelids earlier than the noisy type. The quiet type may simply be the one most able to profit from the nutrition supplied by the mother both before and after birth and therefore the most contented, with little cause to vocalize in frustration/complaint.

3. In their third week when S1's (Emy's) kittens began to come to the front of the box uttering mhrns in response to the human companion's greeting voice, it was the noisy type kitten that immediately left off suckling to approach the human companion. The quiet type kitten showed great concentration on its natural mother and suckling, while the noisy kitten responded quickly and vocally to the human companion. Thus, noisy–quiet differences revealed themselves to be quick/noisy–slow/quiet differences.

4. The quick/noisy type kitten (L20A, Gerrit; L21B, Gilda; L22A, Grinetta) is the explorer of the litter. It makes the longest and most frequent excursions away from the mother and other home spots. The quick/noisy kitten is usually the leader in all activity. The lead kitten tends to attract the human observer's attention more than the others do. The quick/noisy lead kitten may be male or female and born in any order in the litter. (Litters 20–22 are not a large enough sample to demonstrate this fact.)

5. The basic difference between kittens at the quick/noisy and slow/quiet ends of the continuum is in their response to new stimuli. In the nest stage the human companion supplied the new stimuli, and the difference therefore appeared to be that the quick/noisy type is most responsive to the human companion, while the slow/quiet type is most responsive to the natural mother. After the kittens were able to run about in the larger space of the kitchen, the familiar human companion became almost as much a home site as the natural mother. It then could be recognized that the difference was new versus old, familiar versus unfamiliar stimulus, not human versus feline mother figure. The quick/noisy kitten is quick to approach and accept new experiences, while the slow/quiet kitten is not easily diverted from a familiar activity and focus. The most important consequence of the quick/noisy type kitten's readiness to respond acceptingly to new stimuli is that it can lap milk supplied in a dish by a human caretaker as early as 2w6d (L22A, Grinetta), while the slow/quiet type kitten does not begin to lap milk from a dish in any quantity until 5w0d (L22B, Juniper).

6. Middle type kittens can be placed near either the quick/noisy or slow/quiet ends of the continuum as well as anywhere in between or right in the middle. Middle type kittens seldom attracted the observer's attention by being the first to do anything, but when a middle type kitten is mentioned specifically by name in the notes, it is likely to be found responding to whichever littermate is in action rather than to either the mother or the human companion. This characteristic helps to keep the litter together as a group.

7. When the quick/noisy kitten is not the leader of the litter as a whole or in some particular activity, there is likely to be a revealing explanation. There can be a physical handicap such as small size from birth (L23A, Nichette). An especially strong single experience can result in a middle kitten taking the lead in some activity. (Middle L21A, Peleg, after noticing quick/noisy L21B, Gilda,

being picked up first by O at 2w1d, thereafter made sure that he reached O first and was picked up first. Middle L20C, Joella, from 5w3d on jumped out of the cellar box as soon as the light was clicked on at the bottom of the stairs and was also the first of her litter to jump up into O's lap.) Or an entire litter can have a transforming single experience. (L20A–C first climbed out of the cellar box at the very early age of 4w0d when they had had to be returned to the box earlier than usual from play in the kitchen.)

8.   Except for the fact that the quick/noisy type kitten, quick to respond favorably to new stimuli, is able to lap milk from a dish efficiently two weeks or more earlier than the slow/quiet kitten, which is most strongly focused on the natural mother and suckling, quick–slow differences are a matter of the speed and quantity of performance rather than a difference in the age at which a particular kind of behavior appears. At 2w4d when the mother was away from the box, slow/quiet L20B (Hilary) cried, looking up, and came toward O's hand and voice while uttering mhrns. But on that same day when the mother was back in the box and all three kittens were suckling, it was quick/noisy L20A (Gerrit) that immediately left off suckling and came forward, uttering mhrns, in response to O's greeting voice at the front of the box, while middle L20C (Joella) merely lifted her head to look (probably at Gerrit in motion) without leaving her mother, and slow/quiet L20B (Hilary) did not interrupt his suckling at all. In the absence of the stronger attraction of the mother and suckling, slow/quiet Hilary had been able to make the same mhrn response to the human companion as his quick/noisy littermate Gerrit did when the mother was present. Thus, there are no differences in the development of the use of mhrn although the quick/noisy type kitten directs a larger number of mhrns toward the familiar human companion. At 4w2d slow/quiet L22B (Juniper) was described as being able to do everything his quick/noisy littermate L22A (Grinetta) did, and as well and as quickly as she, but he was very slow about starting any activity and, once started, stopped sooner than she did. At a later age it can be seen that the quick/noisy type seems to enjoy motion for the sake of motion while the slow/quiet type does not waste its energy unless a very realistic element of successful prey capture (food) is present.

9.   The human companion–observer (O) needed to be on guard against favoring the quick/noisy type kitten and thinking of it as more "intelligent" than the others because at an early age it interacted most quickly, most easily, and most vocally with herself. Under human care there is, of course, great survival value in being able to ingest food supplied by human caretakers at an early age as the quick/noisy kitten can, and this leaves more mother's milk for the less quick kittens to suckle, but in the wild under the care of only the natural mother the slow/quiet type kitten with its strong focus on the mother may very well have the advantage in many ways over kittens that are much more easily distracted. In studying the development of kittens the behavior of the slow/quiet type kitten is

probably the best clue to behavior in the wild because it is less easily influenced by human companions. It was therefore fortunate that S15 (Rennie) proved to be a slow/quiet type kitten about three days slower than slow/quiet L22B (Juniper) in first being able to lap milk adequately from a dish.

10. In conclusion, the seemingly infinite variation in kittens is made up of (a) physical variation in color and length of hair, shape of body, and other physical characteristics, (b) quick/noisy–slow/quiet differences in temperament, (c) variation in litter experience and conditioning, and (d) variation in individual experience and conditioning.

F. Focus on and Interaction with the Natural Mother and with the Familiar Human Companion

In the 0–4w4d period the kitten is entirely focused on its mother, nest, and suckling, which are generally all one and the same thing. Although, as motor skills and the amount of time spent awake increase, daily exploratory and play excursions keep increasing, at 4w4d these excursions away from and back to the mother and/or nest are still very brief in time and distance. Furthermore, the kitten's radius of awareness at any one moment seems to be nest size, perhaps 3 ft. The technique by which the kitten interacts with its mother and its entire world in the 0–4w4d period is the suckling one of touch–seek and cry–find and suckle–paddle and purr, which in the absence of hunger shortens to touch–paddle (play) and purr, or touch–paddle (play), or touch–purr. Except when the kitten throws itself up at its mother's head in right-side-up play, it is focused on the milk bag bulk of its mother's body.

In the third week when eyesight becomes operative, those of S1's (Emy's) kittens that were reared in the high-sided raised box in the cellar began looking up and out in the right direction for their mother's return. There was only one route from the cellar box to the kitchen that the mother followed, and when kittens were able to climb out of the box, there was never any doubt about what direction they would take. The kitten must seek its food from birth by pressing forward to locate a milk-giving nipple on its mother's body, and its primary task is always to maintain awareness of the location of its mother, the source of its food, as far as it is able.

The mother offers all four types of friendly approach behavior to her kittens and evokes all four types from her kittens. She offers touch and mhrn greetings and often purrs as she nurses them. Her encircling body gives the whole-body contact comfort that the kitten's later rubbing against her momentarily restores, and she creates a cosy nest in which the kittens relax and play, paddle and purr, as well as suckle and sleep. The mother does not play with her kittens in the nest stage, no matter how provocatively they behave, but her body serves as an attractive play target for her kittens' right-side-up body-throwing and batting play.

S1's (Emy's) general aim seemed to be to have her kittens suckling or sleeping after she had cleansed them thoroughly and nursed them. Any other activity in the 0–4w4d period seems to be on the kitten's initiative, with the mother adapting as best she can and sometimes even disapproving, except for the mother's asking to have the kittens brought up into the kitchen at times from the fourth week on. The mother, of course, also chooses and maintains a suitable nest for her kittens. If the nest becomes unsatisfactory to her, she will move the kittens to a new nest, carrying them one at a time by the nape of the neck. The mother is as focused on her kittens as they are on her, but she has to leave the nest at times to obtain food, to relieve herself, to maintain her relations with her human companions, and otherwise carry on a life away from as well as in the nest. Although the mother stays away from the nest for correspondingly longer periods as kittens become able to sleep and stay awake and play without becoming hungry for increasing lengths of time, at 4w4d S1 (Emy) was still spending much time sleeping and resting in the cellar nest or in the kitchen easychair with her kittens.

Unlike the natural mother the human companion preferred to find kittens awake and playful during any time she could spend with them, and was often disappointed to find them merely sleeping. The familiar human companion (O) also offered S1's (Emy's) kittens touch (petting) and murmuring vocal greetings, the cosy body contact of her hand or lap, and the play target of her wriggling fingers, but she tried not to awaken kittens if she found them sleeping at a time when their mother was away from the nest and not likely to return immediately. In general kittens responded to the familiar human companion in the same way as they responded to their natural mother. In the 0–4w4d nest-stage kittens do not distinguish between their mother and their familiar human companion. Under the influence of the suckling technique kittens either expect to find what they need (milk and mother) immediately in front of their noses or expect it to drop down on them from above in short order—as indeed it usually did in the form of their mother. What the human companion could not offer was milk to suckle, but if kittens happened to be on O's lap when they became hungry, they nuzzled for milk on O's arm.

However, the human companion always has a certain advantage over the feline mother, especially in the earliest weeks, because of being able to distinguish among the kittens as individuals. Even if O had not named kittens, she would have thought of them as "the tiger one," "the black one," etc. A human can respond to an individual kitten's need at a particular moment, while the mother distributes her tongue licking and mhrns in a very broadcast manner although eventually all kittens get taken care of. Furthermore, the human companion could lift and move kittens with warm hands supporting all four feet in a manner that did not interfere with the kitten's comfort and functioning, unlike the mother's nape of the neck method, which turns a kitten into an immobile ball. No kitten ever asked to be picked up by its mother, but from their third week on kittens came to the front of the box and cried to be lifted up again by O. As long as they were not hungry, all kittens seemed to enjoy the change of odor and scenery found in the substitute nest of O's hand or lap. If a kitten cried as O held it, she returned it to the box at once, thus permitting the kitten to develop some sense of control over her lifting service. Kittens came to expect to be picked up by O and to suckle at once in their natural mother's presence, but their awareness of the two mother figures was very blended and diffuse.

Because only the mother supplied body milk for kittens to suckle they experienced the high excitement of actually obtaining food only in immediate association with her. They therefore responded at higher levels of tension (excitement) to their mother than to their familiar human companion. This had the result that O received low-tension greeting responses from the kittens earlier than their mother. Held in O's hand, petted and murmured to, L26A (Jemmie) purred at 2d, a day or two earlier than spontaneous purring while suckling was heard. In the third and fourth weeks all litters and almost all kittens gave mhrn responses to O's greeting voice at the front of the box, but it was not until 4w1d that one kitten responded once with a "mhrn" to the mother's greeting.

Inside the high-sided cellar box S1's (Emy's) kittens could not see their mother approaching, and Emy had a habit, whether she was jumping up into the nest box or up into a human lap, of not starting to utter a "mhrn" until after she had begun the jump. Her kittens therefore experienced her chiefly as touch. The human companion, however, was experienced as a voice outside the front of the box before she touched or picked up a kitten. As a result box-reared kittens were always very responsive to O's voice but not very responsive to that of their mother. In the 3-ft square box kittens were not far enough away from their mother for her 'mhrns to act as calls, but O's voice at the outside of the front of the box gave them something to move toward. While L21A–C (Peleg, Gilda, Jeda) were playing in the kitchen at 4w1d, O returned to the kitchen after a brief absence and found no kittens in sight. O said something like, "Well, where have all the babies gone?" Instantly all three kittens quickly came running to her out from under the coal stove. These kittens had not yet been given any food by O. They were described as coming to O very readily when she called them, but as paying no attention to their mother's calls. On this same 4w1d evening middle L21A (Peleg), which had formed the unusual habit of making sure he was the first to reach O, cried while looking up. O lifted him up on her lap, where he settled down contentedly. Slow/quiet L21C (Jeda) noticed Peleg being picked up, and also came over to O, crying, but in a somewhat less purposeful manner. O lifted Jeda up to her lap. A feeling of aloneness now seemed to descend upon quick/noisy L21B (Gilda) in her distant wanderings. She stopped and cried in a bewildered tone. O called her. Gilda started earnestly running toward O at once, twice improving her direction on the way at the sound of O's voice. It would be 5w0d before a kitten—and that slow/quiet L21C (Jeda)—would focus in on the mother's mhrns in a similar manner.

At 5w5d L22A–B (Grinetta, Juniper), the kittens reared in an open nest on the floor of a nearly empty coal bin, were described as being unusual because they came as readily when their mother called them as they did when O called them. It was the openness of their nest that was unusual for S1's (Emy's) kittens. Wilson and Weston (1947) describe kittens reared in a low basket as responding to their mother's "very throaty 'prrp prrrp' " first by crying, then, when older, by stumbling toward her, and still later by making a mad dash toward her. In the wild kittens are apparently reared in nests on the ground, and human strangers who approach are responded to with snarls and hisses (Adamson, 1969; Lindemann and Rieck, 1953). There would be no question of kittens not responding to the voice of their mother since she would be the only animal interacting with them in a maternal manner. Even pet housecat males cannot be trusted not to treat the very young of their own kind as prey, and are fiercely driven off by housecat mothers.

In the human household the mother and the familiar human companion are both simply part of the kitten's mothering experience in the 0–4w4d nest stage. Both the mother and the familiar human companion (when she was with the kittens, and to a large extent even when she was not) were strongly focused on the kittens, and the kittens were strongly focused on both mother figures. They reacted to both in the same manner as far as circumstances permitted, but the circumstance of the shape and location of the nest could alter the degree to which kittens could form associations with and so respond to the voice of their feline mother and that of their familiar human companion. Kittens react to their mother figure in the 0–4w4d nest stage with the suckling technique of touch–seek and cry–find and suckle–paddle and purr, which in the absence of hunger shortens to touch–paddle (play) and purr, touch–paddle (play), or touch–purr.

VII. Four Weeks Five Days–Seven Weeks Six Days:
Expansion of Focus to Include Small, Movable, Inanimate
Objects as Food and Toys
(Development of Dead Prey Handling Technique)

For the 4w5d–7w6d period there are two sets of data: notes on the behavior of S1's (Emy's) kittens (chiefly L19A, L20A–C, L21A–C, and L22A–B), and more detailed notes on the behavior of S15 (Rennie) beginning at 5w2d, after removal from his mother and littermates to the observer's home. Each set will be reviewed separately.

A.   S1's (Emy's) Kittens

1.   Eye-Paw Coordination, Motor Development, Play, and the Use of Space

In the second half of the fifth week there are no more comments in the notes on S1's (Emy's) kittens about kittens playing only one at a time instead of "cooperatively" with littermates. At 4w5d L21B (Gilda) and L21A (Peleg) played fortress attacker and defender, one inside a low-sided basket keeping the other from getting in, turn and turn about. Kittens now begin to show eye–paw coordination, which enables them to counteract accurately the small motions of small objects such as the paws of littermates. This was recognized most specifically of L21A–C at 5w3d: "The kittens are only beginning to see each other's motion, that is, to follow if one moves, to accept a challenge to play, etc." This change is not in eyesight, for kittens are able to watch motion from their third week (at 2w2d L22A, Grinetta watched O move a finger back and forth), but in the ability to make a countering motion with paw and foreleg in response to the sight of a small object in motion.

Two other new elements appear in the play activities of the 4w5d–5w4d week. One, for some but not yet all kittens, is an interest in playing with small inanimate but movable objects, and the other, more universally, is an interest in squeezing into very small boxes such as a shoe box or the even smaller box for an electric iron. When S15 (Rennie) first encountered what was to him the enormous expanse of his new home at 5w2d, he found for himself and hid in a tiny hollow on the floor formed by furniture, thus revealing the significance of the interest in playing in very small boxes from 4w5d on. It is a new ability to take the initiative in quickly finding a small space to hide in and feel safe in because the body touches four or more sides of a snug, nestlike hollow. At 8w2d when L26A (Jorrie) from the kitchen side of the screen door heard his mother fiercely driving another cat out of their yard, Jorrie immediately dashed behind a broom, which stood in a corner near the door, and stayed quietly hidden there for a long time after the battle had ended. In the wild a kitten no doubt must be able to find hiding places for itself before it can safely follow its mother about for any distance.

Play with small movable objects like a paper ball or a pair of large buttons tied to a string is related to the eating of solid food and will be discussed in that connection below.

After 4w5d the running and jumping abilities of kittens increase rapidly, and consequently the amount of space that the kitten uses increases rapidly in three, not just two dimensions. In the 4w5d–5w4d week litters of S1's (Emy's) kittens varied considerably in their ability to climb and jump and play with small movable objects. L22A–B (Grinetta, Juniper), reared in the open space of the coal bin from 2w6d, were the most advanced. L20A–C (Gerrit, Hilary, Joella), which climbed out of the high-sided cellar box at 4w0d, at 5w3d could climb in and out of the cellar box, could easily climb all the way up the cellar stairs, and could climb in and out of the kitchen easy chair. L21A–C (Peleg, Gilda, Jeda), which had to be coaxed by O to climb out of the cellar box at 5w0d so they could be given supplementary milk, at 5w3d could not yet climb in and out of the box or all the way up the cellar stairs very easily, and they could jump down from the kitchen easy chair but not up into it.

However, what kittens can do only with great effort one day they can perform with ease a few days later in this period. Skill in jumping, running, and playing with small movable objects increases even more rapidly after 5w5d so that what only certain kittens or litters could do in the 4w5d–5w4d week all kittens can easily do by 6w0d. At 6w1d L21A–C (Peleg, Gilda, Jeda), the slowest litter, could very easily jump up into O's lap when O was sitting in a kitchen chair higher than the easy chair. Being able to jump up into the human companion's lap and up into the kitchen easy chair that their mother made her headquarters, as well as being able to run quickly, enabled the kittens to follow their mother and familiar human companion wherever either went.

In the 6w5d–7w6d week kittens reach a peak of skill and speed in running, jumping, and playing with small movable objects as well as with their littermates. and mother.

2. Supplementary Food (Milk by the Dish): Response to More and More Food Signals and to More and More Distant Signals

In the wild kittens would be weaned to eating dry solid food such as a dead mouse at about six weeks of age. Human caretakers, however, can supply a range of intermediate liquid and soft, moist, cooked food by the dish, which kittens can suck up at an earlier age.

Normally, with adequate milk from the mother, supplementary food and independent urination would not have been part of kitten behavior until the 4w5d–5w4d week. Because what eventually proved to be a mammary tumor made the amount of S1's (Emy's) milk too scanty for more than one kitten in the latter part of the nest stage during the period under study, warm milk by the dish was given to L20A–C from 4w2d since they could climb out of the cellar box from 4w0d, and to L22A–B from 2w6d, after their mother moved them to the large coal bin. It was then discovered that, if a kitten consumed food other than its mother's body milk on or after 4w2d, it began urinating independently of its mother's licking, but if a kitten ate supplemetary food before 4w2d, it did not. In the wild kittens could walk a considerable distance from their nest spot beginning at 4w2d, and so could keep it clean, although it is not likely that they would eat any supplementary food before 6w0d, when they could cover a great deal more ground much more easily. Sl's (Emy's) cellar-box-reared kittens had to be able to climb out of and back into the high-sided box to use a litter box on the floor before they could be given supplementary milk, but this was no problem at the normal age of 5w0d. Once a kitten was urinating independently, use of the litter box for bowel evacuation followed sooner or later without arousing any attention, and the mother's licking services gradually were no longer needed for removal of the kitten's waste products.

At 4w2d L20A–C (Gerrit, Hilary, Joella) were first allowed to drink warm milk from a dish with their mother because they were able to get out of their cellar box, because their mother was not behaving "very graciously" toward them, and because they kept running after O and crying. O first dipped slow/quiet Hilary's chin into the milk. He backed away and had to have his chin dipped again before he drank. Middle Joella drank as soon as her chin was dipped into the milk. By this time quick/noisy Gerrit had come over to the dish and begun to drink milk on his own. A little later that day when the mother gave the empty dish a few noisy licks as she passed by, quick/noisy Gerrit immediately dashed over and licked the dish, too. Middle Joella then came bounding over, uttering a by-product "mhrn" on the way. She was reacting to Gerrit's motion, not to the sound of the dish. Slow/quiet Hilary paid no attention. The more independent and successful the dish milk experience was, the more quickly kittens responded to a sound associated with the dish, and before 5w0d it was the quick/noisy kitten that was by far the most successful.

Although L22A–B (Grinetta, Juniper) were first given a dish of milk on 2w6d, and daily from 3w1d on, it was noted on 4w2d that the mother usually drank most of the milk given to the kittens, quick/noisy Grinetta drank to the last drop, but slow/quiet Juniper, who still had to have his chin dipped into the milk, chewed at that first sip for so long that the dish was empty before he got around to trying to take a second sip. It was not until 5w0d, as recorded on 5w2d, that slow/quiet Juniper finally began lapping up his share of the dish milk. On 5w5d it was noted that these L22A–B kittens not only came running very readily when O called them, but surprisingly came running when their mother called them with 'mhrns, but they did not recognize the sound of eating in others or the sound of the dish as food signals. Apparently they had begun to drink dish milk with their mother at too early an age to be able to respond to dish-associated sounds although in their open, floor-level nest they had had the opportunity to learn to come in response to their mother's 'mhrns, while box-reared kittens did not. Not until 7w2d, as recorded on 7w4d, did L22A–B began to respond to the sound of the dish and to the sound of others' eating as food signals.

Before 4w5d–5w0d there are wide quick/noisy–slow/quiet differences in the ability of kittens to lap milk from a dish. After 4w5d the adaptation to dish milk is much more rapid, and much easier. L21A–C (Peleg, Gilda, Jeda) had to be coaxed by O on 5w0d to begin climbing out of their high-sided cellar box so that on 5w1d they could be given their first dish of warm milk, which they drank with their mother. On 5w2d, as soon as these kittens heard their mother start to lap milk, all three kittens came running to join her in lapping up the milk, without any quick/noisy–slow quiet differences. On 5w3d two of these kittens came running as soon as they heard their third littermate chew on the edge of the empty dish.

From 5w0d on kitten ability to respond to dish food signals increases rapidly. On 6w1d some very interesting and important food behavior was displayed in both the L20A–C and L21A–C litters. All three L21A–C kittens (Peleg, Gilda, Jeda) were on O's lap in the kitchen easy chair on 6w1d when their mother started to lap water from their water dish across the room. Instantly all three kittens jumped off O's lap and dashed across the room to the water dish. Although for a few days after they had started to drink milk they also drank water as if it were milk, by 6w1d they were no longer fooled by water and would not drink it when they reached the dish and found it contained only water. Yet the food signal noise of their mother's lapping had drawn them instantly to the dish from which she was lapping water, across the room from where their food was served. L20C (Joella) displayed similar involuntary behavior at 6w1d when she returned to the food dish three separate times upon hearing her mother still eating even though Joella presumably had eaten all that she wanted when she left the dish the first time.

The first part of the new dead prey handling behavior, which dominates kitten behavior after 6w0d, is instant high-tension response to food signal. Quick/noisy–slow/quiet differences disappear in the instant high-tension response (speedy running approach) to a food signal, and all kittens thus have an equal chance at the dead prey or object food.

The second part of the dead prey handling technique is instant high-tension response to more and more food signals and to more and more distant food signals. Although dishes of warm milk are not available in the wild, this was the food served regularly to S1's (Emy's) kittens in the morning, in the evening, and at bedtime after 5w0d, and kitten response to an increasing number of food signals and to increasingly distant signals can be seen most clearly in a series of notes on the response of L20A–C to a dish of warm milk served in the cellar to make them willing to return to the cellar after a play period in the kitchen.

O always called to the L20A–C kittens (Gerrit, Hilary, Joella) as she walked down the cellar stairs carrying the dish of milk, followed by the mother, which, when she had kittens in her care, was always interested in drinking milk. At 5w4d slow/quiet Hilary followed O down the steps for the first time. Gerrit and Joella as usual walked along the trap door opening to the deep end and had to be picked up and carried down by O. On the morning of 5w5d Hilary again followed O down the stairs. The other two kittens, after much coaxing from O, also walked down the stairs. When they heard and saw their mother and Hilary lapping milk, they made a "flash beeline dash" for the dish. At bedtime on 5w5d

O opened the cellar trap door before picking up the milk dish, and she found Hilary waiting for her on the top step. The other two kittens also followed O down the steps after a little coaxing from O. On the morning of 5w6d when O opened the cellar trap door to take milk down to the cellar, all three kittens popped up into the kitchen. Hilary followed O down to the cellar at once. Gerrit and Joella did not start down the stairs until Hilary was already at the bottom, but it was noted that quick/noisy Gerrit and middle Joella always ran all around the kitchen as soon as the door was opened, but slow/quiet Hilary was slower than the others and never got so far away from the door in the first place. On 6w0d when O rolled a cart off the cellar trap door when the L20A-C kittens were in the kitchen, all three kittens ran down the stairs ahead of O. Without the prospect of a dish of milk no kitten ever voluntarily returned to the cellar from the more attractive kitchen. (There is other evidence that at 6w0d/1d kittens show a new ability to anticipate a result that will come some time after a preliminary signal. At 6w0d S15, Rennie, for the first time voluntarily went to his nighttime sleeping place when O began to run bath water for herself.)

At 6w3d a second human companion began to appear in the notes on L20A-C (Gerrit, Hilary, Joella), namely O's father (FA), who enjoyed watching kittens play and who took care of them when O was not at home. On 6w3d when the L20A-C kittens were playing in the kitchen at a time when they were not accustomed to receiving food, FA rolled the cart off the cellar door to take some jars of fruit down to the cellar. All three kittens came running and went to the cellar with him. On 7w3d FA rattled a saucer on the floor, and all three L20A-C kittens came running over to him eagerly. On 7w5d quick/noisy L20A (Gerrit) followed FA around in the kitchen and then sat at the cellar door as if hinting that he hoped FA was going to take milk to the cellar for him. Late in the evening on 7w5d, as these kittens slept in the kitchen easy chair with their mother, slow/quiet L20B (Hilary) raised his head and looked lazily across the room at O. O murmured a greeting to him. Hilary then jumped down to the floor, went to the cellar door, sat down there, and cried for his bedtime milk.

As the kittens responded to more and more food signals, all of S1's (Emy's) litters of kittens came to respond to pretty much the same signals in their eighth week, catching up on any they had failed to respond to earlier. At 7w4d when O got home late at night and found L22A-B (Grinetta, Juniper) in the kitchen, the kittens greeted her by crying for something to eat and, like 20A-C (Gerrit, Hilary, Joella), by going to sit on the cellar trap door, expecting their bedtime dish of milk. These kittens also came running, expecting milk to be served in the cellar, whenever the cart was rolled off the cellar trap door. At 7w4d L25A (Jorrie), which was reared in a small box behind the kitchen stove, initiated demands for food by going to the spot where his food would be served and "yelling" for it.

Kitten behavior changes through anticipation. At first the L20A-C kittens followed O and the milk dish down to the cellar only because O's familiar voice called them. Then they followed O and the milk dish on their own. Then they preceded O and the milk dish. Then finally they initiated demands for milk to be served in the cellar by sitting at the cellar door with or without crying before any human companion had begun to make a move toward preparing warm milk for them. In their eighth week kittens are focused on food as an object in itself, separate from the mother figure, and react to any signals associated with the food. It does not matter who feeds them.

3. *Weaning to Solid Food and Play with Small, Movable,*
   Inanimate Objects (Development of the Dead Prey Handling
   Technique of Instant Arousal of High-Tension Response to Food
   or Play Signal–Seize–Run off with–Eat or Play with Object)

In the wild there would be no intermediate dishes of warm milk or soft cooked food, and the kitten would be weaned to eating a single item of dead prey such as a whole mouse. There seems to be no accurately dated detailed information concerning the age at which kittens begin to eat wild dead prey brought by the mother. Wilson and Weston (1947) report that one housecat mother, which had been tamed from a feral state as a kitten, moved her first litter outdoors to a woodpile when the kittens were "six weeks old" and brought them a large dead rat, which one kitten ate in its entirety, preventing its littermates from seizing it. S1 (Emy) twice brought a small dead mouse to early litters of her kittens that had not yet eaten anything but milk (at an age unfortunately not recorded), and in each case one kitten of the litter ate the whole mouse.

Only one small item of dead prey at a time is brought to the litter by the mother in her mouth, and only one kitten can succeed in keeping it and eating it. This is an extremely competitive situation, requiring highly skilled, high-tension behavior for success. The single item of prey is seized, carried off, and fiercely defended from the encroachment of littermates, and finally eaten by whichever kitten can succeed in doing so. The technique for handling dead prey is instant arousal of high-tension response to food signal (and to more and more food signals and to more and more distant food signals)—seize—run off with—defend—eat. In contrast to suckling, which however competitive in the seek-and-cry stage, always ends peacefully with "find and suckle–paddle and purr" for all kittens of the litter at the same time since the mother has more milk-producing nipples than she has kittens in a litter, success in keeping and eating the single item of dead prey is a fiercely competitive experience that leaves all but one kitten in a very frustrated state while the victorious kitten is tensely defensive.

Human caretakers of animals attempt to see that all get an equal share of food so that all survive and develop equally well. When a deep saucer of milk was no longer enough for S1 (Emy) and her ever-growing kittens, a larger pie plate was used. Soft mushy food fills a dish nearly as easily as liquids with equal access for all, and solid food was cut or broken into small pieces and doled out equally to all. The competitive "seize–run off with–defend" aspect of the dead prey handling technique is therefore not often aroused by food supplied by human beings. The high-tension response most often visible in the human home is the "speedy beeline dash" to the source of a food signal.

Kittens (and adult cats) have only two active ways of interacting with a small, inanimate, movable object. They can play with it or they can eat it. There is a very close relationship between play with such objects and the kitten's ability to eat solid food. The dead prey handling technique can be described as a means of handling small, inanimate, movable objects and as consisting in full of instant high-tension response to play or food signal (and to more and more signals and to more and more distant signals)—seize—run off with—play with or eat object.

S1 (Emy) was given small (about ½-inch-long) pieces of dry bread as a snack whenever she asked for it at the bread box in the evening on the premise that, if she wanted to eat such low-preference food as bread, she must really still be hungry. A kitten's ability to eat such pieces of dry bread seems to come closest to its ability to eat dead prey. When O rolled a paper ball toward L21B (Gilda) at 5w3d, instead of chasing it Gilda picked it up in her mouth, walked off with it, and then tried to eat it. On 5w3d neither the L20A–C nor the L21A–C kittens could eat small pieces of dry bread even though they were aware that their mother was eating and they were trying to get a share of what she was eating. L22A–B, the kittens reared in the open nest in the coal bin, were not only advanced in play with objects such as buttons on a string but also were eating dry bread or bread soaked in milk, beets in milk, and scrambled eggs by 5w5d.

Although unable to eat half-inch-long pieces of dry bread on 5w3d, L21A–C (Peleg, Gilda, Jeda) were able to suck up tiny crumbs of cake that dropped from their mother's mouth as she ate small pieces of cake given to her one at a time. As they sucked up the cake crumbs, the kittens "grew very nasty" toward any other who approached, growling and slapping out for all they were worth. On 6w1d these kittens were not satisfied with a dish of milk but kept jumping up into O's lap and trying to get at O's food, and so O gave them small pieces of dry bread. Each took its piece and ran off alone, growling and slapping out if any other came near. On 7w1d these L21A–C kittens, which had become a little too hungry before they were coaxed by O to climb out of their cellar box on 5w0d so that they could be given supplementary milk, were described as always begging for more food no matter how far their stomachs were visibly bulging with the food that they had already eaten. L21A (Peleg) had a habit of running off growling as soon as he got any solid food in his mouth. Although this would have been a successful technique if the food had been a whole dead mouse, it was not very useful when what he ran off with was a single small kernel of rice (6w6d) or a single kernel of tapioca (7w0d) from a dish full of milky pudding while his littermates remained at the dish eating steadily.

At about 7w0d kittens are found locating small objects to eat for themselves. What they are apt to find lying on the floor of a human house is string or cloth from a sewing project or something equally indigestible, and it becomes a problem to keep kittens of that age from swallowing something dangerously inedible. At 6w6d L21B (Gilda) found a small piece of cloth on the kitchen floor. She carried it around in her mouth, played with it, worried it, and then settled down to eat it, all quite as if it were a mouse. It was noted that L19A (Zeb) at 7w0d "like all kittens" chewed on thread and string whenever he found any. At 7w0d L21A (Peleg) is described as trying to eat bits of wood, and whenever O heard growling, she immediately investigated to see what he was trying to eat that he ought not to. This 7w0d initiative with respect to locating small objects to eat seems to mark the point at which all kittens are thoroughly ready to eat solid food and cope with a dead mouse, although some may be able to do so earlier. It is also the point at which kittens have become very skillful at running, jumping, and playing with small movable objects. L19A (Zeb) would pay no attention to a paper ball on a string until he was 5w0d old, but at that age he still had no interest in playing with the heavier, less destructable version of this toy, buttons hanging from a string or fine chain. At 7w0d, however, he played with the dangling buttons "alone, as he should." That is, he took the initiative in playing with them.

One more point remains to be made with respect to the 4w5d–7w6d ability to eat solid food and play with toys. The kitten shifts from expecting food to come from the mother figure's body to expecting the human companion to supply food, to expecting food to be an object in itself with an origin and destination of its own. When L21A–C (Peleg, Gilda, Jeda) were 7w3d old, S1 (Emy), with Jeda following her, walked over to a cupboard where meat left from dinner was stored and stood looking up, silent because there was no hope of her being given a piece of the meat so soon after dinner. The other two kittens then ran over to their mother and Jeda, and all three kittens stood alternately looking up at the cupboard and down at the floor as if expecting food to appear from the cupboard and land on the floor. Similarly when L25A (Jorrie) at 7w4d was not given a share of his mother's regular Sunday morning treat of a piece of raw meat as FA prepared a roast, Jorrie kept looking and crying up at the table where the pan of meat was resting, fully aware of just where the meat that his mother had received had come from.

The very great change in the kitten's apprehension of food that takes place in only three weeks between 4w5d and 7w6d because of the development of focus on small, movable, inanimate objects as food and toys can be seen in the contrast between two notes made on the behavior of L25A (Jorrie), who was reared in a small box behind the kitchen coal stove. In order to catch an early train on the day Jorrie was 4w4d old, O got up at 3 A.M., an hour that S1 (Emy) and Jorrie seemed to find very satisfactory for starting their day. Emy insisted on having her breakfast then, too, but Jorrie licked his paws and "got to work playing." A singleton kitten getting plenty of milk by suckling, he was not yet interested at 4w4d in the milk or other food eaten by his mother only a few feet away from a low nest box he had been able to climb out of for a whole week. Three weeks later at 7w4d it was quite another story. When preparations were only just being started by human companions to heat milk, Jorrie went to the spot on the floor where the milk would be served and cried loudly for it. When his mother was given her meat dinner, Jorrie had to be held out of the way for a while or else he would have eaten it all, and she would have let him do so. This was also the day on which he was well aware that his mother had been given a piece of meat from the pan on the table. Between 4w4d and 7w4d the kitten becomes able to focus on food as an object in itself with a trajectory of its own.

4. Quick/Noisy–Slow/Quiet Differences

Before the dead prey handling technique becomes dominant at 6w1d/3d, some quick/noisy–slow/quiet differences in behavior are still apparent, although at about 5w0d all kittens become equally able to lap milk from a dish. After the dead prey handling technique becomes dominant, its instant high-tension (speedy-running) response to more and more food signals causes all kittens to respond equally rapidly to the same signals, and so all have an equal chance at solid food, at seizing the single item of dead prey that the mother would bring to her kittens in the wild.

After 6w0d there is only one reference to quick/noisy–slow/quiet differences. At 6w4d when slow/quiet L22B (Juniper) came across a long narrow, black woolen belt lying on the floor for the first time, he treated it as if it were a snake, dragging it out from under the stove, backing off from it cautiously, and then, whenever he gathered courage enough, slapping out at it with the same kind of quick, snappy, forceful strokes his mother used toward large insects on the ground (and which S15, Rennie, in adulthood used toward a real snake). Quick/noisy L22A (Grinetta), on the other hand, simply treated the belt as something to play with right from the start. This is an illustration of how the slow/quiet type requires a strong element of prey (food) even in a toy before it expends its energy, while the quick–noisy type enjoys engaging in motion for the sake of motion. This difference in style continues into adulthood, as does the quick/noisy type cat's greater readiness to explore and welcome new experiences.

5. Focus on and Interaction with Mother and with Familiar Human Companion

In the wild where the natural mother would be the only supplier of solid food (dead prey) as she is of milk to suckle, the development of focus on small, movable, inanimate objects and the ability to snatch and defend and eat dead prey would mean that the kitten would be focusing on its mother in an additional, very intense manner. It would be paying attention to and attempting to seize any object that she might be carrying in her mouth or batting with her paws. She probably would allow her kittens to take prey away from her, as S1 (Emy) allowed L26A (Jorrie) to eat all her meat dinner, but the kitten also would be able to snatch it away whether she wanted it to or not. Since being followed about by playful kittens would be no help to the mother in stalking and capturing prey, the kittens would probably stay quietly hidden perhaps in some nearby spot until their mother, like the cheetah Pippa with her prrps to her cubs (Adamson, 1969), released them with her 'mhrns.

In the human household, where all object food is supplied to the mother as well as to her kittens by human companions, the result of the development of focus on small, inanimate, movable objects as food and toys and the dead prey handling technique is more complex. Solid food is not only an object separable from the mother's body, but much more separated from her. It is kept in various storage places and supplied not only by the familiar human companion but eventually by any human companion.

At 4w5d the kitten has only begun to follow its mother out of the nest for a short distance. By 6w0d it can follow its mother and familiar human companion "all around" for very considerable distances over what in the wild might be very rough ground. The two mother figures become mobile home centers, and at this point kittens can maintain awareness over a much larger area of space than in the nest stage. As S1's (Emy's) kittens became able to jump up into the kitchen easy chair to their mother and up into O's lap, they also could easily run back and forth between the two. They came to expect dish food from the human companion on demand but could approach their mother silently to suckle, and thus the two became more differentiated. The kittens no longer needed O's lifting service, but they obtained dishes of food from her by following her around and crying at her when they were hungry. In the sixth week the sound of the mother lapping milk became a food signal to which most kittens responded instantly. The mother became one among many food signals.

As the kittens became more and more skillful at running and jumping and playing with small objects, their human companions found watching kittens play and playing with kittens very entertaining. The mother responded playfully to her kitten's play after 4w5d and enjoyed wrestling with them, although they might bite her hard enough at times to cause her to cry out. But no matter how many new activities kittens engaged in in the 4w5d-7w6d period and however eagerly they ate dish food, all periods of activity and eating object food still ended with suckling and sleeping with their mother in the kitchen easy chair or in the cellar nest box. When S1's (Emy's) kittens wished to suckle, they were responsive to her 'mhrn calls; when they preferred to play or explore, they were not. However, almost anything the human companion said or did at any time was reacted to as if it were a call to food. The human companion was the major food supplier, the supplier of the most exciting food, and the mother became a secondary supplier of always available and therefore less exciting food.

In their eighth week kittens were focused on food as an object in itself with an origin and destination of its own. They could take the initiative in demanding specific food by crying at the spot where the food would be served or at the spot where it originated instead of at the mother figure. They could also take the initiative in finding something to play with almost anywhere. Since human beings are very good suppliers of food and toy objects and kittens are strongly focused on such objects and easily form associations with them, the end of the eighth week is a good time to transfer a kitten to a new home under human care only. If kittens are removed one at a time so that there is no uncomfortable buildup of unused milk, the mother accepts the loss of her kittens with none of the clinging to the familiar human companion or heightened use of mhrns that occurs if she loses a litter in the first week.

At the end of its eighth week the kitten has two techniques for interacting with its mother, its human companions, and its world in general. One is the suckling technique of touch–seek and cry–find and suckle–paddle and purr, and the other is the dead prey handling technique of instant arousal of high-tension response to food or play signal–seize–run off with–defend–eat or play with object. Running, jumping, and play abilities have reached a peak of skillfulness. At 7w4d/6d the kitten lives in a much larger, more complex, more object-filled world than it does at 4w4d.

6. Friendly Approach Behavior

a. Mhrn. In the 4w5d–7w6d period it becomes clear that the main line of mhrn murmur development is not directly through mhrns of greeting response but through the accidental, nonsocialized mhrns uttered as kittens run around in a state of playful relaxation, that is, by-product of motion mhrns.

There were no mhrn responses to the human companion's greeting voice in the 4w5d–7w6d period. The only instance of mhrn use noted with respect to the human companion came from L19A (Zeb) at 5w1d. Zeb cried to be lifted up on O's lap. When O said, "All right, do you want to come up?," and tapped her fingers on the floor to show where she could reach him, he came to her hand uttering a "mrn." This was probably merely a by-product of his forward motion rather than greeting response or affirmative response to an offer of service.

All mhrn response to the human companion's greeting voice ceased after 3w6d. After that time kittens were too eager to be carried up into the kitchen and the human companion became too associated with the exciting new object food that kittens were more and more able to eat for them to be able to respond with calm mhrns to O's approach to their cellar box. After 5w0d all kittens were able to climb out of the cellar box. Thus, even before the dead prey handling technique with its instant high-tension response to food or play signal became dominant at 6w1d/3d, the kittens were responding to the human companion at too high a level of tension for a calm mhrn of greeting response.

At first glance mhrn response to the natural mother's greeting mhrns seems to have developed in the 4w5d–7w6d period, but a closer look reveals that all instances relate to a habit of mhrn exchange that developed accidentally from a crucial experience between a single slow/quiet kitten, L21C (Jeda), and her mother. At 5w0d quick/noisy L21B (Gilda) hopped out of the basket to the kitchen floor uttering a (by-product) "mhrn." Her mother called "'mhrnhrn" from the easy chair. Quick/noisy explorer type Gilda ran off to play, paying no attention to her mother's call, but slow/quiet L21C (Jeda) responded with a "mhrn" and ran in the direction of the chair. Jeda lost her way. Her mother called again from the chair. Jeda then looked up at her mother, and O lifted Jeda up to her in the chair, thus completing the transaction on a calm mhrn level. On 6w6d S1 (Emy), in the kitchen easy chair, called the L21A–C kittens with rich-sounding 'mhrnhrns. It was only L21C (Jeda) that ran toward the chair uttering answering mhrnhrns, not so deep or full-bodied as her mother's. On 7w0d when S1 (Emy), seated on the kittens' carpet piece on the floor, greeted L21C (Jeda) with "mhrn," Jeda responded with the same sound. If Jeda's 5w0d and 6w6d mhrn responses to her mother were possibly more by-products of forward motion than greeting response, the 7w0d instance seems to have been genuine greeting response.

This habit of mhrn exchange between Jeda and her mother in the 4w5d–7w6d period is an illustration of how, given the right low-tension circumstances, by-product mhrn utterance can develop rather easily into a habit of mhrn greeting exchange. Since kittens are able to utter the independent mhrn sound from the beginning of their third week, the absence of mhrns is a matter of the absence of favoring low-tension circumstances, not a matter of whether the kitten can or cannot utter the sound.

There were only a very few other instances of mhrn use noted in the 4w5d–7w6d period, and all of these were of the accidental, unsocialized by-product of forward motion variety. There was L21B's (Gilda's) by-product mhrn as she hopped out of the basket at 5w0d. On 5w3d L20B (Hilary) uttered "happy mhrn sounds" as he ran around by himself, and again on 6w1d it was noted that Hilary uttered mhrn sounds as he ran around in the kitchen. At 6w1d all three L21 kittens (Peleg, Gilda, Jeda) often uttered mhrn sounds as they ran around, especially Jeda, who kept it up for long stretches. Although some kittens within a litter utter by-product mhrns much more often than others do, the data are not sufficient to determine whether this is due to temperament or some accidental conditioning or individually favorable circumstances.

In the latter part of the 4w5d–7w6d period tense mhrn and mhrn-plus-vowel-pattern cries of complaint begin to appear. These can be analyzed better after 8w0d when their use multiplies, but they begin to occur after 6w0d when the kitten has more awareness of and more expectations about objects in a larger radius of environment and so sometimes finds that objects are not always present or do not always behave as expected. The surprise at finding absence or different behavior apparently slows the kitten down enough for a closed-mouth murmur to be uttered.

At 7w4d O heard L22A (Grinetta) uttering "odd little worried mhrnhrn sounds" when she could not find the litter box in its usual location. These mhrns were tense enough to be described as "worried." On 7w5d L20B (Hilary) was heard uttering mhrn sounds, and when O looked, he was pawing at the hem of her housecoat to come up into her lap. At this age kittens do not yet use mhrns of greeting initiative or request, so it seems likely that these were not-yet-tense mhrns of frustration because O was sitting too close to the table for the kitten to be able to jump up into her lap as he wanted to. There is an area of overlapping where "tense" mhrn is no more tense than an ordinary "relaxed" mhrn, and the observer must judge by sight whether the kitten is moving forward freely doing what it wants to do or is somehow unable to move forward to do what it wants to do. The halted mhrn, however light to begin with, will quickly be followed by vowel cries, while the happy, relaxed mhrn occurs during play and is followed by continued play.

Unlike the situation in the earliest weeks when the kitten grunts as it pushes strenuously forward toward its mother against a physical barrier that is largely a result of its own lack of motor skill, no amount of physical effort (and grunting accompaniment) would solve the more subtle problem of an object being absent or otherwise not as expected. A different kind of awareness of the environment is involved, and the occurence of more or less tense mhrns and mhrn-plus-vowel-pattern cries is a clue to the development of this new stage of cognition. It is an intermediate stage. After awareness of the details of the mother figure's behavior has developed, complaint cries from older kittens and adult pet cats tend to be complaints because the human companion has not done some specific thing that the cat wants him or her to do. In the 6w0d–7w6d stage kittens focus on inanimate but movable objects.

Although there is awareness of specific object food after 6w0d, the dominance of the dead prey handling technique with its instant arousal of high-tension energy in response to food signals causes the kitten's demand cry for food to become louder and more prolonged with too much energy put into the crying for the vocalizing to start from a closed-mouth mhrn murmur.

Every mhrn noted in the 4w5d–7w6d period for S1's (Emy's) kittens has been mentioned in this discussion. The number of mhrn utterances was very small in this period, smaller than the number uttered in the 2w0d–3w6d nest stage and much smaller than at any time after 8w0d.

b. Purring.   Purring while suckling continues to be heard every day as does purring in response to being petted by the human companion when the kitten is in a quiet mood. The use of purring was not being studied in the behavior of S1's (Emy's) kittens at the time the notes were made, and there is only one incidental reference to purring in the 4w5d-7w6d notes.

   c. *Play as Greeting.*   In the eighth week full speed, highly skilled play becomes the dominant non-food-seeking behavior of kittens. They play with everything and anything. At 7w6d when L18A (Jankin) was taken to a new home, L18B (Minette) did not seem to miss him at all because she enjoyed all her toys and games thoroughly with no littermate to interrupt. The notes on the play of S1's (Emy's) kittens were not detailed enough to pick up specific examples of the greeting play that undoubtedly occured in a period when kittens initiate a great deal of play with mother, littermates, and human companions.

   d. *Rubbing.*   There are no references to rubbing in the 4w5d-7w6d notes on S1's (Emy's) kittens, but rubbing behavior was not being looked for at the time. Since the more detailed notes on the 9w0d-11w1d behavior of L19A (Zeb) include several references to well advanced uses of rubbing, some rubbing against mother or furniture undoubtedly occurred during the 4w5d-7w6d period.

   Helen Budinski Moelk reports (1976, personal communication) of her pet cat's second litter of kittens, born during the night of August 3/4, that on the evening of September 9 (5w1d/2d), when the mother came in from outdoors, four of the five kittens rubbed against her at the same time, and it seemed as if, by this rubbing, the kittens were seeking "love or comfort" from their mother.

B.   S15 (RENNIE), FIVE WEEKS TWO DAYS–SEVEN WEEKS SIX DAYS

   S15 (Rennie) was transferred to O's care at an age reported to be 6w5d. Although this was a week earlier than the ideal age for such transfer, he was said to have begun eating from his mother's dish of commercial cat food, and it seemed avisable not to let him become accustomed to eating meat since he was to be reared as a vegetarian in a vegetarian household in which milk and eggs but no meat or fish was used, and therefore no commercial cat food.

   A kitten once observed by O as it was spending a day in a college library office en route to a new home became completely comfortable in its strange surroundings simply by playing with a paper ball tied to a string. This kitten (O27), was said to be "six weeks" old, but was probably nearer 7w0d than 6w0d. In any case, it had made the change to focus on small movable objects. When first seen by O, O27 was being held and petted by one after another of a group of girls and women in a library office. The kitten shrank away from them all and tried to get away. Set down on the floor, it sniffed the wooden floor. A sweater sleeve was dragged along the floor to attract its attention, and after a while it slapped once or twice at the sleeve and then went back to sniffing the floor. O then brought in a small paper wad tied to the end of a string and hung the string from the back of a chair so that the paper ball dangled just above the floor. The kitten immediately began playing with this paper ball, batting at it from all angles, and performing acrobatically on the chair rungs to reach it as it swung. Newspaper was placed on the floor in a far corner of the room, and the top sheet rumpled a bit. Placed on the paper, the kitten began to play with it, hiding behind bits that stuck up and jumping out. Soon the kitten walked quickly back to the chair at the other end of the room to play with the dangling paper ball again. It was very businesslike now with no need to sniff in exploration any more. Its business was play.

Simply being free to play with a dangling paper ball changed a new, frightening environment into a place that a seventh-week kitten could cope with and enjoy in a very efficient manner through play. At the reported age of 6w5d it was expected that S15 (Rennie) would, like O27, adjust to his new environment through acrobatic play with a paper ball hanging from a string. That is, it was expected that he would have made the shift to focus on small movable objects as toys and food.

. However, when S15 (Rennie) was set down on the floor of his new home, nothing went as expected. The small living room and kitchen of the—to human eyes—small house, the two small plastic bins (15 x 8 x 7 inches) that had been prepared as toy and sleeping boxes, and the paper balls made from crumpled 3 x 5-inch slips of paper were all overwhelmingly too large for this kitten. He could not really be induced to play with a paper ball dangling from a string. Given a small amount of warm milk in a dish after he cried, he drank the milk and then used a litter box when he was placed on it. On his own initiative he located a tiny cosy space on the living room rug between a bookcase and a magazine stand and curled up to sleep in this small hollow. Given more milk later, he mostly chewed at it, and not until the following evening was he able to lap milk continuously enough to dispel a worry that he could not take in enough milk by lapping to sustain himself. Until word came back five days later that a ten-day error had been made in the first report of his birth date, O feared that she had a very retarded kitten. After she learned that he had been born on August 12, not on August 2 as first reported, and therefore had been only 5w2d, not 6w5d, old when transferred to her care, S15 (Rennie) could be recognized as a slow/quiet type kitten about three days slower in lapping milk efficiently from a dish than S1's (Emy's) slow/quiet kitten L22B (Juniper) had been 30 years earlier. S15's (Rennie's) behavior then fell into place as normal for his age and type. A quick/noisy type kitten with its ready acceptance of new experiences and human companions undoubtedly would have adjusted much more quickly to being under human care only, but it could easily have been predicted from the notes on S1's (Emy's) kittens that a slow/quiet type kitten with its strong focus on the natural mother and suckling and its lack of easy divertability by new stimuli would have difficulty in adapting to human care only at the too early age of 5w2d, before the development of focus on small, movable, inanimate objects as toys and food. At the end of their 4w5d–5w4d week S1's (Emy's) kittens, and especially her slow/quiet type kittens, were still very much dependent on their mother for suckling and other body contact services even though they had begun eating from a dish and using a litter box. At 5w4d a kitten is still almost entirely in the nest stage of focus on mother, suckling and a cosy nest.

Although by 5w2d S15 (Rennie) had undoubtedly acquired eye–paw coordination for specifically counteracting play with littermates, all the rest of his shift from focus only on the bulk of the mother's body to focus also on small, movable, inanimate objects as toys and food was still to come. His human companion–mother substitute (O) not only had the opportunity to observe this shift step by step in his external behavior but also experienced each step within herself as a release from a nest-stage body contact service demand from the kitten, which was impossible for a human companion to supply fully. Only being allowed to sleep and play on a human lap almost continuously and having milk to suckle at will would entirely have satisfied S15 (Rennie) at 5w2d and for some days thereafter.

S15 (Rennie) first made friends with his new human mother substitute (O) by playfully batting at her fingers from a cosy hollow. O had lifted him up on the sofa where she was lying, and he found a good hollow in the center of a large pillow at her head. When O reached wriggling fingers up toward him, he played with the fingers, batting at them for a long time. O then got out a long shoe string without a paper ball tied to it, and Rennie played intently with the shoestring, sometimes even growling over it, as O dragged it about on the sofa for him. He then wandered over to the other end of the sofa and became a bit lost. O called him from the middle. He came back to her and purred when she petted him. He then fell off the sofa to the floor, wandered over to his milk dish (then under a living room chair), and drank some milk. O called him into the kitchen, and he came to her. O placed him on his litter box, and he used it. Back in the living room he could not jump up on the low sofa, and O lifted him up. From the hollow of the pillow he played again with O's fingers and the shoe string, but a bit sleepily. Twice when O petted him, he purred. By this time O felt that he was familiar with her and accepted her as his tension reducer, his mother figure. He walked into O's lap and went to sleep there. When a bit of noise woke him up once or twice, he purred as soon as O petted and talked to him, and then went back to sleep.

O, of course, talked to the kitten a good deal in a murmuring, soothing tone of voice. S15 (Rennie) never had any difficulty in focusing on his new human companion–mother substitute or in responding to her voice. He eagerly expected O to supply all his needs and to respond to his cries. He was still in the stage at which the kitten expects all food, comfort, and other necessary mothering to descend upon it from above whenever it feels a need for this and cries.

1. Motor Development and Use of Space

   a. 5w2d–5w4d. At 5w2d S15 (Rennie) followed O about the house at a fast walk, but he did not run. He had been reared on a small enclosed porch, which did not provide much open space. It was not until the morning of 5w3d that Rennie began to do a little running. Then as he ran away from his litter box, he uttered a "hrrn" and purred two short, loud series of purrs as he ran, thus showing that he felt sufficiently relaxed and comfortable in his new home to have energy to spare for uttering by-product mhrn and purring vocalization while running. Judged by the standard of a 6w5d-old kitten, he seemed much too light and limp, not positive and confident enough in forward motion. He had a tendency to chase his own tail or shadow in a circle and to engage in clutching and grappling play rather than in running play. That is, he still preferred playing in a small, cosy, nest-sized space.

   When O came home for lunch on 5w3d and sat down, Rennie climbed up into her lap. When O returned home from work in the evening, he came to meet her at the door, walking along the sofa seat to the seat of a chair next to the door, thus revealing that he now could get up on the sofa on his own.

   Because the interior space of the house had seemed to overwhelm the kitten at 5w2d, all doors were kept closed—those to a utility room off the kitchen and those to bedroom, den, and bathroom off a small hall. This left a 16 × 15-ft living room with a 3 × 10-ft hallway off one end and a small kitchen with about 3 × 10-ft of open space at the other end, a total length of 35 ft of open space. By the afternoon of 5w4d, a Saturday, with O home all day, Rennie had mastered this floor space well enough so that all interior doors could be left open without making him uncomfortable. O also carried him outdoors through the living room door in the middle of the long front of the house, set him down, and had him follow her across the front of the house to the ground-level porch entrance to the kitchen at the north end, but he did not enjoy this new experience, although he followed O readily enough. On the evening of 5w3d every little noise within the house caused Rennie to be disturbed and look up, but on the evening of 5w4d he paid no attention even to the loud sound of the furnace going on and off.

   b. 5w5d–6w4d. On the morning of 5w5d S15 (Rennie) bounded along the floor, doing almost full-force running. He was able to jump up on O's bed. While on the bed he cried twice, jumped down to the floor, ran out into the kitchen (that is, the full length of the house, about 40 ft), and used his litter box all on his own. Later during the morning of 5w5d while playing on the bedroom floor Rennie did a few bucking-bronco runs. When O came home for lunch on 6w0d and left the living room door to the outdoors open, Rennie boldly took the initiative in going out and played with a plant in the doorway. He also enjoyed exploring the utility room off the kitchen, and in the den he jumped up on a sewing machine from a chair and jumped up on other furniture. On the evening of 6w0d he found pleasure in a trip outdoors, following O from the living room door around to the kitchen door and playing along the way and on the kitchen porch. That is, he was now comfortable enough outdoors to play along the way. He was then (6w0d) described as being fully confident in all rooms of the house and almost fully confident outdoors. On the evening of 6w1d Rennie did more exploring both indoors and out. By 6w3d he was climbing from the top of the back cushion of a chair to a bookcase to the arm of the chair. On Saturday (6w4d) he played outdoors with grass and plants for a long time as O gardened.

c. *6w5d–7w6d.* By 6w6d Rennie was running speedily along the floor of the house. By 7w1d he was running very quickly from room to room and making high jumps from the floor to the top of a bookcase. On 7w2d he did very speedy running everywhere both at noon and in the evening. Outdoors on 7w1d Rennie dallied near the living room door, and so O went into the house by way of the kitchen door without him. A minute or two later he came in through the kitchen door. On 7w5d, a Sunday, he spent much time outdoors, staying close to the house wall or the bushes near the house as O gardened at the far end of the yard. He made a quick run back to the kitchen porch at the opposite end of the house on his own. By 7w6d Rennie was doing such speedy running indoors and so often turned up under O's foot when she thought he was still some distance away that the house, which at 5w2d had been overwhelmingly too large for the kitten, now seemed so thoroughly occupied by his speedy running and jumping that there no longer was room enough for his human companion to move about in it freely. Outdoors O could be sure that Rennie would stay in the shrubbery near the house and could return from a distance to the kitchen porch and door at a speedy run. He could make use of both house and yard in a very independent manner and move at a speed far above that which any adult human being would use in a house or yard.

2. Food Behavior

a. *5w2d–5w4d.* Although S15 (Rennie) had been reported as having begun to eat from his mother's food dish, continuous lapping of milk was so difficult for him at 5w2d that it was not until the evening of 5w3d, when he continued lapping for as many as 35 laps at a time, that O could feel certain that he was going to be able to take in enough milk by lapping to sustain himself. On the evening of 5w3d he was able to shift from licking O's finger to licking a piece of cheese that O held for him, and he consumed the entire piece of cheese by licking it like an ice cream cone. A little while afterwards on O's lap he kept searching along O's arm and cried when he failed to find milk to suckle.

At noon on 5w4d Rennie again licked a small piece of cheese as O held it for him, and then he bit it. But he could not manage to eat a piece of dry bread, and he could not have eaten the cheese if O had not held it for him. In spite of all the biting and chewing that he did in play, and chewing on O's fingers, on a long shoestring, and on a small corduroy covered sofa pillow, which O moved around for him, was the play activity he engaged in most intently and confidently, there seemed to be no carryover from biting and chewing in play to biting and chewing solid food. He did not, of course, have competition from a mother and littermates to help him focus on the object food. On the evening of 5w4d Rennie would not eat strained baby food vegetables, which later became a favorite food, but he readily emptied a dish of corn chowder made with crushed corn, milk, eggs, and butter.

  b. 5w5d–6w4d.   On the evening of 5w5d S15 (Rennie) managed to eat a small piece of wholewheat bread soaked in milk as O held it for him. Although he cried and even banged at her closed door, he was not allowed to stay in O's bedroom during the night, as he would have liked, and so he continued to use as a sleeping place the tiny hollow between furniture on the living room rug, which he had first discovered for himself on 5w2d. O checked on him at least once during each night and gave him a dish of warm milk, which he drank. During the first reunion of the day at 4:30 A.M. on 5w5d and later while O prepared breakfast, O could recognize from the way Rennie rubbed and clutched at her ankles and was not very eager for his milk that he had more need for body contact than for food after his night alone. On 6w0d when O first opened her bedroom door, Rennie came to her from the afghan on the sofa, where he had evidently spent the night. After breakfast O busied herself putting various papers in order instead of sitting down and holding the kitten in her lap. When she finished, she found Rennie flat against the rumpled, loosely knitted woolen afghan on the sofa, suckling on the afghan and paddling his feet all as if it were his mother allowing him to suckle. There was milk left in his dish. O heated fresh milk and had to lift him from the afghan to the milk dish since he did not come at her call. (Never thereafter, however, did he prefer the afghan to real food.) Two nights later, 6w1d/2d, Rennie spent the night on a second afghan on a couch in the den next to O's bedroom. This then became his regular nighttime sleeping place, and in the two afghans Rennie had, as it were, two patient substitute mothers that allowed him to suckle whenever and for as long as he pleased, night or day.

  For an hour before bedtime on 6w0d O stayed in the bedroom to allow Rennie to play there. When O began to prepare for bed and ran bath water, he voluntarily went to the living room sofa and stayed there on the afghan (his then nighttime sleeping place). For the first time O did not have to give him food during the night, although she continued to check on him and greet him during the night. From this time on he was content to spend the night alone without food.

  Although Rennie being allowed to play in the bedroom for a while in the morning and in the evening and his discovery that he could suckle on afghans undoubtedly made the nighttime separation from his mother figure more endurable to him, by 6w1d he was already beginning to enter into a stage of more independence from body contact with the mother in any case. By 6w1d the development of focus on small movable objects as toys and food had begun within the kitten, and his intense need for frequent body contact was therefore beginning to diminish. At noon on 6w1d after drinking milk in the kitchen Rennie went directly to the sofa afghan to suckle without seeking to sit on O's lap. On the evening of 6w1d he emptied one dish of corn chowder in one try and a second helping in two tries, and for the first time ate a piece of cheese from his food mat on his own, without O having to hold it for him. On 6w1d his play also began to include crouching and stalking and pouncing attacks on O. This was the beginning of his need for rough play with his mother figure, and the need for rough play began to replace his earlier need for cosy body contact. As O prepared breakfast on 6w2d, Rennie ate two pieces of cheese by himself from his food mat. On the morning of 6w3d as O prepared breakfast, Rennie again ate small pieces of cheese from his food mat. He twice had to have it pointed out that he still had some cheese left, but otherwise he focused well on the food. As O prepared the evening meal on 6w3d, Rennie for the first time cried at his dish and at the refrigerator (when O went to the refrigerator) instead of just crying at O, and he did almost no clutching at O's ankles. This was the beginning of his recognition of food as an object separate from the body of his mother figure. On the evening of 6w4d Rennie cried for his food while standing at his dish.

   c. *6w5d–7w6d.* On the evening of 6w5d it was noted that Rennie now cried very loudly for food. As O lay reading on the bed, Rennie was not content to play or rest but kept crying. He ran into the kitchen, and O followed him. O gave him a piece of cheese, which he ate eagerly. Because corn chowder, cheese, and milk were all kept in the same refrigerator, when Rennie cried for food at the refrigerator, O could not tell which food he wanted except as he refused to eat what was offered and continued to cry until he got the one he wanted. That is, by 6w5d the kitten was able to keep a particular kind of food in mind and hold out, refusing other foods he liked, until he got the one he wanted. When O said, "Here," and placed a piece of cheese on his food mat, he could locate the food at once. In the wild the kitten would, of course, have no choice of menu, but under human care kittens seem easily able to make use of more kinds of food, more petting, more admiring vocalization, more of anything so long as it helps the kitten to go in the direction it needs or wants to go in.

   On 7w0d it was noted that no suckling on the afghans had been noticed recently, but a little later that same day after Rennie had been shut out of the bedroom so that O could be free of his playful claws as she dressed, he was found suckling and purring on the sofa afghan. Just as in the first instance of his suckling on an afghan on 6w0d, this suckling seemed to be compensation for his not having been permitted to make the body contact he wanted with his mother figure, but by 7w0d the body contact he wanted was rough wrestling play, not the nestlike contact of her lap as before 6w3d.

   On 7w1d, after Rennie had eaten his own dinner, he kept trying to get at O's dish, and he ate pieces of vegetables that she placed on a paper napkin on her lap.

On 7w2d when O worked at the kitchen sink and his dish was empty, Rennie cried very loudly for food. At breakfast on 7w3d he did not quite finish all the milk in his dish, and he would not eat the corn chowder he usually gobbled up, but kept crying until O offered him pieces of cheese. By 7w4d Rennie was going to his empty dish and crying when O was not already in the kitchen, and his crying for food was very loud and prolonged. When O put food in his dish, he tended to throw himself into the dish, and once he accidentally stuck a foot into his nearby water dish in his rush. That is, he used what seemed an excessive amount of energy in crying for food and in approaching food. Although Rennie did not display the middle portion of the dead prey handling technique (seize–run off with–defend) with respect to food, he did show the instant arousal of high-tension aspect in his loud, prolonged crying for food and in his headlong rushes to his dish.

In the eighth week S15 (Rennie), like S1's (Emy's) kittens, could initiate demands for food by crying either at his dish, that is, where the food would end within his reach, or at the refrigerator, the source from which food came. Food was now an object separate from his mother figure's body and had an origin and destination of its own.

3. Play with Small, Movable, Inanimate Objects (Full Dead Prey Handling Technique)

The development of the ability to focus on small, movable, inanimate objects as toys paralleled the development of focus on small, movable, inanimate objects as food in S15's (Rennie's) behavior and included the full dead prey handling technique of instant arousal of high-tension response to signal–seize–run off with–defend–play with or eat object.

a. 5w2d–5w4d. On 5w2d Rennie batted a little at a half-sized paper ball hanging from a string that O dangled for him, but he was not really interested in it. He initiated chewing on lamp cords and played intently with a long shoe string without a paper ball tied to it. This type of play is probably the equivalent of chewing on the mother's legs and tail. (Slow/quiet L20B, Hilary, showed himself livelier than his quicker littermates only in playing with long strips of yarn at 4w2d.) It was from the hollow of a sofa pillow that Rennie batted playfully at O's wriggling fingers and thus made his acquaintance with her. On the morning of 5w3d he ran about a little and enjoyed kicking, biting, and wrestling, that is, whole body play, with a small, round, corduroy-covered pillow, which O moved about for him as a substitute for wrestling with his mother and and littermates. While on O's lap he played a little with a small paper ball that O dangled for him, but chewing on the plain shoe string was still what he did best. On O's lap he purred easily when petted, especially when he was lying on his back and kicking and clutching at O's hand in play. He had a tendency to chase his own tail or shadow in a circle and to enjoy only clutch–kick–grappling play in cosy hollows, that is, whole body play. At lunch time on 5w3d when O came home and sat down, he climbed up into her lap and played with a paper ball tied to the arm of the chair when O dangled the ball for him.

   b.   *5w5d–6w4d.*   At bedtime on 5w5d O tossed a rather large paper ball from the bedroom into the hallway to get Rennie to leave the room, and he bounced along after it uttering a slight "hrn." This was probably the equivalent of running after a littermate or the mother rather than after a dead mouse. (Successful toys for kittens, depending on their size, mass, speed, and motion, seem to serve as substitute littermates or substitute prey.) On the morning of 6w0d Rennie began to show less need for body contact with his mother figure. After one excursion up on the bed and up on a nearby table cart, he did not need any further body contact with O and instead played on the floor alone. At noon on 6w0d when O entered the house through the living room door, Rennie went outdoors and batted a plant near the doorway. He explored the utility room off the kitchen and jumped up on furniture in the den. On the evening of 6w0d he was fully confident in all rooms of the house and nearly fully confident outdoors on a trip from door to door, playing along the way and on the porch. On the morning of 6w1d Rennie's play began to include edging along a wall, crouching, and stalking. At noon and in the evening he did much rough jumping up at O and crouching and stalking with O as the target. This is a type of action that could be used to snatch dead prey from the mother.

   Play with small, movable, inanimate objects became thoroughly satisfying to S15 (Rennie) at 6w3d. After he had eaten breakfast on that day he played on O's lap as she sat in a living room chair and ate her breakfast or wrote. He did some biting and clutching at O's hand but was easily diverted to playing with paper balls hanging from the arms of the chair, which he could reach from O's lap. He clutched the string in his mouth and hauled the balls up to himself. That is, he showed more focus on objects and less need for body contact even when playing on O's lap. As O prepared the evening meal on 6w3d, Rennie for the first time cried at his dish and at the refrigerator with almost no clutching at O's ankles. (That is, he focused on food as an object in itself.) After dinner he very definitely located and played with his toys much more independently. Play with objects (paper balls, paper balls on strings, shoe strings, yarn ball, and O's slippers) was completely satisfying to him now. He came up on O's lap now and then to play, but he did not need to stay or to seek body contact. Thus, at 6w3d S15 (Rennie) was able to focus on and interact with food and toys as objects in themselves, completely separate from the body of his mother figure.

   When O opened her bedroom door on the morning of 6w4d, Rennie came in and played on the floor with only an occasional excursion up on the bed to O until he was ready to take a nap. He located and played with small objects—a drapery tieback and a 1½-inch-long piece of cellophane—which O had not realized were lying on the floor. After breakfast as O sat in a living room chair, Rennie played on the floor, only occasionally coming up into her lap until he became sleepy. Then he played on O's lap a little, washed himself, and went to sleep.

  c. *6w5d–7w6d.*   On 6w6d when O came home for lunch, Rennie's greeting to her was play. After lunch he played quietly from the floor with a paper ball hanging from a string and jumped up into O's lap only once, briefly. In the evening it was noted that running and jumping high up into the air at O's hand or just up at O were now favorite activities. He jumped up on the back of the sofa and looked out the window briefly from time to time without apparently seeing anything. He ran speedily along the floor. He rolled around with a large yarn ball and played more on his own initiative with smaller paper balls tied to strings. He wrestled strenuously with O's hand, which was encased in a terry oven glove for protection. He grabbed a 2-ft long piece of drapery cord in his mouth and ran about with it, even jumping up on the sofa with it. He brought a dry leaf indoors over the porch and through the kitchen door. He slept as much or more by himself on a chair or on the sofa as on O's lap. On 7w0d Rennie did not climb into O's lap as she ate breakfast but played on the floor instead. In the evening he carried a narrow foot-long drapery tieback from room to room and brought three different dry leaves indoors to play with. On the morning of 7w1d Rennie's greeting to O was play at once, lying down and clutching at her ankle, and playing with her slippers. He engaged in wild wrestling play, pouncing on O, making high jumps with his forelegs outstretched. He located his leaf, or tieback, or small paper ball on his own and played with them. He ran very rapidly from room to room and made high jumps to the top of a bookcase. When O came home for lunch on 7w2d, Rennie for the first (and nearly only) time did not come to meet her. He was found playing with his tieback in a hollow formed by rolls of carpeting. In the evening he was given a small round wicker basket to play with. O put his toys in the basket, and he took them out, running off with them in his mouth one at a time. He jumped up on a table and helped himself to a tiny stem from a dish of small Seckel pears and chewed on the stem as if preparing to swallow it. When O tried to take the stem away from him, he slapped out and slightly growled at her, but not in a very tense manner. On the morning of 7w3d O noticed that Rennie was chewing on something in the doorway of the den. She took it away from him and found that it was just a half-inch long twist of dust. A little later he cried at the same spot where the dust twist had been taken away from him, apparently remembering the loss of his toy at that spot. At 7w3d Rennie ran about very speedily, finding his paper ball or tieback on the floor for himself and carrying the toy away in his mouth as he ran. On 7w4d he was carrying his paper balls and tieback for long distances in the house. On 7w5d he had a very active, prolonged chase going, up and down the two chairs that had paper balls hanging from their arms. His most intimate greeting to his human companion was now a scratchy clutch and an effort to bite, that is, a play greeting. Early on the morning of 7w6d Rennie was very lively, playing in the chairs that had paper balls tied to them. He worked up to very fast, intense play. When O thought he was still playing 5 ft away, he was under her foot as she stepped through a doorway. On the day before she had nearly stepped on him or closed a door on him because he moved more quickly than she had thought possible. On 7w6d Rennie played on O's lap as she sat in a living room chair writing letters. He seized a tiny pad of mailing labels in his mouth, and, when O tried to take it away from him, he quickly clawed tensely out at her with his tail lashing back and forth. It was noted that he put much energy into protecting toys, slapping out and spitting in an aggressive manner even when O was not trying to take them away from him.

Thus, even with no competition whatsoever at the food dish, with a large number of toys, including things in infinite supply such as dry leaves, which he provided for himself, and with no competition for toys except when he seemed about to eat some inedible object or played with an object that his human companion needed to use, S15 (Rennie) in his 6w5d–7w6d week managed to get in a great deal of practice of the dead prey handling technique in speedily running to, seizing, and carrying off toys, defending them by growling and slapping out even when no one was trying to take them away, and finally playing with them or attempting to eat them. He was thoroughly prepared to hold his own in competition with littermates at snatching away the single item of dead prey that would have been provided by his natural mother in the wild.

4. Focus on Mother (Human Mother Substitute)

At 5w2d S15 (Rennie) was still completely focused on mother and nest. He had no difficulty in responding to his new human companion–mother substitute's voice or in following her about. He took for granted the fact that his cries would be responded to and his needs supplied from above. The change in mother figure did not seem to matter as long as his needs were met, although the human mother substitute was not completely satisfactory in that she could not supply built-in running milk to suckle or allow him to stay in the preferred nest of her lap as continuously as he would have liked. O sought to keep the kitten "happy," which meant not only supplying his physical needs but also keeping him relaxed enough to play and purr and utter mhrns as much as possible.

The first of the major shifts in kitten behavior that brought relief to Rennie's human companion from nest-stage demands, which a human companion cannot completely fulfill, were those relating to food. When Rennie finally continued lapping milk for as many as 35 consecutive laps on 5w3d, O was relieved of a 24-hr worry that he could not take in enough dish milk to sustain himself. From the night of 6w0d/1d on Rennie was content to sleep all night without feeding, and O no longer had to heat milk for him in the middle of the night. On the evening of 6w1d Rennie was first able to eat solid food (cheese) on his own without O having to hold it for him to lick, and from then on he was able to eat his food from the dish or floor as an adult cat does. The amount of soft cooked food Rennie ate had also been increasing along with the ease with which he ate it, and from 6w1d on O could consider him to be "full of food" with no more worries that he could not eat enough food without suckling. On the evening of 6w3d Rennie cried at his dish rather than at O, and this was the point at which he became focused on food as an object separate from the mother figure's body. From 6w5d on Rennie cried very loudly for food and could lead O to the refrigerator when he wanted some. Moreover, he could keep crying until he got the particular kind of food he had in mind, refusing other food that ordinarily he enjoyed. From this point on O's only problem with Rennie and food was to decide, as with an adult cat, at what point he had had as many extra snacks as it was wise for him to have. Rearing Rennie as a vegetarian in a vegetarian household was never a problem. Rennie probably enjoyed eating more different kinds of food than nonvegetarian cats do, and there proved to be some surprising ones (cucumbers, canteloupes) that he found as exciting as S1 (Emy) had raw meat.

The fact that everything about a very small house was so overwhelming to S15 (Rennie) at 5w2d that all interior doors had to be kept shut to make him comfortable was also a strain on his human companion. On 5w3d Rennie began to be able to jump up on the sofa and into O's lap on his own. On the afternoon of 5w4d all interior doors could be left open without disturbing him. By 6w0d he could enjoy a trip outdoors from living room to kitchen door and showed his enjoyment by playing along the way. He was then fully self-confident indoors and almost fully self-confident outdoors, and his human companion could relax and enjoy his company without having constantly to be on guard to keep the space about him constricted. After 6w0d Rennie's running and jumping skills increased rapidly until by the end of the eighth week O began to feel crowded out of her own house because the kitten seemed to be everywhere at once. Kitten and human companion then were making about the same range of use of house and garden, with the kitten easily able to get up on chairs and beds indoors and staying within the borders of the yard or near the house outdoors.

At 5w2d S15 (Rennie) had a nest-stage need to rest or sleep or play in the cosy nest of his human companion's lap almost continuously. He needed long and frequent body contact with his mother figure, more than any human being with other duties would have time to supply. (That this need did not result from his being in a new environment is proved by the behavior of O27, the "six-week-old" kitten spending a day among human strangers in the much less satisfying strange environment of a library office, which was able to master this new environment and feel at home simply by batting at a paper ball hanging from a string.) It was especially hard on S15 (Rennie) that O did not allow him to sleep in her bedroom during the night, which would have meant on her bed, and it was 5w5d before O recognized that his need was more for body contact than for food after a night alone. She then began allowing him to spend time in the bedroom before she got up in the morning or went to bed at night. Although he spent most of this time in play, it seemed to help, as did his own discovery on the morning of 6w0d that he could suckle on a woolen afghan. (Presumably the amount of suckling Rennie did on afghans thereafter was no more than the amount of suckling that kittens which remain in their mother's care continue to do after they have begun to eat solid food.) But by 6w0d the kitten's need for continuous body contact with the mother had also begun to diminish. He soon began to sleep as much or more by himself as on O's lap during the daytime. From 6w0d on he was able to spend the night alone comfortably without milk. In the wild this is probably the point at which the natural mother is free to spend more time away from her kittens hunting prey to bring back to them, since this is also the time at which kittens begin to eat solid food, dead prey. It was on the evening of 6w1d that Rennie first managed to eat a piece of cheese from his floor mat without O having to hold it for him to lick.

On the morning of 6w1d Rennie no longer clutched at O's ankles (seeking to sit in her lap) as O prepared breakfast. After breakfast on this 6w1d morning Rennie's play first began to include edging along a wall, crouching and stalking, and at noon this had become stalking O and jumping high up at her. Thus, the type of body contact the kitten needed from his mother figure began to shift from the whole body nest contact of suckling to the type of attack play contact that would enable a kitten to snatch dead prey away from the mother. On 6w2d Rennie engaged in much stalking play, chiefly stalking O and trying to wrestle her feet and legs. Before breakfast on the morning of 6w3d Rennie clutched at O's hands and feet in a manner indicating that he wanted to be wrestled with, but he did not need to sit on O's lap or to have any special body contact. After breakfast he did some clutching at O's hands but was easily diverted to playing with paper balls hanging from the arms of the chair. This was also the day that he became thoroughly absorbed in playing with toys he located for himself, and it could be recognized that he was showing much more focus on objects and much less need of body contact. Play with small movable objects was thoroughly satisfying to him, and he came up on O's lap only occasionally until he was ready to sleep. During the afternoon of 6w4d he slept alone on a living room chair, and in the evening, when O sat in one living room chair, he climbed up into and slept in another chair. O now was free of the kitten's earlier need to have her serve as a nest almost 24 hr a day.

Human beings can easily supply toy objects and food objects and cushions to sleep on. It is very much easier for a human companion–mother substitute to supply a kitten's needs after the kitten has made the shift to focus on small movable objects as food and toys after 6w3d and especially after 7w5d when the kitten has developed still more initiative with respect to food and toys as objects in themselves than it is to try to supply its nest stage need for prolonged body contact and suckling.

5. Friendly Approach Behavior

a. 5w2d–5w3d. At 5w2d during his first evening in his new home, from about 8 P.M. to 11 P.M., S15 (Rennie) became familiar with his new human companion–mother substitute (O) by lying in the cosy hollow of a sofa pillow and batting playfully at the wriggling fingers she reached up near him and by playing with a long shoestring she pulled about for him. After that Rennie purred when O petted him and spoke to him. Thus, he displayed both play and purring greeting responses during his first 3 hr with his new mother substitute.

On the following morning, 5w3d, between 4:15 A.M. and 5:15 A.M. Rennie twice uttered a murmur ("hrrn" and a fast "hrhrhrn") as he ran away from his litter box, thus revealing that he was comfortable enough in his new environment to have energy to spare for uttering a relaxed by-product mhrn. Between 5:15 A.M. and 5:30 A.M. as O sat in a living room chair, Rennie rubbed against a wastebasket, the legs of a table, and the legs of a chair nearby. He could not yet get up into O's lap on his own, and the rubbing seemed to be an indication, if not a direct request, that he wanted to be lifted up into her lap.

Thus, at 5w2d/5w3d, within his first 9¼ hr in his new home under human care only, within less than 4¼ waking hours of interaction, S15 (Rennie) displayed some form of all four types of friendly approach behavior: play, purring, mhrn, and rubbing. Batting or rolling play and purring occurred frequently, mhrn only twice, and rubbing once.

b. Purring. S15 (Rennie) purred every day, and descriptions of his purring were written down almost every day. He purred in response to being petted, that is gently stroked, by O. On 5w3d while lying on O's lap and licking his fur, he looked up at O and started purring on his own initiative. On the evening of 5w3d O held a piece of cheese for Rennie to lick, and he purred as he started to lick the cheese. Later on that evening he purred loudly and cried at about the same time in O's lap. O had the impression that he still hoped to find milk to suckle in her lap, hence the anticipatory purring, but crying on failing to find any. Purring and crying at what seemed to be the same time was a habit with Rennie for some time, and the ability to do both at the same moment seems to indicate that purring through the nose and crying through the mouth are two completely separate systems of vocalizing. On 5w4d Rennie again cried through his mouth and purred through his nose at about the same time. On 5w5d he cried from the bedroom floor, and, when O spoke to him from the bed, he jumped up on the bed to her and purred. In the afternoon as O lay on the bed, Rennie jumped up on the bed uttering a faint by-product "hrhr" and started to purr when he reached O. Purring was his form of vocal greeting at this time. Thus, by the age of 5w5d

Rennie already had purred in response to O's touch and in response to her voice, and had initiated purring on sight of O.

After breakfast on 6w0d Rennie discovered that he could suckle on a loosely knitted woolen afghan. As he did so, he moved his feet in a paddling motion and purred, all as if he were suckling from his mother. Thus, Rennie managed to reproduce the original paddling and purring as part of the suckling situation, without the mother's milk, licking, and mhrns. On 6w2d it was noted that since Rennie could then eat dish food easily, he did less crying and clutching at O (for body contact), but perhaps he did less purring also. He now did more sleeping by himself and less on O's lap. On 6w3d, the day that Rennie could first focus on and play with toys with total absorption, it was noted that with the lessening of his need for reassuring body contact with his mother figure, he purred less and played more with objects. On 6w4d Rennie purred at the sight of O placing his dish of milk on the floor for him. While sleeping on O's lap, Rennie twice started to purr when O spoke to him without petting him. He looked up, moved his head toward the cards on which O was writing, and purred sleepily. On 6w5d, when O got up, Rennie came out of the den, stretched, and purred. Although O always spoke to him and petted him in greeting, purring on his own initiative also became part of early morning greeting, including play greeting. On 7w1d it was noted that, when Rennie was quiet (and on 7w5d when he was sleepy), he usually purred when O petted him. When more alert his response was now play instead of purring. He continued to initiate purring while approaching food. Since he had no littermates or mother to compete with at the food dish, he was able to continue the association of purring with food.

In his 5w2d–7w6d period of first living alone with a human companion-mother substitute S15 (Rennie) purred (1) in response to touch greeting (petting), (2) in response to voice only greeting, (3) to initiate greeting on sight, (4) while suckling and paddling on an afghan, (5) while starting to eat food held or supplied by his human mother substitute, (6) in anticipation of locating or receiving food, that is, when approaching food or being approached by food, and (7) while initiating touch play greeting to the mother figure.

Purring reaches its fullest and most frequent use before 6w3d, before the dead prey handling technique with its instant arousal of high-tension response becomes dominant in kitten behavior and replaces much of the need for suckling type body contact with a need for rough attack play contact of the type that would enable the kitten to snatch dead prey from its natural mother.

c. Rubbing. On the morning of 5w3d as O sat in a living room chair S15 (Rennie) rubbed his side against a wastebasket and other furniture. He could not yet get up into O's lap on his own and the rubbing seemed to be a seeking of, even a request for, body contact rather than casual greeting. On the morning of 5w5d Rennie rubbed or nearly rubbed against O's ankles, and O realized that he needed body contact more than food after a night alone. By 7w0d, however, when O got up, Rennie at once set to playing around her feet. "There was no rubbing greeting, only play."

What little rubbing behavior S15 (Rennie) displayed in his 5w2d–7w6d period was a seeking of body contact or a greeting seeking body contact, and it occurred before the 6w3d shift to focus on small, movable, inanimate objects as toys and food. After 6w3d play greeting became dominant and replaced whatever rubbing greeting had been occurring.

Rubbing, like batting play, first occurs on the kitten's own initiative, unlike purring, rolling play, and mhrn, which first occur in response to the mother's touch or voice.

d. Play as Greeting. When O arose on the morning of 6w3d, the day when focus on small, movable, inanimate objects as toys became all absorbing and dominant, Rennie came into the bedroom and played. He did some clutching at O's feet and especially her hands, wanting to be wrestled with, but he did not need to sit on O's lap or have any special body contact. When O came home for lunch on 6w6d, Rennie's greeting to her was play. After eating his lunch Rennie quietly played on his own from the floor with paper balls instead of playing on O's lap. When O got up on 7w0d Rennie set at once to playing around her feet; there was no rubbing greeting, only play. On 7w1d Rennie's greeting to O was play at once, lying down and clutching at her ankles, and playing with her slippers or shoes. During the initial stages of greeting play he purred. On 7w5d "Rennie's most intimate greeting to O is a scratchy clutch and effort to bite," that is, play. On 7w6d it was noted that Rennie seemed to need to jump up and clutch at O's hands and feet. Anything O used to keep her skin at a distance from his claws seemed not so satisfactory to him.

After 6w3d, when the dead prey handling technique became dominant with its instant high-tension response aspect making it difficult for the kitten to exhibit low-tension behavior unless he was sleepy, play greeting dominated in greeting situations just as play with objects dominated all non-food-seeking waking behavior.

e. Mhrn. S15 (Rennie) had a very high, light voice. His murmur vocalization sounded like "hrn" rather than "mhrn," and his vowel cries were very high and squeaky. Not until near the end of his 12w0d–19w6d period did O hear and write his murmurs as "mhrn," but even in adulthood his initial "m-" sound is not as strong as that of many cats. The initial "m-" sound results from starting vocalization with a strong explosive force. In writing down animal vocalization one must be careful not to place too much emphasis on initial and final consonants. A duck does not quite say "quack," but these initial and final consonants help convey the type of vowel actually uttered. "Mhrn" and "hrn" are the same type of vocalization with the same significance. It is the pattern of utterance, not the surface details, that matters. In this article "mhrn" is used generically to refer to murmur vocalization and "hrn" when S15 (Rennie) is being directly quoted.

Notes were made on S15's (Rennie's) mhrn vocalization on 5w3d, 5w5d, 6w0d, 6w1d, 6w4d, 6w5d, 7w3d, 7w5d, and 7w6d.

Most of Rennie's 5w3d–7w6d mhrn murmurs were of the by-product of motion variety, neither giving nor expecting response. On 5w3d as he ran away from his litter box without have used it, he uttered a long but not very distinct "hrrn" and purred two short, loud series of purrs as he ran. Later that day after he used his litter box, he uttered a faint "hrhrhrn" as he ran away. On 5w5d Rennie uttered a faint "hrhr" as he jumped up on the bed to O, but purred when he reached O. At that age jumping up on the bed was not yet simultaneous with reaching O, and purring, not mhrn, was his vocal greeting initiative. At bedtime on 5w5d O tossed a paper ball out of the bedroom doorway into the hall to get Rennie to leave the room, and he uttered a slight "hrn" as he bounded along after the ball. On 6w1d Rennie uttered "hrhrhr" trills several times when moving forward in play. At bedtime on 7w3d O arranged the afghan on the couch in the den, Rennie's nighttime sleeping place by this time, and patted the edge of the afghan to get him to notice that it was back on the couch. As he jumped up on it, he uttered a short "hrn." On 7w5d when Rennie had a very active, prolonged chase going in the living room up and down the two easy chairs with paper balls hanging from them, he twice uttered a trilled "hrhrn" while running or jumping. His murmur sound was a very birdlike trill. After he jumped down from a chair he uttered a "hrhrn" as he ran forward. A little later he uttered a "hrhrn" as he tried to jump up into a chair from the side. Still later on 7w5d after running around in the living room in play Rennie came running into the kitchen uttering a "hrhrn." O was working at the kitchen sink, but it was noted that this murmur was not a greeting to O but merely a by-product of the running in relaxed play. It was also noted that Rennie did not utter a mhrn greeting when jumping up on the sofa to O. At this time his greeting initiative to O had become a scratchy clutch and an effort to bite, that is, rough play.

S15 (Rennie) rather often uttered mhrns when sleeping on or near O. On 6w5d after he had been sleeping on O's lap for some time, he raised his head, uttered a trilled "hrhrn," and went back to sleep. A short time later he uttered another shorter "hrhrn" when he woke up momentarily. On 7w3d while asleep on O's lap Rennie purred when O petted him. He then stretched a bit, and as he readjusted his position with his eyes still closed, he uttered a very short, light "hrn" several times. On 7w5d while lying on O's lap in a very sleepy state, Rennie uttered a "hrhrn" as he moved his head to curl up more. On 7w6d as he slept at O's shoulder on the bed, he stretched a bit while still sleeping and uttered tiny "hrhrn" sounds. Then he purred a couple of short rounds of purring.

Such mhrns uttered while asleep or nearly asleep were not noticed in the behavior of S1's (Emy's) kittens, but Emy herself displayed them. On May 30,

Year 14, as S1 (Emy) was curled up "fast asleep" on O's bed, she twitched her paws, etc. as if dreaming of catching something, quieted, and then, still asleep, curled her nose up in the air and started making mhrn sounds. The sounds apparently woke her up, for she lifted her head high, uttered a "mhrn" clearly, and immediately went back to sleep.

The uttering of mhrns during certain stages of sleep may simply be something that cats do. In the case of S15 (Rennie) the mhrns were uttered as he moved his head in a situation (drowsiness or light sleep) in which he was likely to purr, so they seem perhaps a combination of purring thickened to mhrn and mhrn as a by-product of motion under relaxed conditions. On 6w0d it was noted that while Rennie was purring after having jumped up on the bed to O, some strokes of his purring thickened nearly to "hrhrhrn."

After 6w3d kittens have an awareness of objects in their environment that sometimes leads to uttering more or less tense mhrns of complaint and mhrn-plus-vowel-pattern cries of complaint because the objects are not present or otherwise behaving as expected. On 6w4d while playing on the bedroom floor, Rennie made a noise nearly like a complaining, that is a tense, mhrnhrn, but O could not see what was troubling him. Otherwise he had not been heard to utter a grunt or complaint type murmur up to that time.

However, just as Rennie could purr and cry at the same or almost the same time, on 5w5d he cried and uttered a "hrrhn" as he tried but failed to climb up on the sofa and once a "hrrr" at the end of a cry. In this case the murmurs seemed merely less tense versions of the cries and were sounds of frustration rather than of more specific complaint. On 7w3d while standing in the doorway of the den Rennie uttered "hrn-(plus squeaky vowel pattern), hrrn-(plus squeaky vowel pattern)" cries of complaint for no apparent reason except that this was the spot on the floor where O had taken a twist of dust which he had been playing with away from him a little while before. He could remember that he had played with a toy there and complained because it was no longer there to be played with. The initial murmurs probably occur because the kitten is slowed down enough by the surprise of things not being as expected for vocalization to start from a low tension closed-mouth murmur rather than from a more vigorous open-mouth vowel cry.

S15's (Rennie's) crying for food became very loud after 6w3d under the influence of the high-tension aspect of the dead prey handling technique, and it never started from a mhrn murmur. In fact, Rennie's demand crying never even started from a simple "m-" or "w-" sound as the cries of S1's (Emy's) kittens frequently did, but always directly from the stressed vowel.

It should be assumed that S15's (Rennie's) human companion (O) always responded with at least a murmur if not further vocalization and petting and other action to every sound made by the kitten, just as the feline mother does.

Although S15's (Rennie's) use of mhrns was increasing in his eighth week, the quantity of mhrns of any sort that he uttered in this 5w2d–7w6d period was extremely small compared with the number of mhrns uttered in later periods. Most significantly, there were as yet no mhrns of greeting response or greeting initiative at all, and no mhrn-plus-vowel-pattern cries of demand.

VIII. SUMMARY OF BIRTH–SEVEN-WEEK SIX-DAY KITTEN DEVELOPMENT AND FRIENDLY APPROACH BEHAVIOR

The behavior of housecat kittens and the development of their behavior is organized by the manner in which the kitten would be obtaining its food in the wild at various ages—first, by suckling only, then also by snatching the single item of dead prey brought to the litter at a time by the mother against competition from littermates, and finally by capturing and killing live prey in the prey's own territory.

The kitten starts out at birth with the ability to drag itself forward a few inches and move its head from side to side, usually crying, until it locates a milk-producing nipple on its mother's body. The kitten also starts out with the ability to vocalize not only with its mouth held open but also with it closing and closed in a whole continuum of sound from strained-intensity open-mouth vowel cries to less tense vowel pattern cries to closed-mouth grunting, which with further relaxation can become purring. When the mother returns to her kittens after having left them asleep in the nest, she greets them with mrhns, calls them with 'mhrns, and licks them before lying down to permit them to suckle. Thus, she offers her kittens voice and touch greeting as well as milk and other body comforts. When comfortably suckling kittens tend to purr and paddle their feet against the mother's body. By the end of their first week kittens display the full suckling technique of touch–seek and cry–find and suckle–paddle and purr. When they are awake but not hungry, this shortens to touch–paddle (play) and purr, or touch–paddle (play), or touch–purr. In their third week kittens can sit up and utter independent mhrn murmurs, and in their fourth week they begin to engage in right-side-up pawing and batting play. By the end of the nest stage in the first half of their fifth week when kittens are able to run about outside their nest in play, kittens are able to respond to touch or voice greeting from their mother with mhrn, rolling play, or a bit of purring, and to take the initiative in greeting their mother with batting play or a rub.

After 4w5d kittens develop eye–paw coordination, which enables them to counteract specifically the small motions of small objects, and their running and jumping skills increase rapidly. At 6w1d/3d the kitten becomes able to eat solid food and focus with absorption on small, movable, inanimate objects as food and toys, and it changes from needing to seek a suckling type of frequent and prolonged whole body contact with its mother to needing to direct at her rough attack play of a sort which would enable the kitten to snatch away from its mother the single item of dead prey that the mother would bring at a time to her litter in the wild and only one kitten can succeed in seizing, keeping, and eating. The kitten then is under the dominance of the dead prey handling technique, which consists of instant high-tension response to food or play signal (and to more and more signals and to more and more distant signals)–seize–run off with–defend–eat or play with object.

The instant arousal of the high-tension energy aspect of the dead prey handling technique makes it largely impossible for low-tension greeting behavior (purring, mhrn, rubbing) to surface. Play greeting then becomes dominant as does an intense interest in object food and play with toys. It is not until after the dead prey handling technique no longer has to be constantly practiced and the kitten then comes under the dominance of a new focus involving the much less tense technique of quiet observant or cautious behavior that lower-tension friendly approach behavior, especially the mhrn murmur, will have the opportunity to surface again and expand in socialized use.

References

Adamson, J. 1960. "Born Free: A Lioness of Two Worlds." Pantheon Books, New York.
Adamson, J. 1961. "Living Free: The Story of Elsa and Her Cubs." Harcourt, New York.
Adamson, J. 1964. "Forever Free." MacFadden-Bartell, New York.
Adamson, J. 1969. "The Spotted Sphinx." Harcourt, New York.
Escalona, S. K. 1968. "The Roots of Individuality: Normal Patterns of Development in Infancy." Aldine, Chicago, Illinois.
Leyhausen, P. 1956a. Das Verhalten der Katzen. *Handb. Zool.* **10**, 1–34.
Leyhausen, P. 1956b. "Verhaltensstudien an Katzen." Parey, Berlin.
Lindemann, W., and Rieck, W. 1953. Beobachtungen bei der Aufzucht von Wildkatzen. *Z. Tierpsychol.* **10**, 92–109.
Moelk, M. 1944. Vocalization in the house-cat: A phonetic and functional study. *Am. J. Psychol.* **57**, 184–205.
Rosenblatt, J. S., and Aronson, L. R. 1958. The decline of sexual behavior in male cats after castration with special reference to the role of prior sexual experience. *Behavior* **12**, 285–338.
Rosenblatt, J. S., and Schneirla, T. C. 1962. The behaviour of cats. *In* "The Behaviour of Domestic Animals" (E. S. E. Hafez, ed.), pp. 453–488. Ballière, London.
Schaller, G. B. 1972. "The Serengeti Lion: A Study of Predator–Prey Relations." Univ. of Chicago Press, Chicago, Illinois.
Schaller, G. B. 1973. "Golden Shadows, Flying Hooves." Knopf, New York.
Schneirla, T. C., Rosenblatt, J. S., and Tobach, E. 1963. Maternal behavior in the cat. *In* "Maternal Behavior in Mammals" (H. L. Rheingold, Ed.), pp. 122–168. Wiley, New York.
Spitz, R. A. 1957. "No and Yes." International Univ. Press, New York.
Spitz, R. A., with Gobliner, W. G. 1965. "The First Year of Life: A Psychoanalytic Study of Normal and Deviant Development of Object Relations." International Univ. Press, New York.
Wilson, C., and Weston, E. 1947. "The Cats of Wildcat Hill." Duell, Sloan & Pearce, New York.

Progress in the Study of Maternal Behavior in the Rat: Hormonal, Nonhormonal, Sensory, and Developmental Aspects

JAY S. ROSENBLATT, HAROLD I. SIEGEL, AND ANNE D. MAYER

INSTITUTE OF ANIMAL BEHAVIOR

RUTGERS—THE STATE UNIVERSITY OF NEW JERSEY

NEWARK, NEW JERSEY

Copyright © 1979 by Academic Press, Inc.
All rights of reproduction in any form reserved.
ISBN 0-12-004510-9

I. Introduction

This chapter will review the progress that has been made in the study of maternal behavior in the rat with special focus on the research that has been carried out at the Institute of Animal Behavior. Our approach has been a developmental one in which hormonal and nonhormonal (that is, sensory stimuli, emotional states, etc.) factors that regulate maternal behavior have been studied at each stage of the maternal behavior cycle. This research has its roots in earlier research on maternal behavior in the rat and other small mammals with altricial young, the mouse and rabbit, which has been reviewed elsewhere (see Lehrman, 1961; Noirot, 1972; Richards, 1967).

The study of maternal behavior in the rat has a relatively long history as one of the earliest natural patterns of behavior to be studied in the laboratory. The earliest approach was descriptive–functional (Small, 1899; Sturman-Hulbe and Stone, 1929) but there was also analytical research on the motivational and physiological bases of maternal behavior (Kinder, 1927; Nissen, 1930; Richter, 1927; Stone, 1925; Sturman-Hulbe and Stone, 1929). A landmark was the report by Wiesner and Sheard (1933) of their extensive studies on the hormonal and motivational basis of maternal behavior. This volume, modern in its tone, methodology, and thinking, remains the richest source of ideas in the study of maternal behavior. It was followed by the extensive studies of Riddle et al. (1942) on the hormonal basis of maternal behavior and by Beach (1937) on the neural substrate. Following this period of active investigation, there was practically no further research on maternal behavior in the rat for nearly 15 years.

Interest was revived when maternal behavior in the rat became one of the principal theoretic battling grounds for the nature–nurture controversy that raged in American comparative psychology around the end of the 1940s and early 1950s. In the course of this theoretic controversy the developmental approach that we have adopted emerged in the writings of Schneirla (1952), Lehrman (1956), and Birch (1956). A similar approach was adopted by Lehrman (1965) and Hinde (1965) in their studies of the reproductive cycles of the ring dove and canary, respectively.

Rather than existing preformed and on an endogenous basis alone, the maternal behavior cycle of the rat, like the reproductive cycles of the ring dove and canary, develops during the course of the cycle on the basis of the interaction of hormonal and external stimulation. The early sections of this chapter are devoted to the problem of how maternal behavior arises from the hormonal conditions that exist during pregnancy, particularly around parturition when maternal behavior normally begins. Later sections deal with postpartum stimulus factors, which regulate maternal behavior and are involved in its maintenance and eventual decline.

Schneirla (1952) and Birch (1956) were first to point to the pivotal role of parturition in the transition from pre- to postpartum maternal behavior. Leblond (1940), Leblond and Nelson (1937), and Koller (1956) had proposed that the onset of maternal behavior in the mouse was based upon hormones and that its maintenance was independent of hormones (Richards, 1967). The former authors proposed, in effect, that the transition between these two principal phases of the cycle in the rat might be mediated by the trophallactic (reciprocal stimulation) relationship that forms between the mother and the young. The evidence on this issue will be presented and discussed in this chapter.

Finally, in our studies, we have extended the developmental approach to the investigation of the ontogeny of maternal behavior. These studies are presented in the final sections of this chapter. We consider that knowledge of the background of hormonal stimulation and experience is essential for an understanding of the maternal behavior cycle in the adult.

II. Onset of Maternal Behavior

A. Behavioral Observations

Until recently, studies on the onset of maternal behavior indicated that retrieving, nursing, licking, and nest construction did not begin until parturition which normally takes place in the light phase of the 23rd day after mating. Wiesner and Sheard (1933) and Rosenblatt and Lehrman (1963), among others, tested late-pregnant rats by presenting them with pups during the last few days before parturition and found that they did not exhibit any of the above items of maternal behavior until they had given birth. Carrying nest material has been observed to increase as parturition approaches, but, most often, a nest was not constructed. Recently Slotnick et al. (1973) and Rosenblatt and Siegel (1975) found that the onset of maternal behavior occurs before parturition. If pups are presented to late-pregnant females either continuously or at regular short intervals (for example, 2 hr) and left with them for at least 15 min at each of these periods, nest building can be observed beginning around 34 hr prepartum, and nests are completed (while pups are present) by all of the females 2 hr before parturition begins. Retrieving is initiated by some animals as early as 28 hr prepartum, and

75–80% of the females retrieve pups by 4 hr prepartum. There are strain differences in the timing of the prepartum onset of these aspects of maternal behavior: Slotnick *et al.* (1973) found that they did not begin until 14 hr prepartum in their Wistar rats as compared with the earlier onsets in our Sprague–Dawley (Charles River CD) rats, and only 45% of their females exhibited retrieving prepartum compared with 75% in our strain. Bridges *et al.* (1978a) have recently found that nursing–crouching also begins about 24 hr before parturition.

It is likely that, as additional maternal responses are discovered, they too will be found to begin before parturition. This is already the case with respect to maternal responsiveness to pup calls, which Koranyi *et al.* (1976) studied by testing females in late pregnancy for their approaches to the calls of hidden pups. They found that approaches to these calls are low early in pregnancy but increase in frequency and reach a sustained high level just 24 hr before the females give birth. Food pellet hoarding, which has been studied apart from maternal behavior, has recently been found by Stern (unpublished observations) to be correlated with the maternal behavior cycle and to begin about 24 hr before parturition.

Maternal aggression toward male intruders or a stick moved toward the female as she crouches over young may also begin before parturition. Sadowsky (unpublished observations) found that maternal aggression toward a stick first appeared at about the same time as the prepartum onset of maternal behavior. On the other hand, Erskine (1978) did not find a prepartum onset of maternal aggression in 18- and 22-day-pregnant Long–Evans females presented with male intruders in their home cages. The females did not have pups in their cages, and she has shown that even postpartum mothers that are known to be aggressive show a rapid decline in aggression when their pups have been removed for as little as 4 hr. This, however, is not likely to have been the cause for the absence of prepartum aggression among the females since even when pups are present during the first 5 days postpartum, maternal aggression remains low and is not significantly greater than during the last 5 days before parturition. The only hint of a prepartum increase in maternal aggression is in the intruder male's adoption of a subordinate posture. This occurs more rapidly and for longer durations when males are faced with late pregnant females than with nonpregnant females, both of which have had equal lengths of residence in their cages and therefore equal opportunity to establish home territories (Erskine, 1978).

B. HORMONAL STIMULATION OF MATERNAL BEHAVIOR

1. Humoral Factors around Parturition

Behavioral observations indicating that the onset of maternal behavior occurs immediately prepartum or during and after parturition have led to the search for a basis in hormonal changes occurring at or shortly before parturition. The studies of Terkel and Rosenblatt (1968, 1972) in which plasma or blood from newly parturient females was either injected intravenously into nonpregnant females or cross-transfused via a chronically implanted heart catheter (Terkel, 1972) provided the first clear evidence that at this time there is a humoral factor(s) capable of inducing maternal behavior. Maternal blood injected into intact nonpregnant females stimulated maternal behavior (nest building, crouching over young, retrieving, and pup licking) with an average latency of 2.25 days (from the start of pup exposure at the time of blood injection) compared with 4–6 days required when plasma taken from a female in proestrus or saline was injected. Cross transfusion, which permitted larger amounts of blood to be introduced into the nonpregnant intact female and enabled the maintenance of up to 50% mixing of the blood of the donor and recipient for more than 6 hr continuously, was even more effective than injecting plasma and 88% (7/8) of the females receiving the blood exhibited all components of maternal behavior, except nest building, with an average latency of 14.5 hr. Nest building was displayed subsequently at the end of 24 hr but whether it had been stimulated by the exposure to pups or was a delayed response to the transfused blood was not determined.

The findings with plasma injection have been confirmed by Koranyi *et al.* (1976), but the mean latency for the onset of maternal behavior obtained by these investigators was 18 hr, which is comparable to that obtained by Terkel and Rosenblatt (1972) using cross transfusion of blood.

In contrast to Stone (1925), who joined females in parabiosis, our procedures thus succeeded in demonstrating a humoral (that is, hormonal) basis for the onset of maternal behavior. However they did not define the effective agent(s). The cross-transfusion study was able to define the period during which humoral factors were present in the donor mother that could stimulate maternal behavior in the recipient cycling female. Two additional groups were studied: one received blood from prepartum females about 27 hr before they gave birth and the other from postpartum females 24 hr after delivery. Only 13% (1/8) of the females receiving 24-hr postpartum blood and 34% (3/8) of the females receiving prepartum blood exhibited maternal behavior with latencies shorter than two days: most had longer latencies (that is, between 5 and 6 days) equal to those of the controls that had received blood from estrous cycling donors. The humoral factor(s) is therefore not yet present in sufficient concentration (or is not transfused to the recipient in a sufficient amount) during the 6-hr transfusion starting 27 hr prepartum and it has already disappeared 24 hr after parturition (Terkel and Rosenblatt, 1972).

2. Hormone Assays during Pregnancy

The development over the past 15 years of radioimmunoassay procedures for measuring circulating levels of ovarian and pituitary hormones and their application to the study of pregnancy in the rat has provided us with measurements of what are believed to be the most important hormones for the onset of maternal behavior, namely, ovarian progesterone, estradiol, and pituitary prolactin. The concentrations of these hormones in the circulation of pregnant females at various times during pregnancy have been compiled from various sources and are shown in Fig. 1. Circulating levels of follicle-stimulating hormone and luteinizing hormone are initially high after mating and then fall and remain relatively stable until postpartum estrus when both gonadotrophins increase dramatically (Bast and Melampy, 1973; Linkie and Niswender, 1972; Mori et al., 1974).

Important to note are the decline in circulating levels of progesterone, which occurs about 30 hr before parturition, the slow and subsequent rapid rise in estradiol, and the sharp increase in prolactin shortly after estradiol has risen. In effect the prepartum period is characterized by rapid changes in the spectrum of hormones compared with that characterizing the earlier period of pregnancy: specifically, progesterone declines, estrogen rises, and shortly thereafter prolactin rises.

3. Stimulation of Maternal Behavior by Hormone Administration

On the basis of the hormonal picture during pregnancy and particularly at its termination, several investigators have attempted to stimulate maternal behavior

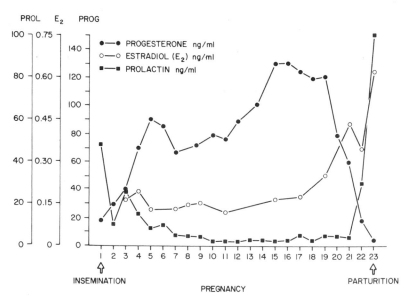

FIG. 1. Circulating levels of progesterone, estradiol (E₂), and prolactin during pregnancy in the rat. (From various sources: progesterone, Pepe and Rothchild, 1974; estradiol, Shaikh, 1971; prolactin, Morishige et al., 1973.)

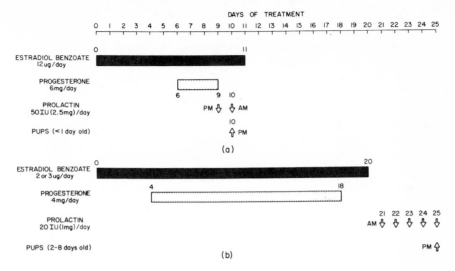

FIG. 2. Hormone treatments for stimulating maternal behavior in ovariectomized rats. [(a) From Moltz et al., 1970; (b) from Zarrow et al., 1971.)

by administering progesterone, estradiol benzoate, and prolactin, thereby simulating the hormonal changes during pregnancy. Moltz et al. (1970) administered the hormone treatment shown in Fig. 2 to ovariectomized nulliparous females and found that females were maternally responsive to pups, with latencies of 35–40 hr in all cases. The females exhibited nest building, retrieving, nursing–crouching, and licking. Using a somewhat similar injection schedule of the same three hormones (Fig. 2), Zarrow et al. (1971) were also successful in stimulating maternal behavior in nearly all ovariectomized nulliparous females. Although Zarrow et al. reported the almost immediate appearance of maternal behavior when pups were presented at the end of the prolactin treatments, if latencies are calculated from the second prolactin injection, as in the Moltz et al. study, then they are nearly equal in the two studies and range from 1½ to 3 days.

Moltz et al. (1970) and Moltz (1972) attributed specific roles for each of the three hormones and for the order and timing of their administration. Of particular importance was the short period of progesterone administration between Days 6 and 9 and the continuation of estradiol benzoate beyond this time, simulating the progesterone withdrawal and estradiol rise at the end of pregnancy. These investigators proposed that progesterone withdrawal causes a lowering of the neural threshold for estradiol benzoate and prolactin stimulation of maternal behavior. Under normal conditions, exogenous estradiol stimulated the release of endogenous prolactin (Blake et al., 1972; Caligaris et al., 1974; Kalra et al., 1973; Vermouth and Deis, 1974) but in the Moltz et al. study, additional prolactin was administered shortly after progesterone withdrawal. In view of this, one finding is not easily understood: one group of females was given estradiol benzoate and progesterone in the same sequence as the group described above but was not given prolactin. The latencies for the onset of maternal behavior in this group were only slightly shorter than those in control animals given hormone vehicles only. Yet if estradiol benzoate stimulated the release of endogenous prolactin, this treatment should have been as effective as when prolactin was given; however, without direct measurement of circulating hormones it is difficult to interpret these particular findings. A subsequent study by Moltz *et al.* (1971) substituted phenothiazine for the progesterone in the estradiol benzoate and prolactin treatment schedule, with equally rapid stimulation of maternal behavior. A control group lacking prolactin did not have short latencies for maternal behavior yet here again phenothiazine is a potent releaser of pituitary prolactin (Ben-David *et al.*, 1971). The question is why administered prolactin had an effect but endogenously released prolactin did not.

4. Pregnancy Termination Studies: Estrogen Stimulation of Maternal Behavior

There are several practical difficulties involved in directly studying the effects of normal changes in hormone secretion at the end of pregnancy on the onset of maternal behavior. This is exemplified by a study reported by Moltz *et al.* (1969) in which progesterone was administered to primiparous pregnant females duirng the last 4 days prior to parturition and was found to prevent the onset of maternal behavior in 50% of the animals. The animals were delivered by Caesarean section to avoid the complications that arise from progesterone block of parturition. Although the investigators interpreted their findings as evidence of progesterone inhibition of the action of prolactin, progesterone administration has many effects on hormone secretion at this time, which might have affected the results. It is likely that the normal rise in estradiol did not occur and in addition that prolactin was not released as normally (Rothchild, 1962). Since vaginal smears were not taken to provide at least a minimum indication of the hormonal status of the females before and after Caesarean-section delivery, the results cannot be interpreted solely with respect to the administered progesterone. A similar difficulty exists in attempting to interpret studies by Herrenkohl (1971) and Herrenkohl and Lisk (1973) in which it was reported that progesterone had no effect on maternal nursing behavior in intact females when given daily from the 18th to the 23rd day of pregnancy. In these animals, however, parturition was delayed for 3 days after termination of progesterone treatment, and in the interim the females may have undergone hormonal changes that nullified the effect of the previously administered progesterone: progesterone may have declined sufficiently to permit the display of maternal care. Almost all prepartum hormone treatments are likely to have multiple effects that are difficult to disentangle from one another in order to identify which effects are specifically responsible for the onset of maternal behavior. One reason for this is that the hormonal changes at the end of pregnancy not only play a role in the onset of maternal behavior but also are important for triggering the beginning of parturition and the onset of lactation and milk letdown.

A few studies have employed ovariectomy shortly before parturition to eliminate the principal source of estradiol and progesterone. Moltz and Wiener (1966) reported that ovariectomy 2 days before parturition prevented the onset of maternal behavior in 50% of primiparous females, measured in part by cannibalism and also by failure of the young to survive (see also Jost, 1959). However, Terkel (1970) and Catala and Deis (1973) found that females ovariectomized on Days 20 and 21 of pregnancy exhibited maternal behavior in nearly all cases, but there were many among them that were unable to lactate and their young starved to death. The failure of lactation could be traced to failure of suckling-induced release of oxytocin necessary for the milk-ejection reflex, and this, in turn, may be attributed to the lack of estrogen during the immediately prepartum period (Catala and Deis, 1973). It is not clear, therefore, what percentage of the females designated as nonmaternal in the earlier studies were actually nonmaternal and what percentage lost their young because of failure of lactation.

An alternative approach that we have adopted is to terminate pregnancy under controlled conditions that simulate those occurring at the normal termination: knowing the starting point of the hormonal changes that follow and measuring the effects of these changes by introducing pups to females to measure the latencies for the onset of maternal behavior has enabled us to make considerable progress in understanding the hormonal basis for the onset of maternal behavior (Rosenblatt and Siegel, 1975).

The procedure of artificially terminating pregnancy and observing the effects on maternal behavior was originally used by Wiesner and Sheard (1933), who found that deliveries by Caesarean section performed on 19-day-pregnant females resulted in the accelerated appearance of maternal behavior 1 or 2 days later, a finding that has been confirmed by other investigators (Labriola, 1953; Moltz, et al, 1966; Moltz and Wiener, 1966). Zarrow et al. (1962) employed this procedure to study the hormonal basis for the onset of maternal behavior in the rabbit.

Our studies of pregnancy termination have been reported elsewhere in detail (Rosenblatt and Siegel, 1975; Siegel and Rosenblatt, 1975a, b); therefore, only the main results will be presented here. Hysterectomies were performed on pregnant females at different times during pregnancy (Fig. 3) and pups were presented to females 48 hr later and remained with them continuously from then on with daily exchanges of fresh pups for those that had been with the nonlactating females overnight. During the interim between surgery and testing, females, therefore, were not influenced by pup stimulation, a potent stimulus for maternal

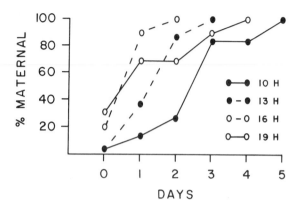

FIG. 3. Cumulative percentage of females showing maternal behavior following hysterectomy (H) on the 10th, 13th, 16th, and 19th days of pregnancy. First test (O) at 48 hr after surgery. (From Rosenblatt and Siegel, 1975a.)

behavior even in cycling females (Cosnier and Couturier, 1966; Rosenblatt, 1967).

A larger percentage of the late-pregnant females (that is, 16- and 19-day-pregnant females) initiated maternal behavior during the first 24 hr following exposure to pups (that is, 48–72 hr postoperatively) than those pregnant for only half the duration of pregnancy (10- and 13-day-pregnant females). These females, in turn, initiated maternal behavior sooner than intact and hysterectomized nonpregnant females. These results indicate, first, that hysterectomy during the second half of pregnancy induces changes in hormone secretion that stimulate the onset of maternal behavior, and second, that the rate at which these changes occur is more rapid the later in pregnancy the hysterectomy is performed (see page 238).

It will be shown later that virgin females can be stimulated to exhibit maternal behavior simply by exposure to pups in 5–7 days (Cosnier and Couturier, 1966; Rosenblatt, 1967). This maternal behavior, when it appears, however, is not based upon hormones (see Section V,A). The reduction in latencies resulting from hysterectomy performed on the pregnant females during the second half of pregnancy, which was about 2–4 days, can therefore be attributed to the hormonal changes caused by the termination of pregnancy in this manner. Parturient females have even shorter latencies following the natural termination of pregnancy, and this may be considered a continuation of the trend we have already noted.

Although all previous work had indicated that pregnant females do not exhibit maternal behavior when exposed to pups (Rosenblatt, 1965; Rosenblatt and Lehrman, 1963; Wiesner and Sheard, 1933), the fact that nonpregnant females can be stimulated to display maternal behavior suggested that further study of pregnant females would reveal a similar responsiveness with prolonged pup exposure. This, in fact, proved to be the case (Rosenblatt and Siegel, 1975): starting around 13 days of pregnancy females exposed to pups exhibited shorter latencies for maternal behavior than nonpregnant females. However, these latencies were not as short as those of hysterectomized females at the same stage of pregnancy.

Progressively shorter latencies as pregnancy advances were not found in a recent study (Bridges *et al.*, 1977) in which pregnant females of the same strain were sham hysterectomized on Days 8, 13, and 17 and given pups starting 24 hr later. The 17-day-pregnant females did have shorter latencies than virgin females. It is possible, as we have found in other studies, that sham operation leaves animals ill for at least 24 hr, and this may have delayed the onset of maternal behavior, uniformly, in all of the groups.

The 16-day-pregnant hysterectomized females have provided us with an excellent group to study the hormonal changes responsible for the onset of maternal behavior: we know when these changes begin, since they are initiated by the hysterectomy, and we know how long they must be present to stimulate maternal behavior—the problem is to identify them.

Normally females remain in diestrus during pregnancy and they do not undergo estrous cycling until the postpartum estrus, which occurs during the first 24 or 48 hr after parturition (Johnson, 1972; Ying and Greep, 1973; Ying *et al.*, 1973). Hysterectomy, however, interrupts diestrus and initiates an estrous cycle at whatever stage during the second half of pregnancy that it is performed. Figure 4 depicts the relationship between the onset of maternal behavior following hysterectomy and the latency to proestrus as reported by Morishige and Rothchild (1974) and Johnson (1972). There is a fairly close correspondence between the latencies of maternal behavior and the phase of the estrous cycle called proestrus during which estradiol levels in plasma are high (except at the 8th day when maternal behavior is initiated after 7 days of pup stimulation, whereas proestrus does not occur for more than 11 days). In all stages of pregnancy from the 10th day onward there is a shortening of the latencies for both the onset of the estrous cycle and of maternal behavior; on the 8th day of pregnancy, after hysterectomy, pups alone can stimulate maternal behavior in 7 days, and this is earlier than hormonal stimulation, which is not present until around 11 days when the first estrous cycle occurs.

The suggestion that hysterectomy stimulates an estrous cycle when performed after midpregnancy is confirmed by additional measures that we have made after hysterectomy, including vaginal smears, sexual receptivity, and ovulation, as shown in Fig. 5 (Siegel and Rosenblatt, 1977). These findings imply that following hysterectomy there is a decline in ovarian secretion of progesterone and a rise in estradiol secretion, which is followed, once again, by a rise in progesterone secretion with ovulation occurring shortly afterward. This sequence of hormonal

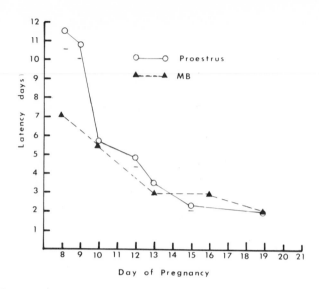

FIG. 4. Comparison of latencies from surgery for the onset of maternal behavior (MB) and for the appearance of proestrus in females hysterectomized on the day of pregnancy shown. [Median latencies for maternal behavior and mean latencies and standard errors (dash below mean) for proestrus. Latencies for proestrus taken from Morishige and Rothchild, 1974, and Johnson (1972).] (From Rosenblatt and Siegel, 1975.)

changes and the associated release of gonadotrophic hormones would account for the vaginal changes, sexual receptivity, and ovulation. Thus far, we have only measured changes in the levels of circulating progesterone following hysterectomy at the four stages of pregnancy, and they conform to the predicted pattern (Fig. 6; Siegel and Rosenblatt, 1977). The results with respect to the initial decline in progesterone following hysterectomy are in good agreement with those reported by Pepe and Rothchild (1972), Rothchild et al. (1973), Takayama and Greenwald (1973), and with those found in another study at the Institute of Animal Behavior (Bridges et al., 1978b). Assays of circulating levels of estradiol are currently in progress using blood obtained at various times after hysterectomy from females at different stages of pregnancy.

Our experimental studies (as contrasted with correlational studies described above) have been aimed first at the role of the ovary in the onset of posthysterectomy maternal behavior in pregnant females. When females were both hysterectomized and ovariectomized, on the 10th, 13th, 16th, or 19th day, their latencies were longer than when the ovaries were retained following hysterectomy (Fig. 7; Rosenblatt and Siegel, 1975). Both ovariectomized and nonovariectomized females undergo a decline in progesterone following hysterectomy but only the females with ovaries secrete estradiol, and their latencies, in tests started at 48 hr after surgery, were shorter than those of the ovariectomized females. The differences in latencies between the ovariectomized and nonovariectomized–hysterectomized females suggested therefore that the hormone which stimulates maternal behavior may be estradiol, once it has reached a certain level and after progesterone has declined sufficiently.

A series of studies has confirmed that estradiol (that is, estradiol benzoate) can stimulate short-latency (that is, 2 days or less) maternal behavior in hysterectomized–ovariectomized pregnant females during the second half of pregnancy (Fig. 8–11; Siegel and Rosenblatt, 1975a). The injection (estradiol benzoate 100 or 20 μg/kg, sc) was given at the same time as hysterectomy-ovariectomy, and pups were introduced 48 hr later to test for maternal behavior. They remained with the females continuously from then on as described earlier. [Initially progesterone (0.5 mg/rat) was given 44 hr after estradiol benzoate to stimulate sexual behavior in 16-day-pregnant animals that were hysterectomized and ovariectomized, but we have found it is unnecessary for stimulating maternal behavior (Fig. 10; Siegel and Rosenblatt, 1975a)].

Among the pregnant females hysterectomized–ovariectomized between 10 and 19 days, there were no differences in responsiveness to 100 μg/kg estradiol

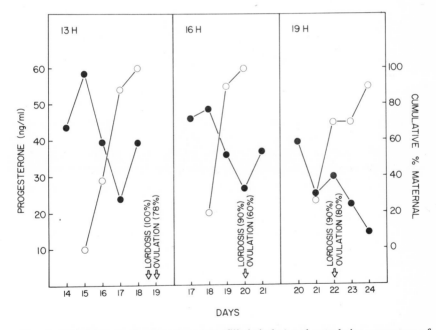

FIG. 5. Circulating levels of progesterone (filled circles) and cumulative percentage of females showing maternal behavior (open circles) following hysterectomy (H) at the 13th, 16th, and 19th days of pregnancy. Time of lordosis and ovulation also shown. Maternal behavior tests started two days after surgery. (From Siegel and Rosenblatt, 1977.)

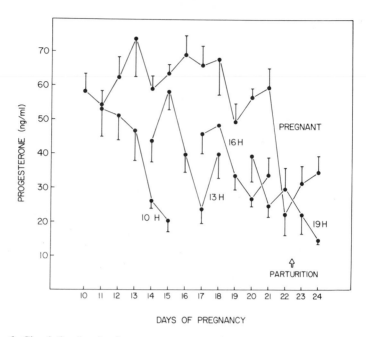

FIG. 6. Circulating levels of serum progesterone in intact pregnant and pregnant hysterectomized (H) rats on Days 10, 13, 16, and 19. Each point presents the mean ±sem of six decapitated animals. (From Siegel and Rosenblatt, 1977.)

benzoate, which is contrary to differences in onset latencies resulting from hysterectomy alone. When 20 μg/kg of estradiol benzoate was given to 10–19-day-pregnant operated females, the 13–19-day-pregnant females were more responsive than the 10-day females and the 13-day females were slightly less responsive than 16- and 19-day females. Referring to our earlier finding of more rapid onset of maternal behavior following hysterectomy as pregnancy advances (see page 234), we can say that there may be two explanations: as we have seen, it takes longer for the onset of the estrous cycle earlier in pregnancy, following hysterectomy, than later, and therefore estrogen stimulation may occur somewhat later. Second, there may be some increase in responsiveness to estrogen as pregnancy advances. This has already been shown using 20 μg/kg of estradiol benzoate and may become more evident when even lower doses of estrogen are used.

5. Prolactin

There is difficulty in attributing the onset of maternal behavior following estradiol benzoate treatment to the steroid hormone alone, however, since estrogen has been found to cause the release of prolactin in pregnant and nonpregnant females (Blake *et al.*, 1972; Caligaris *et al.*, 1974; Kalra *et al.*, 1973; Vermouth

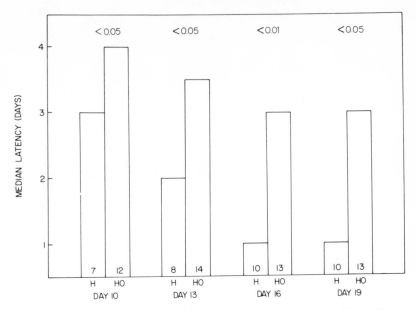

FIG. 7. Comparison of the median latencies to the onset of maternal behavior in hysterectomized (H) and hysterectomized–ovariectomized (HO) rats on Days 10, 13, 16, and 19 of pregnancy. Maternal behavior tests started 48 hr postoperatively. Numbers in bars represent sample sizes. (From Siegel and Rosenblatt, 1977.)

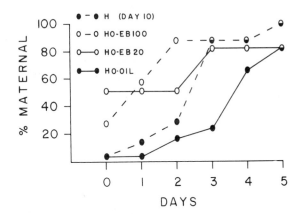

FIG 8. Cumulative percentage of 10-day pregnant rats showing maternal behavior after hysterectomy (H) or hysterectomy–ovariectomy (HO) plus estradiol benzoate (EB) treatment of 100 or 20 μg/kg. Maternal behavior tests started 48 hr after surgery and EB injection. (From Siegel and Rosenblatt, 1975a.)

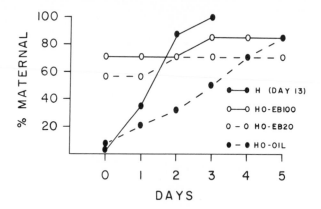

FIG. 9. Cumulative percentage of 13-day pregnant rats showing maternal behavior after hysterectomy (H) or hysterectomy–ovariectomy (HO) plus estradiol benzoate (EB) treatment of 100 or 20 μg/kg. Maternal behavior tests started 48 hr after surgery and EB injection. (From Siegel and Rosenblatt, 1975a.)

and Deis, 1974). Our findings could be attributed, therefore, to either a combined estrogen–prolactin effect or a prolactin effect alone. We have examined this possibility in two separate studies using 16-day-pregnant hysterectomized-ovariectomized females that were given estradiol benzoate and pharmacological agents that block the release of prolactin (that is, ergocornine hydrogen maleate, CB 154-2 bromo-α-ergokryptine, or apomorphine hydrochloride; Numan *et al.*, 1977; Rodriguez-Sierra and Rosenblatt, 1977). Both studies showed that block-

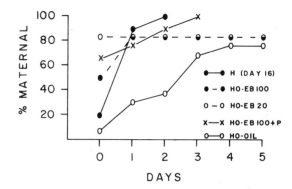

FIG. 10. Cumulative percentage of 16-day pregnant rats showing maternal behavior after hysterectomy (H) or hysterectomy-ovariectomy (HO) plus estradiol benzoate (EB) treatment of 100 or 20 μg/kg and progesterone (P) at 0.5 mg. Maternal behavior tests started 48 hr after surgery and EB injection. P injected 4 hr prior to first test. (From Siegel and Rosenblatt, 1975a.)

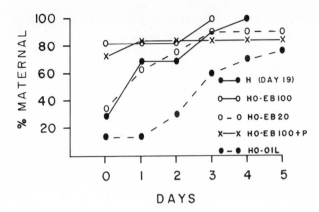

FIG. 11. Cumulative percentage of 19-day pregnant rats showing maternal behavior after hysterectomy (H) or hysterectomy–ovariectomy (HO) plus estradiol benzoate (EB) treatment of 100 or 20 μg/kg and progesterone (P) at 0.5 mg. Maternal behavior tests started 48 hr after surgery and EB injection. P injected 4 hr prior to first test. (From Siegel and Rosenblatt, 1975a.)

ing the release of prolactin did not affect the ability of estradiol benzoate to stimulate short-latency maternal behavior. Zarrow *et al.* (1971) and Stern (1977) were also unable to prevent the onset of maternal behavior at parturition by blocking the release of prolactin for 4 or more days before parturition.

6. *Estrogen Stimulation of Maternal Behavior in Nonpregnant Females*

Although we were able to stimulate maternal behavior in pregnancy-terminated ovariectomized females with a single injection of estradiol benzoate, all of the females had a background of hormonal stimulation of nearly half the length of gestation or longer. Our treatment might, therefore, be considered a prolonged one similar to those of Moltz *et al.* (1970) and Zarrow *et al.* (1971), which lasted for 11 and 25 days. Yet we already knew from our cross-transfusion and blood plasma injection studies (Terkel and Rosenblatt, 1968, 1972) that a short treatment with blood plasma was effective in stimulating maternal behavior in 48 hr without the background of hormonal stimulation typical of pregnancy. It was necessary, nevertheless, to show that estrogen would also be effective in nonpregnant ovariectomized females lacking such a background of hormone stimulation, that is, a chronic low level of estrogen and high level of progesterone followed by progesterone withdrawal.

Our first attempt to induce maternal behavior in ovariectomized nonpregnant females using 100 μg/kg estradiol benzoate failed: the treated females had latencies that were no shorter than those of oil-treated controls (Siegel and Rosenblatt, 1975b). We realized, however, that we were not using the same procedure that we had used with the pregnant females; they had been hysterectomized to terminate their pregnancies—a procedure that we had not thought in itself contributed to the effectiveness of estrogen on maternal behavior. We therefore hysterectomized and ovariectomized the nonpregnant females and gave then estradiol benzoate, and this time we successfully induced short-latency maternal behavior (Fig. 12; Siegel and Rosenblatt, 1975b). This effect of the uterus in reducing the effectiveness of estradiol benzoate was unexpected and will be discussed shortly.

Our demonstration that estradiol benzoate resulted in a significant reduction in tbe latency to the onset of maternal behavior in nonpregnant rats is important for a number of reasons. First, using a single injection of one hormone, we accomplished what other investigators obtained using 11–22 or more days of treatment with three different hormones. Second, in certain respects, we confirmed the Terkel and Rosenblatt studies on the effectiveness of a short-term treatment, and the data suggested the possibility that estrogen was a key element in the transfusion experiment. Finally, we were able to show that a period of progesterone stimulation followed by its withdrawal is not essential for the display of maternal care.

The same problem of interpretation involving the estradiol benzoate-induced release of prolactin is present in the studieś using virgin animals as in those using pregnancy-terminated animals. Estradiol benzoate-treated virgin rats that are either hypophysectomized or injected with prolactin blockers have not been studied, and hence, we cannot rule out the possibility of a combined hormonal facilitation. However, multiple injections (Beach and Wilson, 1963; Lott and

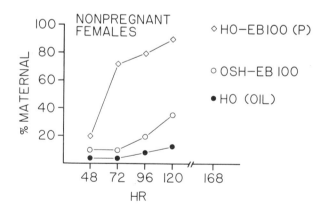

FIG. 12. Cumulative percentage of nonpregnant rats showing maternal behavior after hysterectomy–ovariectomy (HO) or ovariectomy–sham hysterectomy (OSH). Groups treated with 100 μg/kg estradiol benzoate (EB) or oil at surgery. Progesterone (P) was given 44 hr after EB. Maternal behavior tests started 48 hr postoperatively. (From Siegel and Rosenblatt, 1975b.)

Fuchs, 1962) and the chronic release of high levels of prolactin accomplished by pituitary transplants (Baum, 1978) have been shown to have no effect on the latency to maternal behavior in nonpregnant females that are exposed continuously to young pups.

Returning to the question of the uterus, the problem is more serious in pregnant animals because in removing the uteri we may have created an artifactually effective treatment that has no relationship to the normal situation before parturition in which the uteri are present. One way to test this possibility is to terminate pregnancy in such a way that the uteri do not have to be removed. This can be done using prostaglandin $F_{2\alpha}$ ($PGF_{2\alpha}$), a naturally occurring uterine secretion that is involved in normal parturition in the rat (Buckle and Nathanielsz, 1975; Chester et al., 1972; Strauss et al., 1975). To match the groups that were hysterectomized on those days $PGF_{2\alpha}$ was administered to 16- and 19-day-pregnant females (Rodriguez-Sierra and Rosenblatt, unpublished observations). Pregnancies were terminated within 36 hr and females exhibited short-latency maternal behavior starting at 48 hr. The presence of the uteri during late pregnancy therefore does not prevent the rapid onset of maternal behavior after pregnancy termination. This indicates perhaps that the uterine factor, which is normally present in nonpregnant, recently ovariectomized, females and which counteracts the effect of estrogen on maternal behavior, is absent at the end of pregnancy.

We have explored the effect of the uterus on estrogen-dependent behavior patterns in nonpregnant females in one further study on maternal behavior and in one experiment on female sexual behavior. Nonpregnant females were ovariectomized but were not treated with estrogen until 8 weeks later. At the end of 8 weeks, they were given estradiol benzoate without removing the uteri, and they exhibited short-latency maternal behavior similar to that of hysterectomized females. Evidently the absence of the ovaries for 8 weeks altered the uteri in such a way that they no longer exerted their effect.

In our study on sexual behavior, lordosis quotients were measured in hysterectomized–ovariectomized and ovariectomized–sham hysterectomized rats injected with several hormone treatments (Siegel et al., 1979). Hysterectomy resulted in a facilitation of sexual receptivity in groups treated with either a single injection of 5 μg/kg of estradiol benzoate or daily injections of 3-μg/kg dosage, providing progesterone (0.1 or 0.5 mg) was also administered.

The uterus has an effect on maternal latencies and lordosis quotients only in animals treated with estradiol benzoate and estradiol benzoate plus progesterone, respectively. Hysterectomized, hysterectomized–ovariectomized, and sham operated virgin rats exposed to pups all have identical latencies to the onset of maternal care in the absence of estradiol benzoate. Without exogenous hormones, ovariectomized animals that were either hysterectomized or sham operated displayed lordosis in response to manual stimulation and cervical probing, but the groups did not differ in the intensity of the behavior (Rodriguez-Sierra, 1977). Our attention has therefore been focused on the possibility that the uterus competes with other estrogen-sensitive tissues for circulating levels of estradiol. This type of competition has already been found between peripheral tissues (uterus and mammary gland; Kuo *et al.*, 1974) and may also exist between peripheral and central structures.

An important question is whether the quality or intensity of maternal behavior induced by estradiol benzoate in 16-day-pregnant hysterectomized–ovariectomized females is similar to the maternal behavior displayed by postparturient females. While we have been concerned mainly with onset latencies and simply the appearance of retrieving, nursing–crouching, nest building, and pup licking, noting that in the main they appear identical (except for actual lactation) to the same behavioral items performed after parturition, Krehbiel and LeRoy (1978) have examined in greater detail the retrieving, nest building, and responses to male intruders of such females. Estradiol benzoate-treated females (16-day pregnant hysterectomized–ovariectomized) performed identically to postparturient females with respect to the speed of retrieving pups, the size of the nest built, and the intensity of responses to male intruders.

We have recently studied the effects of adrenalectomy on the maternal responsiveness of pregnancy-terminated animals to determine if adrenal estrogen and progesterone could influence the postoperative latencies to the onset of maternal behavior in hysterectomized, hysterectomized–ovariectomized, and hysterectomized–ovariectomized plus estradiol benzoate-treated females. In this study half of the animals in each of these three groups were adrenalectomized and the other half were sham operated (Siegel and Rosenblatt, 1978b). The data clearly showed that the adrenals have little if any effect on the onset of maternal care, and these results are in agreement with those of Numan (personal communication), who also examined pregnancy-terminated animals, and of Thoman and Levine (1970), who studied postpartum maternal behavior in animals adrenalectomized prior to mating. Apparently, adrenal corticoids also are not involved in the onset of ma-maternal behavior although Leon *et al.* (1978) have shown that they do play a role in certain aspects of maternal behavior postpartum.

7. Progesterone: Inhibition of Maternal Behavior

Normally progesterone declines at the end of pregnancy, and its decline may enable the rising levels of circulating estrogen to be effective. The studies cited before (see Section I,B,4) on the inhibition of initiation of maternal behavior by progesterone suggest this. In several studies with nonpregnant and pregnant hysterectomized females we have investigated this problem (Siegel and Rosenblatt, 1975c, 1978a, unpublished data). Using *nonpregnant* hysterectomized–ovariectomized females we administered progesterone at a high (5.0-mg/animal) or low (0.5-mg/animal) dose at three time intervals in relation to treatment with estradiol benzoate and tested all animals 48 hr postoperatively. Different groups received progesterone at the same time as the estrogen treatment, 24 hr later, or 44 hr later. At the high dose, all animals were inhibited in the display of short-latency maternal behavior, but at the low dose, only those given progesterone 24 hr after the estrogen were inhibited (Fig. 13). Those given progesterone at the same time as estrogen were initially delayed, but they soon caught up with the control animals that had been given only estrogen. Progesterone given at 44 hr had no inhibiting effect on maternal behavior.

The results were somewhat similar when 16-day-*pregnant* hysterectomized or hysterectomized and ovariectomized females were tested for the inhibitory effect of progesterone on maternal behavior. Hysterectomized females given a low dose of progesterone (0.5 mg) either at the time of hysterectomy or 24 hr later and tested at 48 hr after surgery were initially delayed in the onset of maternal behavior compared to untreated controls (Fig. 14). However, they soon caught up and at 72 hr, 70–80% of the females had initiated maternal behavior. Paradoxically, at a higher dose of progesterone (2 mg) at the time of surgery females exhibited no delay in onset of maternal behavior, and 70% were maternal at 48 hr upon their first exposure to pups.

When using hysterectomized pregnant females the time at which circulating levels of estrogen begin to rise is not known; therefore, the timing of treatment

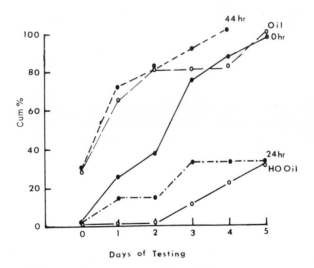

Fig. 13. Cumulative percentage of nonpregnant rats showing maternal behavior after hysterectomy–ovariectomy (HO) plus 100 μg/kg estradiol benzoate (EB) at surgery and 0.5 mg progesterone (P) at 0, 24, and 44 hr after EB. Additional groups received EB or oil at surgery and oil 44 hr later. Maternal behavior tests started 48 hr postoperatively (Day 0). (From Siegel and Rosenblatt, 1975c.)

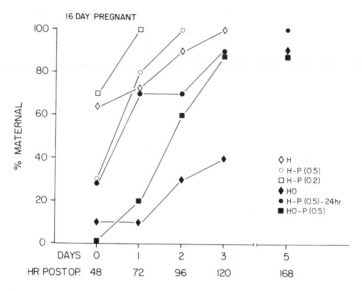

FIG. 14. Cumulative percentage of 16 day pregnant rats showing maternal behavior after hysterectomy (H) or hysterectomy–ovariectomy (HO) plus 0.5 or 2 mg progesterone (P) injected at surgery or 24 hr later. Maternal behavior tests started 48 hr postoperatively. (From Siegel and Rosenblatt, unpublished data.)

with progesterone can only be approximate. To be more precise in our progesterone treatment we have also used 16-day-pregnant hysterectomized-ovariectomized rats that we injected with estrogen (5 μg/kg) at the time of surgery. They were then injected with progesterone (0.5 mg) on the same schedule as previously (at the time of surgery and estrogen treatment, 24, and 44 hr later). The results were very similar to our previous findings (Fig. 15): progesterone given at 0 and 24 hr inhibited, somewhat, estrogen stimulation of maternal behavior in tests started 48 hr after surgery but progesterone given at 44 hr had no effect. Numan (1978) has also studied the effects of progesterone on estrogen stimulation of maternal behavior in 16-day-pregnant hysterectomized-ovariectomized females. He administered 20-μg/kg estradiol benzoate immediately after surgery and progesterone at two dose levels, low (1 mg/day) and high (5 mg/day), using two injections, one immediately after surgery and the second 24 hr later, testing the females starting at 48 hr after estrogen treatment by exposing them continuously to pups thereafter. Females given the lower dose of progesterone were inhibited in maternal behavior and were similar to those not given estradiol benzoate.

Females given the higher dose of progesterone were inhibited in maternal behavior for 24 hr after they were exposed to pups, but then at 48 hr after the last progesterone injection and 96 hr after estradiol benzoate treatment 75% exhibited maternal behavior, a percentage equal to the females given estradiol benzoate alone. These females therefore were ultimately more responsive to the estradiol benzoate treatment than those given the lower dose of progesterone. They were similar to our hysterectomized animals given 2 mg progesterone at surgery that were as responsive at 48 hr as females not given the progesterone. These findings suggest that, in addition to its inhibitory effect on estrogen, progesterone may have a facilitating effect on maternal behavior under certain conditions, a subject we shall return to shortly.

The possibility also exists that it is not only the progesterone which is in the circulation concurrent with the rise in circulating estrogen that can inhibit the action of estrogen but that which has been circulating *before* estrogen has risen. This has been found to be the case in the sexual behavior of the female rat (Blaustein and Wade, 1977). To test this possibility Siegel and Rosenblatt (unpublished observations) gave progesterone at two dose levels (0.5 and 2.0 mg) to 16-day-pregnant hysterectomized–ovariectomized females at surgery. Twenty-four hr later they gave estradiol benzoate (5 μg/kg) to the females. Progesterone given 24 hr before estrogen was *less* effective in inhibiting the action of estrogen than when given at the same time (Fig. 16): at 48 hr 64–70% of the females were maternal compared with 40% maternal when females were given progesterone at the same time as estrogen. At either dose the pretreatment effect of progesterone was short lived, and at 72 hr 80–90% of the females had initiated maternal behavior.

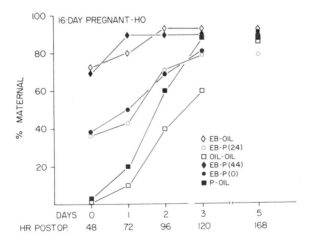

FIG. 15. Cumulative percentage of 16 day pregnant rats showing maternal behavior after hysterectomy–ovariectomy (HO) plus 5 μg/kg estradiol benzoate (EB) at surgery and 0.5 mg progesterone (P) at 0, 24, or 44 hr after EB. Additional groups received either EB, P, or oil at surgery plus oil 44 hr later. Maternal behavior tests started 48 hr postoperatively. (From Siegel and Rosenblatt, 1978a.)

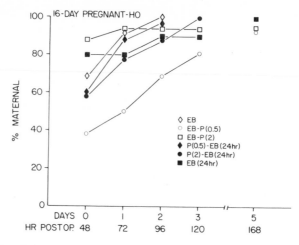

FIG. 16. Cumulative percentage of 16-day pregnant rats showing maternal behavior after hysterectomy–ovariectomy (HO) plus either 5 μg/kg estradiol benzoate (EB) alone or in combination with 0.5 or 2 mg progesterone (P) at surgery. Additional groups received P at surgery plus EB 24 hr later or EB only 24 hr after HO. Maternal behavior tests started 48 hr postoperatively. (From Siegel and Rosenblatt, unpublished data.)

8. Estrogen Stimulation of Maternal Behavior: Timing and Duration

The latency and duration of estrogen effects on maternal behavior have been studied in pregnant and nonpregnant females as a model of what happens at the end of pregnancy when estrogen rises and maternal behavior is initiated. The use of estradiol benzoate instead of estradiol in our studies is justified because of the slow absorption rate of the benzoate, a rate that more closely approximates the continuous secretion of estradiol than would a single or even multiple injections of estradiol. Knowing the time of onset of maternal behavior prepartum, information about the latency and timing of estrogen effects would help us to specify at what point during the prepartum rise in circulating estrogen maternal behavior may be stimulated. This, combined with information about the duration of estrogen effects, would help us to deal with certain problems in the transition between hormonal and nonhormonal regulation of maternal behavior to be discussed later (Section IV).

In 16-day-pregnant females, hysterectomized–ovariectomized, given estradiol benzoate (5 μg/kg) at surgery and tested with pups for the first time at 24, 48, 72, or 96 hr after surgery, the maximal effect is shown at 48 and 72 hr (Fig. 17). At 24 hr and at 96 hr estrogen-treated females do not differ from oil-treated controls. Thus, estrogen has a 48-hr latency for the onset of maternal behavior and it continues to act for at least 72 hr, but its effect appears to wane at 96 hr in late-pregnancy-terminated females.

The latency and duration of estrogen stimulation of maternal behavior in hysterectomized–ovariectomized nonpregnant females has been studied more extensively than in pregnant females. In an initial study (Siegel and Rosenblatt, 1975d), estradiol benzoate (100 μg/kg) given at the time of surgery was ineffective at 24 hr and maximally effective at 48 hr. The effect lasted through 72 hr and was equal between 24 and 48 hr and between 48 and 72 hr, as measured by the number of females that initiated maternal behavior during 24 hr of pup exposure (Fig. 18). These findings in nonpregnant females are therefore similar to those in pregnant females.

In a more recent series of studies (Siegel *et al.*, 1978), a dose–response relationship was found between estradiol benzoate and initiation of maternal behavior at 48 hr and afterward: 200 μg/kg was most effective, 100 μg was slightly less effective, whereas lower doses (2 × 50 μg/kg, 50 μg/kg, and 25 μg/kg) were ineffective in stimulating maternal behavior. In a second study aimed at excluding the influence of pup stimulation on the duration of estrogen action, only the higher doses (200 and 100 μg/kg) were used and females were initially tested at 72 and 96 hr after surgery and estrogen injection. Tests were repeated after 24-hr periods of pup stimulation. At 72 and 96 hr neither dose had

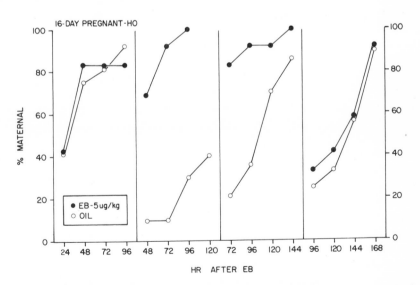

FIG. 17. Cumulative percentage of 16-day pregnant rats showing maternal behavior after hysterectomy–ovariectomy (HO) plus either 5 μg/kg estradiol benzoate (EB) or oil at surgery. Different groups of EB and oil-treated animals were initially tested 24, 48, 72, or 96 hr postoperatively. (From Siegel and Rosenblatt, 1978a.)

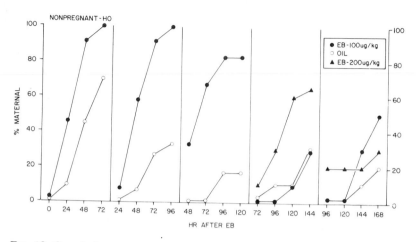

FIG. 18. Cumulative percentage of nonpregnant rats showing maternal behavior after hysterectomy–ovariectomy (HO) plus either 100 or 200 μg/kg estradiol benzoate (EB) or oil at surgery. Different groups of EB and oil-treated animals were initially tested at 0, 24, 48, 72, or 96 hr postoperatively. (From Siegel and Rosenblatt, 1975d; Siegel *et al.*, 1978.)

an *initial* effect, but at 72 hr the 200 μg/kg dose combined with 24 hr of exposure to pups had a significant effect on short-latency maternal behavior.

These studies show that in both late-pregnancy-terminated and nonpregnant females estrogen requires 48 hr to have its maximal effect on maternal behavior, independent of pup stimulation. It is active before 48 hr since presenting pups at the time of estrogen injection increases the percentage of females that initiate maternal behavior at 24 hr and again at 48 hr above that of oil-treated controls (Siegel and Rosenblatt, 1975d). It remains active between 48 and 72 hr after an injection of 100 μg/kg, but only at 200 μg/kg does it remain active between 72 and 96 hr after injection.

9. Facilitation of Maternal Behavior by Progesterone Withdrawal

Progesterone withdrawal refers to the decline in circulating levels of progesterone following hysterectomy, hysterectomy and ovariectomy, and, presumably, ovariectomy alone, during the latter half of pregnancy. The rate at which progesterone declines and reaches levels at which it can no longer be detected in the blood differs according to the stage of pregnancy and the type of surgery. It is most rapid at all stages when the ovaries and fetal placentas (that is, hysterectomy) are removed since the two main sources of progesterone are thereby eliminated. The rate of progesterone withdrawal is less rapid when only hysterectomy is performed as the ovaries decrease their release of progesterone and, presumably, increase the secretion of estradiol. The change in the pattern of ovarian steroid hormone secretion occurs more slowly following hysterectomy around midpregnancy (see Figs. 5 and 6) because the fetal placentas are not yet fully regulating ovarian progesterone secretion. Regulation by anterior pituitary gonadotrophins and prolactin remains after the placentas are removed. By the 16th day of pregnancy the fetal placentas have fully established their regulation of ovarian progesterone secretion, and their removal therefore precipitates a rapid decline in progesterone secretion. Ovariectomy alone undoubtedly causes a rapid decline in circulating progesterone as shown by the abortions that occur after a variable interval, but the rate of decline has not yet been measured.

There is evidence that progesterone *withdrawal* may facilitate maternal behavior independently of the permissive relationship between progesterone and estrogen shown in the previous studies. It will be recalled that *pregnant* hysterectomized–ovariectomized females had shorter latencies for maternal behavior than nonpregnant intact or hysterectomized–ovariectomized females.

Bridges et al. (1977) have investigated this possibility by once again comparing pregnant hysterectomized, hysterectomized–ovariectomized, and ovariectomized females at different times during pregnancy, starting the tests at 22 hr after surgery (with continuous exposure to pups from then on) and testing females at 24-hr intervals thereafter instead of waiting 48 hr as in most of our previous studies. The results at 17 days of pregnancy (Fig. 19) are representative of those at 8 and 13 days: at 24 hr a few hysterectomized–ovariectomized females (25%, 3/12) but only 1 of both the ovariectomized and the hysterectomized females were responsive to the pups. At 48 hr 75–90% of all groups were responsive and the groups did not differ. Earlier Stern and MacKinnon (1976) had also found that 16-day-pregnant hysterectomized–ovariectomized females that are given pups soon after surgery (in this study about 18 hr after surgery) responded to pups in 25–40% of the cases initially, and this increased to 60–80% at 48 hr. We have confirmed in a study cited earlier (Fig. 17, 24-hr group) in which 16-day-pregnant hysterectomized–ovariectomized females were first tested at 24 hr (Siegel and Rosenblatt, 1978a). In all of these studies, therefore, by starting pup stimulation earlier than 48 hr it was shown that hysterectomized–ovariectomized females are as responsive as those that are only hysterectomized.

The two main sources of progesterone, the ovaries and fetal placentas, are absent only in the hysterectomized–ovariectomized females. It has been proposed, therefore, that the higher percentage showing maternal behavior at 24 hr results from the more rapid withdrawal of progesterone. Radioimmunoassays of circulating levels of progesterone are negatively correlated with the appearance of maternal behavior (Bridges et al., 1978b). Moreover, progesterone administered by silastic implants to these hysterectomized–ovariectomized females prevents the initiation of maternal behavior, presumably by preventing progesterone withdrawal effects (Bridges et al., 1978b).

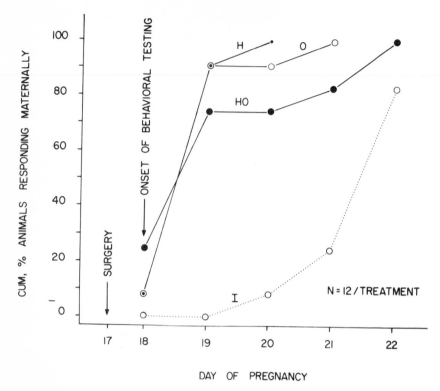

FIG. 19. Cumulative percentage of day 17 pregnant rats showing maternal behavior after hysterectomy (H), ovariectomy (O), hysterectomy–ovariectomy (HO), or sham surgery–intact (I). (From Bridges *et al.*, 1977.)

Another study more clearly defined the time course of the facilitatory effect of progesterone withdrawal on maternal behavior. Pregnant females on the 17th day were hysterectomized, hysterectomized–ovariectomized, or ovariectomized, and different groups of each of these treatments were first given pups at 6, 24, or 48 hr after surgery. The initial performance of each of these groups upon receiving pups is shown in Fig. 20. Among hysterectomized–ovariectomized females endogenous hormonal changes (without pup stimulation) stimulated a brief surge of maternal behavior at 24 hr, which subsided to 0% by 48 hr. Among ovariectomized females there was a less pronounced increase (10%) and none at all among hysterectomized females. At 48 hr, however, nearly 60% of the hysterectomized females initiated maternal behavior without prior pup stimulation, whereas less than 20% of the ovariectomized females became responsive.

In view of the facilitating effect of progesterone withdrawal shown at 24 hr in hysterectomized–ovariectomized pregnant females, it is possible that the hysterectomized females are delayed in exhibiting maternal behavior because they have a slower rate of decline of circulating levels of progesterone (Bridges *et al.*, 1978b) than when the ovaries are also removed. This is an alternative explanation to one that has been offered earlier that it is the rise in estradiol secretion that stimulates maternal behavior at 48 hr and that some hours are required for this rise to begin.

If, however, estrogen played no role in posthysterectomy onset of maternal behavior, then ovariectomizing such females and treating them with estrogen should not affect the timing of maternal behavior; this should be determined solely by the rate at which progesterone declines.

A series of recently completed studies shows that estrogen does have an important effect on the timing of maternal behavior. Following the hysterectomy–ovariectomy of 16-day-pregnant females, the time of onset of maternal behavior was largely determined by the time at which estradiol benzoate was administered. In this study females were hysterectomized and ovariectomized and given either estradiol benzoate or oil, and groups were tested initially at 24, 48, or 72 . hr after surgery. To begin with, at 24 hr estrogen-treated

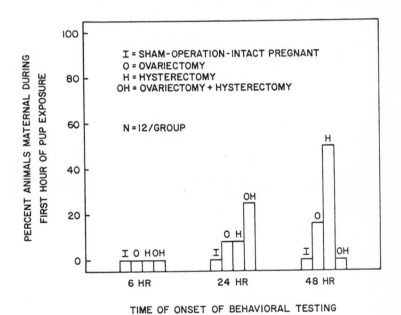

FIG. 20. Comparison of the percentage of animals that were spontaneously maternal (responded during the first hour of exposure to foster young) after surgery at 10:00 A.M. on Day 17 of pregnancy and first presentation of pups at 6 hr, 24 hr, and 48 hr later. (From Bridges *et al.*, 1978a.)

(5 μg/kg) and oil-treated females (given at surgery) were equally responsive (40% of females initiated maternal behavior in each group). More crucial for deciding between the estrogen stimulation versus progesterone withdrawal hypotheses for the initiation of maternal behavior at 48 hr were the findings at 48 hr. Oil-treated females (hysterectomized–ovariectomized) were highly *unresponsive* to pups at this time (10%), whereas estrogen-treated females were highly *responsive* (70%) at their first exposure to pups (see Fig. 16). They were as responsive as hysterectomized females tested initially at 48 hr (64%). Still further, when hysterectomized–ovariectomized females were given progesterone at surgery (2 mg) to prevent the decline in circulating progesterone but then were given estradiol benzoate at 24 hr to stimulate maternal behavior, 60% were maternal at 48 hr. If the earlier progesterone was omitted, then 80% were maternal at 48 hr as a result of the intervening treatment with estrogen at 24 hr.

At 72 hr neither estrogen-treated nor oil-treated females were initially responsive to pups (less than 15% of the females). A higher dose of estrogen would be required to initiate maternal behavior 72 hr after treatment, as we have seen in nonpregnant hysterectomized–ovariectomized females (see Fig. 18), but progesterone withdrawal is maximal at this time, and were it the effective stimulus beyond 24 hr after pregnancy termination it should have stimulated maternal bebavior. The evidence suggests that maternal behavior at 48 hr in hysterectomized females (16 or 17 days pregnant) is stimulated by estradiol rather than by progesterone withdrawal since the same high level of responsiveness can be produced in hysterectomized–ovariectomized females by injected estrogen. Moreover, this high level of responsiveness is not blocked by slowing the decline in circulating levels of progesterone by prior treatment with progesterone (2 mg) at the time of surgery (Fig. 16).

This does not rule out the possibility that progesterone withdrawal alone is responsible for the 40% of females that initiate maternal behavior at 24 hr after 16- or 17-day hysterectomy–ovariectomy during pregnancy. It has not been possible, however, to produce this progesterone-withdrawal effect by administering progesterone for 12 or 16 days via implantation of a series of silastic capsules then withdrawing the progesterone from the nonpregnant hysterectomized–ovariectomized females and testing them for maternal behavior (Bridges and Siegel, unpublished observations; Siegel, Doerr, and Rosenblatt, unpublished observations). These females were no more responsive than oil-treated females or animals in which the progesterone implants were not removed prior to testing.

III. Postpartum Maternal Behavior

The aim of this section is to discuss the regulation of maternal behavior postpartum. We shall not review the many descriptive studies of maternal behavior and mother–young interaction except insofar as it is necessary to deal with the factors that regulate maternal behavior.

A. Postpartum Separation

Chief among the factors regulating maternal behavior are stimuli which the mother receives from the pups. An early study (Rosenblatt and Lehrman, 1963) investigated whether mothers could maintain maternal behavior if pups were removed at parturition. The mothers were tested at weekly intervals after parturition with 5–10-day-old pups for the maintenance of their maternal behavior in the absence of pup stimulation. By the end of the first week only a small percentage of mothers were exhibiting some nest building, but none of them exhibited nursing or pup licking and only one exhibited retrieving. Maternal behavior waned rapidly when pups were removed at parturition: if pups were removed at parturition and returned on the third day, about 30% of the mothers still retrieved them and exhibited other components of maternal behavior, but if the pups were kept away for 4 days and returned early on the fifth day postpartum, none of the mothers exhibited maternal behavior except for nest building, which rose rapidly upon the return of the pups.

These studies suggest that, following the hormonal onset of maternal behavior, pups are required to maintain it. The question arises whether pups maintain maternal behavior by evoking the release of hormones or by nonhormonal means, that is, solely through the sensory stimuli they provide. To answer this question we must discuss the hormones that are secreted during the period of postpartum maternal care and their relationship to the postpartum maintenance of maternal behavior.

B. Postpartum Estrus–Diestrus

During gestation, estrous cycling is suspended, but soon after parturition the female undergoes a single estrous cycle, exhibiting estrous behavior between 8 and 13 hr postpartum and ovulation several hours later (Sachs *et al.,* 1971; Ying and Greep, 1973; Ying *et al.,* 1973). If a male is present, the female may mate and become pregnant again, and she becomes acyclic until the next postpartum estrus (Fig. 21). Implantation of the newly fertilized ova is delayed up to 13 days, depending upon the number of suckling pups in the current litter, and the new litter is not delivered until the present one is about 35 days of age and has terminated its suckling. Whether or not the female is mated during postpartum estrus, she becomes acyclic, but in unmated females lactation diestrus lasts only 2–4 weeks, depending upon the number of suckling pups in the litter.

The suspension of estrous cycling, following the postpartum estrus, is based upon pup suckling stimulation (Rothchild, 1960). The relationship is shown in

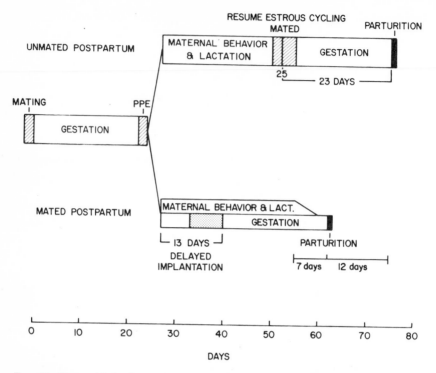

Fig. 21. Maternal behavior cycle of the rat when female is mated or remains unmated at postpartum estrus (PPE). Estrous behavior and estrous cycling (oblique lines), range of days beyond normal implantation (stipple), and parturition (black bars).

Fig. 22, where it can be seen that the interval between parturition and resumption of estrous cycling is shorter for females suckling 1–4 pups than for those suckling 5 or more pups (see also van der Schoot *et al.*, 1978). If pups are removed immediately after parturition (in unmated females), estrous cycling is resumed after 4–5 days in nearly half the females, and the remaining females resume cycling between 10 and 15 days postpartum. Rothchild (1960) and others (Gala, 1970) have shown that suckling prevents estrous cycling by inhibiting the release of follicle-stimulating hormone (FSH) and luteinizing hormone (LH).

The important point to be gathered in relation to pituitary–ovarian activity is the vast hormonal changes that occur during the immediately prepartum period. Gonadotrophic and ovarian hormones undergo rapid increases followed by equally rapid decreases, and then a stable pattern of secretion is established by the anterior pituitary and ovaries in which FSH and LH secretion is inhibited, prolactin is secreted in large amounts, and progesterone secretion is maintained while estradiol is not secreted.

C. NONHORMONAL BASIS OF MATERNAL BEHAVIOR POSTPARTUM

The rise in prolactin prepartum and its maintenance at high levels throughout lactation until weaning have led many to suggest that postpartum maternal behavior may be based upon prolactin (Moltz *et al.*, 1970; Riddle *et al.*, 1942). We have already shown that prolactin does not play a role in the *onset* of maternal behavior, and there is equally compelling evidence that it does not play a basic role in the maintenance of maternal behavior postpartum.

Postpartum administration of ergocornine, blocking the release of prolactin, has no effect on maternal behavior although lactation fails and pups receive little milk (Numan *et al.*, 1972; Stern, 1977; Zarrow *et al.*, 1971). Similarly, failure of release of oxytocin from the posterior pituitary gland, a condition that prevents milk ejection and leads to high mortality among pups, does not prevent the appearance and maintenance of maternal behavior when foster pups replace the mother's own litter to maintain pup stimulation (Catala and Deis, 1973; Terkel, 1970). Finally, hypophysectomy performed during pregnancy does not prevent the appearance of maternal behavior in those females that are able to give birth (Obias, 1957). Moreover, when it is performed between the 1st and 5th day postpartum, maternal behavior continues although without hormone replacement, lactation fails, and the pups eventually die (Bintarningsih *et al*, 1958; Erskine, 1978; Rothchild, 1960).

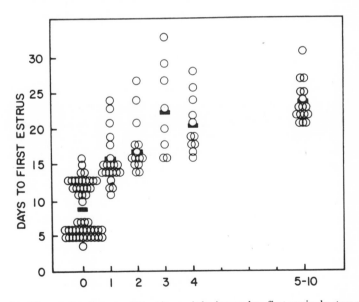

FIG. 22. The relation between litter size and the interval to first vaginal estrus postpartum in intact rats. Individual values are shown by the open circles; means for the group are shown by the solid black bar. (From Rothchild, 1960.)

Since ovarian and adrenocortical hormone secretion undergoes such vast changes immediately after parturition, and, moreover, hypophysectomy does not prevent maternal behavior postpartum, we should expect that ovarian and adrenocortical hormones play a small role in the maintenance of maternal behavior after parturition. This has been shown directly in females that were ovariectomized either prepartum or immediately postpartum (Rosenblatt, unpublished; Terkel, 1970) or adrenalectomized before mating or after parturition (Leon *et al.*, 1978; Thoman and Levine, 1970). Moreover, the administration of progesterone to recently parturient mothers for 4 days has no effect on their maternal behavior, which is in contrast to the strong effect it has in preventing the appearance of maternal behavior when given prepartum (Moltz *et al.*, 1969).

The situation with respect to postpartum maternal aggression is similar to that with respect to maternal care: hypophysectomy when performed on the 5th day after parturition does not affect its maintenance (Erskine, 1978).

There is little evidence, therefore, that maternal behavior, once it has been initiated under the influence of hormones prepartum, requires hormones to maintain it. This is not to say that hormones do not play any role in maternal care since it has already been shown that prolactin and other lactogenic hormones are required to maintain lactation and that oxytocin causes milk ejection. These and perhaps other hormones play a role in modulating various capacities of the lactating mother, but the motber is not dependent upon them for maintenance of her maternal behavior.

D. Pup Stimulation as the Basis for Maintenance of Maternal Behavior Postpartum

Our study in which removal of pups at parturition led to the waning of maternal behavior after 4 days can now be interpreted as evidence that pup stimulation is required to maintain maternal behavior after it has been initiated prepartum by hormonal stimulation (Rosenblatt, 1970, 1975a; Rosenblatt and Lehrman, 1963). A later study added support to this interpretation: pups were removed from mothers for 4 days but now starting some time after maternal behavior had been initiated. The removals occurred either at the 4th, 10th, or 14th day postpartum. When 5–10-day-old pups were returned to these mothers on the 9th, 15th, and 19th days between 40 and 70% exhibited retrieving and nursing almost immediately, and nest building appeared shortly afterward, having declined during the interim without pups. These mothers went on to rear the foster pups to weaning in all cases (Rosenblatt, 1965).

It will be recalled too that cross transfusion of blood from mothers 24 hr after parturition was ineffective in stimulating maternal behavior in nonpregnant females. Mothers from whom the blood was transfused therefore were no longer stimulated by a humoral factor(s) (that is, hormones) in the performance of their maternal behavior (Terkel and Rosenblatt, 1972).

The evidence therefore points to a change in the regulation of maternal behavior from the hormonal onset before parturition to the nonhormonal maintenance after parturition. Moreover, pup stimulation is required immediately after parturition to sustain the maternal behavior that has been established hormonally, and in its absence maternal behavior declines. The decline of hormonally stimulated aggression in females when their pups are removed is even more rapid (Erskine, 1978). On the other hand, once the nonhormonal regulation of maternal behavior has been established, mothers are able to withstand a 4-day separation from their young with less disastrous consequences for their maternal responsiveness than when the separation occurs at parturition.

IV. Transition in the Regulation of Maternal Behavior: Transition Period

The existence of a *prepartum period* during which the onset of maternal behavior is stimulated by hormones and a *postpartum period* during which regulation shifts to a nonhormonal (that is, pup stimulation) basis implies that there is a transition period during which this shift takes place. We shall discuss this transition period in this section. In the writings of Leblond (1940) and Leblond and Nelson (1937), with reference to the persistence of maternal behavior in mice following hypophysectomy, we find an early statement of this concept "... after the hormonic stimulus has once set up the nervous mechanism of maternal behavior, this mechanism would remain after removal of the hormonic stimuli, such as after hypophysectomy" (Leblond and Nelson, 1937, p. 171). Similarly, Koller's (1956) finding that prepartum nest building in the mouse is hormonally stimulated while postpartum nest building is maintained by pup stimulation implies a transition period between the two modes of regulating nest building, but the concept was not clearly formulated by this investigator. A clearer statement is contained in Richards (1967).

Schneirla (1952) and Birch (1956) most clearly proposed a transition between pre- and postpartum maternal behavior in the rat mediated by the mother's licking of the pups and the formation of a behavioral bond with her offspring. They were unaware, however, of the nonhormonal basis of postpartum maternal behavior, which is an important feature of the transition period as proposed in this section. The relationship between the present concept and their formulation will be discussed in a later section (Section VIII).

A. Early Separation and the Waning or Retention of Maternal Behavior

The waning of maternal behavior during the first 4 days postpartum following removal of the pups during parturition has already been cited as evidence of dependence of postpartum maternal behavior upon pup stimulation. It also defines the period during which the transition to nonhormonal maternal behavior occurs and the new basis of regulation becomes established. If females are allowed the first three days of contact with pups before the pups are removed for a period of 4 days, then maternal behavior is retained over the interval in the majority of females and in all of them it is reinstated within 24 hr (Rosenblatt, 1965; Rosenblatt and Lehrman, 1963). By the third day, therefore, the nonhormonal stage of maternal behavior has become established sufficiently to sustain the separation and from then on until maternal behavior declines a 4-day separation from the young does not cause maternal behavior to wane in most females (Rosenblatt, 1965).

The capacity for sustaining maternal behavior over long intervals without pup stimulation is a characteristic of the nonhormonally based maternal behavior rather than the hormonally stimulated onset of maternal behavior; its appearance therefore indicates that this stage has been attained by the postpartum female. Fleming and Rosenblatt (1974b) showed that removal of pups from females that had just given birth (and therefore, one can assume, were hormonally primed to exhibit maternal behavior) prevented the "retention" of maternal behavior two weeks later when the females were exhibiting estrous cycles. Bridges (1977) has also shown that 22-day-pregnant, Caesarean-section delivered females denied access to pups fail to show any effect of the hormonal stimulation in sensitization tests 25 days later, and Fleming and Cummings (personal communication) have shown that sensitization latencies are unaffected even 7 days after Caesarean-section delivery.

Tests of retention of maternal behavior, as measured by reduced latencies for sensitization after 14–25 days following removal of the litter, have been used to define more precisely the amount of contact with pups that is necessary to establish postpartum nonhormonal maternal behavior. Bridges (1975, 1977) has done the most systematic work on this problem and his findings are summarized in Fig. 23. In normally parturient females 48 hr of postpartum contact resulted in significantly shortened latencies and this was also true when females had had only 6–8 hr of postpartum contact with their litters. Most surprising was his finding that contact with the pups only during parturition was highly effective and contact with only half the litter was equally effective. As cited above, lack of contact either in parturitive females or those delivered by Caesarean section resulted in long latencies for sensitization that were equal to those of inexperienced nonpregnant females.

The nature of the experience with pups during the brief contact with them postpartum that enables mothers to exhibit shortened latencies 6 days later was studied by Fleming and Cummings (personal communication). The mothers were delivered by Caesarean section on the 22nd day of pregnancy, and on Day 1 following delivery they were allowed either (1) no contact with pups; (2) contact

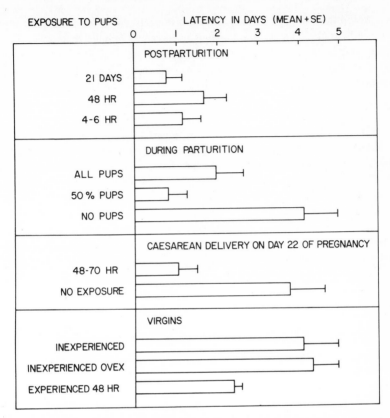

FIG. 23. Effects of parturition and postparturition contact with pups on latencies for induction of maternal behavior 25 days later. Period of exposure to pups shown on ordinate and latencies on the upper abscissa. Caesarean-section delivered females and virgins shown for comparison. (From Bridges, 1975, 1977.)

with the smells and sounds of pups that were enclosed in a metal box with perforations; (3) contact with the smells, sounds, and sight of pups that were enclosed in a Plexiglas box with perforations; or (4) direct contact with pups placed in the mothers' cages. Only mothers that had had direct contact with pups and therefore could interact with them during the first day were immediately responsive to pups presented to them on the 7th day postpartum. Those mothers that had experienced pups enclosed in Plexiglas boxes in their cages had slightly reduced latencies for sensitization but those with either a smaller amount of contact or no contact had latencies that were no shorter than those of inexperienced nonpregnant females.

Perhaps most convincing that the retention is nonhormonally based is the finding by Bridges (1978) that retention was not affected when females were tested during their next pregnancy, which is reminiscent of the finding by Terkel and Rosenblatt (1971) that "spontaneous retrievers," females that exhibit maternal behavior at their first exposure to pups and from then on, also were not affected in their responsiveness to pups by becoming pregnant.

Finally, some of the results in postpartum females can be duplicated in nonpregnant females that have been sensitized. If females were sensitized and then permitted 36 hr of contact with pups, their resensitization latencies two weeks later were significantly reduced (Fleming and Rosenblatt, 1974b; Mayer and Rosenblatt, 1975). Interestingly, the reduction in latencies in these nonpregnant females was the same found in postpartum females allowed 36-hr contact with their litters when they were sensitized for the first time 2 weeks later (Fleming and Rosenblatt, 1974b). A related finding has been demonstrated by Cohen and Bridges (unpublished observations), who sensitized nulliparous females and allowed them 1, 4, or 9 days of pup exposure. Primiparous females also were allowed 1, 4, or 9 days with pups postpartum. When nulliparous and primiparous females were again exposed continuously to pups after a 25-day isolation period, the primiparous females that had received 1 or 4 days of pup exposure reinstated retrieving, grouping of the young, and crouching more rapidly than did the nulliparous animals that had received equal amounts of pup experience. After 9 days of initial pup exposure, both nulliparous and primiparous females had equally short latencies to retrieve and group the pups 25 days later although the nulliparous animals were slower to adopt the nursing posture.

B. HORMONAL AND NEURAL PROCESSES UNDERLYING THE TRANSITION

During the transition there is a period of overlap between the effects of hormonal stimulation that gave rise to maternal behavior and those of pup stimulation which are destined to maintain this behavior. In a previous section (Section II,A,8) it was shown that in both pregnancy-terminated and nonpregnant females, at the higher doses (100–200 μg/kg) of estrogen, effects on maternal behavior reached their maximum at 48–72 hr and could still be detected at 96 hr when the females are also given pup stimulation. If this also applies to normally delivering females, then hormonal effects may still be present while nonhormonal stimulation is growing more effective, and, in fact, the hormonal effects may be potentiated by early pup stimulation.

Different patterns of interaction between these two processes may yield different kinds of transition periods. Slotnick et al. (1973) tested females for prepartum onset of maternal behavior and found that 25% exhibited retrieving, whereas 75% did not. Postpartum the females were allowed only 10 min of contact with pups per day. Only those females that had exhibited maternal behavior prepartum were able to maintain their maternal behavior under the severely restricted conditions of contact with pups, whereas those that had not exhibited maternal behavior showed a rapid waning of responsiveness to pups. Apparently for the latter females the restricted amount of pup contact coming at a time when hormonal stimulation was reaching its maximum was insufficient to mediate the transition to nonhormonally based maternal behavior. Further study of the interrelationships between these two processes would be of value for our ability to predict the outcome of hormonal or behavioral manipulations during the transition period.

Studies on the neural site of action of estrogen and of pup stimulation on maternal behavior may provide a mechanism by means of which the transition from hormonal to nonhormonal regulation occurs during the transition period. Numan et al. (1977) found that estrogen implants into the medial preoptic area (MPOA) in pregnancy-terminated females (hysterectomized and ovariectomized) stimulated short-latency maternal behavior 48 hr later. This same region when lesioned in postpartum females that were exhibiting maternal behavior resulted in the abrupt disappearance of this behavior (Numan, 1974) and when produced in nonpregnant females prevented them from being sensitized. How complete the elimination of maternal behavior is in postpartum females with lesions of the lateral input to the MPOA and the anterior hypothalamic area (AHA), which Numan (1974) found to be as effective as MPOA lesions, has recently been studied by Terkel et al. (1979). Females remained capable of responding passively to the suckling approaches of the young but did not actively retrieve them and they responded to suckling by releasing increased amounts of prolactin.

Although they are incomplete and require further study, these findings suggest that at the onset of maternal behavior, estrogen activates this neural site and as these effects wane, sensory stimulation from the pups maintains the activity of this neural site and therefore maintains maternal behavior. Unlike the situation with nonpregnant females that are not primed with hormones in which prolonged pup stimulation is necessary to stimulate maternal behavior, in the postpartum female pup stimulation is immediately effective in stimulating maternal behavior because of the prior action of estrogen.

V.   MATERNAL BEHAVIOR IN NONPREGNANT FEMALES

Wiesner and Sheard (1933) were partially successful in inducing retreiving and nest building in nonpregnant females (30% of more than 75 females) leading the way for later studies by Cosnier and Couturier (1966) and Rosenblatt (1967), using foster pups maintained in good condition and extending the period of exposure, to show that nearly all females can be induced to show maternal behavior (nest building, retrieving, crouching, pup licking) by exposing them to pups continuously for 5–7 days on the average (that is, sensitization). In nearly all of our studies we have used 3–10-day-old pups to induce maternal behavior in nonpregnant females. Stern and MacKinnon (1978) have shown that 1- and 2-day-olds are more effective than 3–12-day-olds, all of which are equally effective. Older pups, however, are much less effective stimuli for inducing maternal behavior in nonpregnant females. Even without prior hormonal stimulation pup stimuli are able to activate the neural processes mediating maternal behavior, but under these conditions females require a longer period of exposure than is necessary when maternal behavior is initiated by hormonal stimulation.

Nonpregnant female rats induced to show maternal behavior are good preparations for various experimental purposes: they have been used as "aunts" to mouse and rat pups being reared by their mothers (Denenberg et al., 1969) and have served as nonlactating control mothers in infant rat nutritional studies (Slob et al., 1973). They serve as baseline groups for the study of hormonal factors and various other experimental treatments that are intended to alter the latency for the onset of maternal behavior.

In this section we shall confine ourselves to discussing this phenomenon in relation to the problem of hormonal and nonhormonal bases of maternal behavior as successive phases in the regulation of normal maternal behavior during the reproductive cycle of the rat (see Sections IV and VIII). A number of investigators have viewed sensitization as an alternative way of eliciting maternal behavior from female rats (a view originally proposed by one of the authors, Rosenblatt, 1967) to be compared with maternal behavior arising naturally. While this has been useful in bringing out differences between hormonally induced and pup-stimulated maternal behavior as Noirot (1972) has done so well in the mouse, to view this behavior only in this way poses a false problem since in reality pup-stimulated maternal behavior occurs under natural conditions only when it has been first initated by hormones. A number of investigators have adopted the latter position, and their thinking therefore contributes to a better understanding of the specific influence that hormones exert on the initiation of maternal behavior, which is subsequently maintained by pup stimulation (Erskine, 1978; Numan et al., 1972).

A. NONHORMONAL BASIS OF PUP-STIMULATED MATERNAL
    BEHAVIOR IN NONPREGNANT FEMALES

In their original studies Cosnier and Couturier (1966) and Rosenblatt (1967) showed that ovariectomy did not prevent pup-induced maternal behavior, and Rosenblatt (1967) showed that hypophysectomy also had no inhibiting effect. Subsequent research has not required any significant modification of the hypothesis that this behavior is nonhormonally based although there is now evidence that hormones may modulate this behavior.

The cross-transfusion studies of Terkel and Rosenblatt (1972), already cited, support the nonhormonal hypothesis. Moreover, cycling females that exhibit retrieving upon their first contact with pups (spontaneous retrieving) are not subsequently prevented from doing so during gestation despite the change in the pattern of hormones secreted at this time (Terkel and Rosenblatt, 1971). This suggests that spontaneous retrieving is not based upon hormones. Similarly, females with previous sensitization experience, which normally have short latencies for the reinduction of maternal behavior when they are tested several weeks after their initial experience, continue to be easily sensitized even when they are in the middle of pregnancy during the second sensitization (Bridges, 1978).

Rosenblatt (1967) reported that estrous cycles were not altered during pup induction of maternal behavior and that the onset of maternal behavior in intact females was not associated with any particular phase of the estrous cycle either with respect to the start of pup exposure or to the first appearance of maternal behavior. Koranyi et al. (1977) also found no disruption of estrous cycling during and after induction of maternal behavior in a similar strain of rats, and Erskine (1978), using Long-Evans rats, reported similar findings during and for at least 9 days after induction of maternal behavior. Stern and Siegel (1978) also reported estrous cycling among maternal virgins (Sprague-Dawley strain) during the period of induction and subsequent display of maternal behavior. However, Marinari and Moltz (1977) have recently reported that the onset of maternal behavior in their nonpregnant females may be more closely associated with the proestrous and estrous phases of the cycle than diestrous phases and, moreover, that continued exposure to pups produced alterations in regular cycling. The latter effect may be considered to be a consequence of the induced maternal behavior since females that were exposed to pups but did not exhibit maternal behavior did not show the subsequent alterations of estrous cycling. Suspension of estrous cycling in females exhibiting maternal behavior as a consequence of pup stimulation is reminiscent of the suspension of cycling as a consequence of pup suckling. In the postpartum period this is associated with elevated levels of prolactin, which is released by suckling. It is interesting, therefore, that Koranyi et al. (1977) have found that nonpregnant maternal females release more prolactin to pup exposure for 1 hr (following removal of the pups for the preceding 5 hr) 10 and 30 days after onset of maternal behavior than nonmaternal females. The exteroceptive release of prolactin is characteristic of lactating females (Grosvenor, 1965) although the response in lactating females is several times greater than in nonlactating, pup-induced maternal females. Corticosterone release was reduced in nonpregnant maternal females after 1 hr exposure to pups as it is in lactating mothers, whereas in nonmaternal and multiparous females that were no longer exhibiting maternal behavior it was increased, due, very likely to stress. Stern and Siegel (1978) have now found an increase in the release of prolactin to pup stimulation in maternal virgins, which was already evident by the 8th day of the display of maternal behavior.

There are indications that ovarian and adrenocortical hormones may modulate latencies for pup-induced maternal behavior in females (Leon et al., 1973, 1975). Ovariectomized females exposed to pups 8 weeks after surgery had shorter latencies (24 hr) for the onset of maternal behavior than intact controls (120 hr), and, more significantly, than ovariectomized females (120 hr) at 4 weeks after surgery (Leon et al., 1973). The investigators proposed that residual estrogenic effects were responsible for the longer latencies at the shorter post-operative interval, and they supported this by showing that administering the antiestrogen MER-25 reduced latencies (48 hr) at 4 weeks and administering estradiol benzoate (12 μg/animal) for 8 days before exposing females to pups increased latencies (168 hr) at 8 weeks.

Further study showed that adrenal gland secretions were also involved in the postovariectomy effect on maternal behavior in virgins (Leon et al., 1975). Removal of the adrenals in ovariectomized virgins reduced latencies to those of 8-week-ovariectomized females (40 hr) at 4 weeks postsurgery, whereas removal of the adrenals alone had no effect. Either estradiol benzoate (12 μg) or proges-terone (3 mg) given for 7 days prior to pup exposure at 4 weeks reinstated the longer latencies for the onset of maternal behavior of ovariectomized-adrenalectomized females.

The longer latencies exhibited by 4-week compared to 8-week ovariec-tomized females could therefore be based upon adrenocortical release of either estradiol or progesterone following ovariectomy (Leon et al., 1973, 1975), both of which, presumably have inhibitory effects that delay but do not inhibit mater-nal behavior in virgins. At 8 weeks after ovariectomy, removal of the adrenals did not reduce the already shortened latencies so that, presumably, the adrenals are no longer secreting these hormones in effective amounts at this time.

These findings raise several questions that require further investigation. Siegel and Rosenblatt (1975b), using Sprague–Dawley rats, were unable to find reduced latencies 8 weeks after ovariectomy as reported for the Wistar rats. Mayer and Rosenblatt (1979a) found, in fact, that latencies were *increased* after intervals of 8 weeks and longer following ovariectomy at 32, 60, and 90 days. Siegel and Rosenblatt (1975b) and Mayer and Rosenblatt (1979a) also found that estradiol benzoate (25 μg) given in a single injection 1 or 2 days before pup exposure caused a significant *reduction* in latencies (except at 90 days) rather than an increase whereas given immediately after surgery it had no effect in ovariectomized–sham hysterectomized animals (Siegel and Rosenblatt, 1975b).

In a recent study aimed once again at determining the role of estrogen in the sensitization of nonpregnant females we followed the procedures used by Leon et al. (1973) with our strain of rats (Sprague–Dawley), using 90-day ovariec-tomized females, as in the original study (see Section I,B,6). We have found that

7 days of estradiol benzoate injections (12 μg/injection, per day) started 7 weeks after ovariectomy, with exposure to pups beginning at the 8th week, resulted in significantly shorter latencies than among oil-injected females (Mayer and Rosenblatt, 1979a). Our results are consistent, for our strain, with findings that (1) estrogen stimulates short-latency maternal behavior, and (2) that ovariectomy for prolonged periods (3½ weeks and longer) either increases sensitization latencies or has no effect on these latencies (see Section V,D).

B.   SIMILARITIES AND DIFFERENCES IN THE MATERNAL BEHAVIOR
      OF NONPREGNANT AND POSTPARTUM MATERNAL FEMALES

In reviewing the similarities in the performance of maternal behavior of nonpregnant (sensitized) and postpartum maternal females once again the aim is not to present these as alternative methods—that is, hormonal and nonhormonal—for arriving at the "same" endpoint, namely, the performance of maternal behavior. The heuristic value of comparing similarities and differences between pup-induced and normally appearing maternal behavior should not be elevated into a theoretical dispute; it should be viewed as a starting point for evaluating the hormonal and nonhormonal influences upon maternal behavior during the maternal behavior cycle, as Noirot (1972) has done in the mouse.

Fleming and Rosenblatt (1974a) compared the behavior of sensitized mothers and postpartum mothers on a variety of measures during the first 10 days postpartum in the case of lactating females and for 10 days following the onset of maternal behavior in the nonpregnant females. To provide the two groups with similar pup stimulation, after the nonpregnant rats had become maternal, the 5–10-day-old pups used to induce maternal behavior were removed and replaced with a litter of five 1–2-day-old pups and each day thereafter, 1-day-older pups were exchanged for the pups that had been with the female overnight.

The performance of maternal behavior was similar both quantitatively and qualitatively in both groups of mothers over the 10 days of the study except, of course, for actual lactation, which was absent in the nonpregnant mothers (Fig. 24). With respect to various measures of retrieving, time spent in the nest, pup licking (including general body and specifically anogenital licking), nursing–crouching, grooming, and nest building, there were only minor differences that were present usually at the beginning but then disappeared. Nest building occurred for longer periods among lactating females, but pup licking was more frequent among nonpregnant mothers, probably as a result of receiving a new litter of pups each day. In daily tests retrieving was initiated somewhat more slowly by sensitized mothers. The several qualitative differences found in the nursing positions of lactating and nonlactating mothers could be attributed to the difference in lactation.

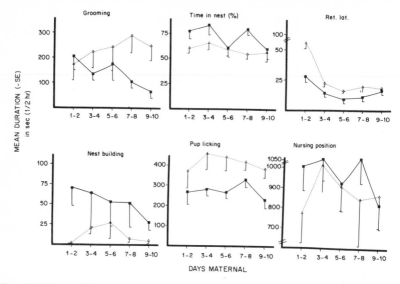

FIG. 24. Maternal behavior in the lactating (solid line) and virgin (dashed line) female rat: mean duration (in seconds) of different maternal components as a function of days maternal. Abbreviations: ret. lat., retrieval latency; lact, lactating. (From Fleming and Rosenblatt, 1974a.)

Observations of lactating and sensitized mothers were extended to 4 weeks in order to include the period of decline in maternal behavior by Reisbick *et al.* (1975). Similar procedures were used, except that pups were rotated among lactating mothers each day to equate for the presentation of new litters to non-pregnant mothers. The results of this study are presented in Fig. 25, which divides maternal behavior into two categories: *maternal care,* represented by retrieving and nursing, and *maternal rejection,* represented by actual rejection and mater-nal withdrawal as well as by excitable behavior (that is, dart, hop, shake) in response to approaches by pups. The decline in maternal care and rise in maternal rejection during the third week of "rearing" pups were similar in sensitized mothers and their lactating counterparts. There were differences, however, and these consisted mainly of an earlier decline of nursing and a more precipitous decline of retrieving in nonpregnant mothers and a corresponding earlier rise in maternal rejection. Both of these differences can be attributed either to an initial, weaker maternal motivation among nonpregnant mothers compared with lactat-ing mothers or to some continuing influence of hormones during lactation that is absent in the nonpregnant mothers.

There was some indication that nonpregnant mothers were basically not less "maternal" than postpartum mothers but that their earlier decline in maternal care and rise in maternal rejection were based upon more severe reactions to the

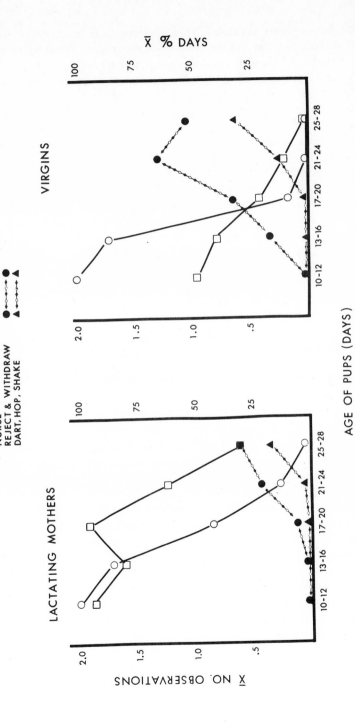

Fig. 25. Comparison of sensitized virgins and lactating mothers in the decline of maternal behavior during tests with live-in pups 10–28 days of age. (Each point represents the mean of 3- or 4-day averages for nine females.) (From Reisbick *et al.*, 1975.)

older pups that they were "rearing." This, of course, is an aspect of the strength of maternal motivation but only one measure of its strength. When both kinds of mothers were tested daily with 4–8-day-old pups throughout the entire 28 days of the study, and their behavior toward these pups was compared with their behavior toward the older pups that they were rearing, there were a number of similarities between the two groups that were not apparent in the tests with older pups. Both groups retrieved pups, adopted a nursing posture, and licked the anogenital regions of the younger pups about equally during the period when these activities were declining with the older pups, and the females did not withdraw or lie with their nipples unavailable to the younger pups to the same extent as with the older pups. One could say, therefore that the maternal "condition" of the two groups was quite similar even during the 3rd and 4th weeks despite the fact that their behavior with their own litters was somewhat different.

What emerges from these two studies is the fact that the main outlines of the postpartum maintenance and decline, and many of the details of maternal behavior during the 4-week postpartum period, are quite similar (provided that they are presented with growing young) among females whose maternal behavior was initiated hormonally and those whose behavior toward pups arose through sensitization.

Differences between sensitized and postpartum mothers that are performing maternal behavior equally well in their home cages can be detected when both are tested for retrieval of pups from a T maze that extends 14 inches from the opening of the home cage (Bridges et al., 1972). Lactating mothers retrieved in 35–90% of the cases over the 6 days of testing, which started immediately after parturition, while sensitized females retrieved pups in fewer than 15% of the cases until the last test, in which the percentage rose to 40%. Sensitized mothers hesitated at the entrance to the arms of the T maze and only entered them on 1.2 days compared with entries on 4.6 days by lactating mothers. Given a choice between a live and a dead pup both kinds of mothers retrieved the live pup more often as their first choice.

These findings suggest that the hormonal onset of maternal behavior confers upon the postpartum mother behavioral characteristics that are absent when maternal behavior arises nonhormonally through pup stimulation alone. Several possible bases for differences were studied by Stern and MacKinnon (1976). First, they considered that they might arise either from the nursing, which is performed by lactating mothers but is absent in nonpregnant mothers, or from the effects of hormones either at the onset or during maintenance of maternal behavior. The T-maze performance of normal lactating mothers was compared with that of thelectomized mothers (that is, without nipples), and these, in turn, were compared to the performance of estrogen-induced maternal females (that is, 16-day-pregnant hysterectomized–ovariectomized females given a single injection of 100 μg/kg estradiol benzoate) and sensitized females.

The three groups that had been hormonally stimulated (intact and thelec-tomized postpartum females and estrogen-treated pregnant hysterectomized–ovariectomized females) had similar T-maze performances in nearly all respects (pup retrieval, maze entry, traversing maze), but sensitized mothers were deficient in all of these respects: no more than 30% of the females retrieved pups from the T maze, and nearly all showed some hesitation about entering the alley of the T maze. Short-latency sensitized females (0–3 days) performed no better than those that had long latencies (4–10 days).

In subsequent studies Stern (1977) and MacKinnon and Stern (1977) have explored further the basis for the differences in T-maze pup retrieval between lactating and sensitized mothers, focusing on the secretion of prolactin during lactation. Thelectomized mothers were given ergocryptine (CB-154) beginning 4 days prepartum continuing through the 11th day postpartum to block the release of prolactin (Stern, 1977). Females were observed for the display of maternal behavior with their own pups, and it was found that neither the prepartum blocking of prolactin release nor thelectomy prevented the onset of maternal behavior. In T-maze retrieving tests the ergocryptine-treated females performed as well as the control mothers: the same percentage of females retrieved pups from the T maze and their latencies were comparable to those of the untreated females. Thus, the difference between lactating and nonpregnant mothers in their T-maze performance is not based upon the continued secretion of prolactin during lactation.

In a second experiment (MacKinnon and Stern, 1977), pregnant–thelectomized females were hysterectomized–ovariectomized at 10, 16, and 21 days to determine the earliest time during pregnancy when hormonal influences on maternal behavior and T-maze retrieving could be detected. Three days after maternal behavior had been established, the animals were tested in the T maze and simultaneously tested in the home cage for retrieving to assure the continuation of this behavior. Females whose pregnancies were terminated after the 10th day, that is, on the 16th and 21st days, had higher percentages of T-maze retrievers than nonpregnant controls of an earlier study (Stern and MacKinnon, 1976). The 21-day-pregnancy-terminated females were nearly as proficient in T-maze retrieving as the sham-operated pregnant females that had given birth prior to testing. These findings suggest that by the 16th day of pregnancy females have been stimulated by hormones sufficiently so that upon terminating their pregnancies by hysterectomy–ovariectomy, a procedure that leads to relatively short-latency maternal behavior, they exhibit T-maze retrieving that approaches that of full-term postpartum females.

An interesting sidelight of this study is the characterization of T-maze behavior of the nonretrievers of the various groups, but particularly the 10- and 16-day-pregnancy-terminated females. These females entered the T maze very slowly or not at all, they hardly explored it once they had entered it, and they spent little time in the T maze after entering it. They displayed a fearfulness in the T maze that probably interfered with their retrieving, but they retrieved quite readily in the home cage. This raises the question of how closely related the behavior seen in T-maze tests is to maternal behavior and what significance with respect to maternal behavior can be attributed to failure to retrieve in this situation.

A recent study by Mayer and Rosenblatt (1979a) provides tentative answers to some of these questions. When nonpregnant females were tested in the T maze without pups present, *prior to* the beginning of sensitization, their emergence behavior (how rapidly and how far they moved into the maze) predicted how rapidly the T maze would be entered after sensitization but not how rapidly they would be sensitized. Those females that were less timid before sensitization also were less timid after sensitization, but timidity per se did not predict maternal responsiveness. Sensitized females that are reported not to retrieve pups in the T maze generally have been tested under conditions in which they must leave the familiar home cage and enter a well-lit maze to retrieve the pups. When, on the other hand, sensitized females were tested in a dimly lit room, we found that nearly all of the females retrieved pups from the T maze. Among these females, timidity, as expressed in emergence behavior prior to sensitization, did not predict T-maze retrievals after sensitization, since nearly all entered the maze and retrieved pups from it. It would appear, therefore, that timidity determines whether retrievals occur in the T maze when the conditions of testing are such that only the least timid females enter the maze and contact the pups in it. Therefore, T-maze retrievals under these conditions may measure an aspect of maternal responsiveness, namely, whether the female is capable of overcoming her normal timidity to enter the maze and retrieve the pup.

Under conditions that arouse less fearfulness, however, T-maze retrieving by sensitized females may measure something else. As an example; we tested sensitized (ovariectomized) females in the T maze in dim light and found that nearly all *entered* the T maze. Under these conditions, therefore, timidity about entering the maze did not interfere with the possibility of retrieving pups, and those females with short sensitization latencies were significantly more likely to *retrieve* pups from the maze than those with long sensitization latencies. Thus, when there is little emotional barrier to entering the T maze, differences in maternal responsiveness (that is, sensitization latencies) may determine what different females do among alternatives such as exploring the maze or retrieving a pup.

Returning to the original question posed by the earlier studies, the conclusions that can be drawn from the T-maze test about maternal responsiveness are limited unless the animals are tested before and after sensitization or the conditions are systematically varied since timidity varies among females and under different test conditions.

VI. Olfaction and Maternal Behavior

A. Olfaction and the Onset of Maternal Behavior

1. Nonpregnant Females

The nonpregnant female has proved to be a good subject for studying the role of olfaction in the onset of maternal behavior. Unlike the postpartum female in which the onset is rapid and is under the influence of hormonal stimulation, the nonpregnant female is almost entirely dependent upon sensory stimulation from the pups, and the onset is protracted over several days. During sensitization the female passes through several distinct phases. These have been described as an *initial avoidance* of pups that may last for several days and is characterized by exploratory sniffing of the pups then withdrawal from them, a growing *tolerance* of them when they contact her (that is, she stiffly remains in place), with occasional licking of them during the latter part of this phase, and a *sudden onset* of extensive licking, nest building, crouching over them, and retrieving if given the opportunity (Cosnier, 1963, 1965; Cosnier and Couturier, 1966; Fleming and Rosenblatt, 1974c; Terkel and Rosenblatt, 1971).

Nonpregnant females make an initial distinction between a pup and a small plastic toy the size of a pup when presented with both in their home cages (Plume *et al.*, 1968; Rosenblatt, 1975b). While both are investigated equally by sniffing, the toy often is bitten and never licked, while the pup is licked from the beginning. The toy elicits playlike behavior involving, in sequence, repeated picking up, biting, dropping, and picking up again, whereas when the pup is picked up, it is carefully carried to a corner, where it is deposited and left (that is, it is retrieved). The responses of nonpregnant females to a single pup indicate a differentiation and appropriateness of behavior that is obscured for a period when a group of 4–5 pups is presented, but which appears after the initial withdrawal subsides.

a. Olfactory Bulbectomy: Effects on the Onset of Maternal Behavior. Since olfactory investigation of newly introduced pups is the most prominent reaction of the nonpregnant female, olfaction has been studied extensively as it contributes to the onset of maternal behavior. Surgical removal of the olfactory bulbs (olfactory bulbectomy) was the first technique used to produce anosmia for this purpose. Schlein *et al.* (1972) removed the olfactory bulbs of inexperienced nonpregnant females and exposed them to 2-day-old pups, with the result that 80% of the females cannibalized the young within 24 hr, and none exhibited retrieving or other aspects of maternal care. Fleming and Rosenblatt (1974b) found that nearly 60% of the inexperienced, nonpregnant females cannibalized young (5–10 days old) when they had been bilaterally bulbectomized in one stage, but when the same operation was done in two stages, with a week intervening between removal of the first and second bulbs, none of the females cannibalized; the same results were obtained whether or not the females were exposed to pups during the week between bulb removals. Unilateral bulbectomy alone did not provoke cannibalism.

A further finding by Fleming and Rosenblatt (1974b) suggested that an alternative response to cannibalism, following olfactory bulbectomy, is the rapid onset of maternal behavior. Among the females that were bilaterally bulbectomized in one stage, nearly 60% displayed maternal behavior instead of (or alternately with) pup killing, and of these 70% initiated maternal behavior after less than 24 hr with the pups. None of the intact females had latencies of 24 hr, but nearly all of the females that were bilaterally bulbectomized in two stages, and exposed to pups after the second bulb was removed, exhibited maternal behavior in less than 24 hr (between 86 and 100%). Even unilaterally bulbectomized females exhibited short-latency maternal behavior in an unusually large number of cases (36%).

Some adult rats of both sexes cannibalize pups when first exposed to them, although they have not been bulbectomized. An analysis of this behavior among intact males by Paul and Kupferschmidt (1975) indicated that young rat pups (1–5 days of age) are regarded as food by male rats that have had no previous experience with pups, and are simply eaten rather than attacked. Ten-day-old pups sometimes are simply eaten, and sometimes are first killed by a well-directed biting attack and then eaten. In contrast, pup killing by olfactory bulbectomized animals has been described as disorganized, with multiple bites directed to all parts of the pup's body (Fleming and Rosenblatt, 1974b).

Pup killing after olfactory bulbectomy resembles mouse killing, which also has been reported to increase in incidence after bilateral bulbectomy in one stage (Didiergeorges et al., 1966; Hull and Homan, 1975; Karli, 1961; Karli et al., 1969; Spector and Hull, 1972; Vergnes and Karli, 1963; 1965). As with cannibalism, mouse killing occurs spontaneously among both males and females, although the percentages of adult rats that will kill mice (and rat pups) varies widely, depending upon the strain of rat and testing conditions. "Natural" mouse killing generally is regarded as predatory behavior, and is characterized by a stereotyped pounce and rapid bites to the back of the neck, which sever the spinal cord (Moyer, 1968); this differs, therefore, from bulbectomy-induced mouse killing. Olfactory bulbectomy has been found to convert a high proportion (50–100%) of animals that previously did not kill mice into killers (Karli, 1956). A number of investigators, however, have reported that the attacks on mice initiated after bulbectomy tend to be disorganized and inefficient, and therefore have proposed that this behavior is not predatory, but rather the result of irritability, which is seen under many circumstances in bulbectomized rats (Bernstein and Moyer, 1970; Cain, 1974; Douglas et al., 1969; Hull and Homan, 1975; Spector and Hull, 1972).

It seemed, therefore, that the cannibalism seen after bilateral olfactory bulbec-tomy also might be a consequence of irritability, whereas the facilitation of maternal responsiveness after bulbectomy might be the result of the sensory loss (anosmia). To test this hypothesis, Fleming and Rosenblatt (1974c) produced permanent anosmia in nonpregnant females by lesioning the lateral olfactory tracts, without removing the olfactory bulbs, and induced temporary anosmia in other females by an intranasal injection of zinc sulfate, which selectively de-stroys much of the olfactory epithelium (Alberts and Galef, 1971). Both treat-ments had a similar effect on the females' behavior; nearly all exhibited maternal behavior after less than 24 hr of exposure to pups, and continued to behave maternally as long as pups remained with them. Although 4 of 11 animals with lateral tract lesions (but no animal treated with zinc sulfate) cannibalized pups on first exposure, only one continued to kill for more than 2 days. These results indicate that the pup killing seen after bilateral olfactory bulbectomy is not simply a consequence of anosmia, since it does not occur in anosmic animals if the olfactory bulbs remain intact, and the data support the hypothesis that the loss of olfaction facilitates the initiation of maternal behavior.

Both the absence of cannibalism and the facilitation of maternal responsive-ness when pup exposure followed intranasal zinc sulfate were confirmed in further studies of nonpregnant animals by Mayer and Rosenblatt (1975, 1977, 1979b) and Mayer et al. (1979). Intranasal zinc sulfate reduced the latencies to become maternal not only of adult virgin females, but also of males (45–60 days of age), adult females ovariectomized before or after puberty, and, less reliably, immature, 30-day-old females.

The zinc sulfate technique, since it interrupts the olfactory system at the level of its receptors, has unique advantages in behavioral studies. However, it also has disadvantages: loss of olfactory function following single intranasal infusions is variable in extent, and recovery often takes only a few days (Mayer and Rosenblatt, 1975; Sieck and Baumbach, 1974; Slotnick and Gutman, 1977). Moreover, under some circumstances treatments may be followed by persistent mouth breathing, weight loss, and lethargy. A fuller discussion of the use of zinc sulfate in behavioral research, particularly in studies of maternal behavior, is included in the Appendix.

b. *The Olfactory Inhibition of Maternal Behavior in Nonpregnant Females.* Reducing olfactory input, then, by means of bilateral bulbectomy (performed in two stages), lesions of the lateral olfactory tracts, or intranasal infusion of zinc sulfate reduced the amount of pup exposure that is needed to elicit maternal be-havior from nonpregnant adult females and (by extension from zinc sulfate studies) from adult males and subadult (45-day-old) females and males. This suggests that the olfactory system plays a major role in inhibiting nonlactating animals from approaching and, in particular, from maintaining close contact with young pups. This inhibition significantly delays the onset of maternal behavior when continuous exposure to pups is enforced experimentally, and that probably precludes the sensitizaton of nonlactating rats under more normal conditions.

While intranasal zinc sulfate significantly reduces the sensitization latencies of subadult animals, adult males, and ovariectomized (adult) females, interrupting olfactory input has its most dramatic impact on the latencies of mature, nonpregnant females. In most studies, nonpregnant females treated with zinc sulfate or surgery to interrupt the olfactory system have required on the average no more than one day of pup exposure before becoming maternal, in contrast to sham-treated females that required 3.5–6 days (Fleming and Rosenblatt, 1974c; Mayer et al., 1979; Mayer and Rosenblatt, 1975, 1977). In a recent study in which we tested for the appearance of maternal behavior at hourly intervals for the first 8 hr after pup exposure was begun, we found that 80% of zinc sulfate-treated females became maternal within 6 hr.

The nature of the olfactory inhibition of close contact with pups is not yet understood. A plausible hypothesis is that nulliparous adult rats respond aversively to some pup odor or odors, and, by withdrawing from this stimulus, are prevented from responding to other attractive pup stimuli. This implies that the process of sensitization is largely a process of adaptation to the initially aversive odors, after which maternal behavior appears rapidly. The hypothesis is consonant with the earlier finding of Terkel and Rosenblatt (1971) that reducing the area of the female's cage will reduce the number of days of pup exposure before maternal behavior appears; that is, forcing the females into closer contact with the pups accelerates the onset of maternal behavior, presumably by enabling them to overcome more rapidly the initial tendency to withdraw. A similar process may be operative in the effect that periodic tail pinch has on sensitization latencies. In response to a series of brief tail pinches, administered over two days, nonpregnant females approach and lick pups, and their sensitization latencies then are shorter than those of non-tail pinched animals (Szechtman et al., 1977).

Despite the seeming plausibility of the hypothesis that the olfactory inhibition is mediated by aversive pup odors, there is as yet only indirect evidence that such odors exist. Certainly they have not been identified. It remains possible, therefore, that the withdrawal of nonpregnant animals from pups is not a response to specific odors, but rather depends upon some critical level of stimulation of the olfactory system or systems, perhaps acting in combination with nonolfactory, novel pup characteristics such as vocalizations or movements.

Recently Fleming et al. (1979) have shown that the vomeronasal or accessory olfactory system, as well as the primary olfactory system, plays a role in delaying the initiation of maternal behavior in nonpregnant females.

The vomeronasal (or Jacobson's) organs are paired tubular, fluid-filled structures lined with sensory epithelium, located within the nasal septum of many vertebrates. In rats the single opening of each organ is a small pore in the floor of the nasal chamber a few millimeters within the external nares. Sensory neurons from the vomeronasal organs join the vomeronasal nerves, which course caudally between the olfactory bulbs, synapsing in the accessory olfactory bulbs. The anatomy of this system in rodents and other mammals, including its pathway and projections within the central nervous system, have been described by Scalia and Winans (1975), who established for the first time that the primary and accessory olfactory systems project to different areas within the pyriform lobe.

Fleming *et al.* (1979) subjected nonpregnant female rats, before exposure to pups, to bilateral dorsoventral cuts into but not through the olfactory bulbs, transections of the vomeronasal nerves, or sham operations; some females sustained both types of lesion. An additive effect was found on subsequent latencies to become maternal. Vomeronasal nerve transections alone and partial olfactory bulb cuts alone significantly reduced latencies from about 8 days to 4 days, but when vomeronasal nerves were cut and lesions were made in the primary olfactory bulbs as well, latencies were reduced to 1–2 days.

In our own laboratory we have been able to occlude the vomeronasal organs by cautery, a procedure that does not impair the functioning of the primary olfactory system (Mayer and Winans, unpublished observations). Transecting the vomeronasal nerves, in contrast, inevitably damages the olfactory bulbs to some extent. Since intranasal zinc sulfate apparently does not damage the vomeronasal organs (see the Appendix), we were able to use the two procedures in combination to study the involvement of both systems in the initiation of maternal behavior (Mayer and Rosenblatt, unpublished observations). We found that latencies to become maternal were significantly reduced in nulliparous females after either vomeronasal occlusion or intranasal zinc sulfate, compared with the latencies of females receiving only sham procedures. It was not clear, however, that one procedure was more effective than the other, and we were particularly interested to find that females that had received both vomeronasal occlusion and intranasal zinc sulfate did not have latencies that were shorter than females treated with zinc sulfate alone; the median latency in each group was 1 day. Further studies in this area are in progress.

2. *Hormonal Onset of Maternal Behavior: Amelioration of*
   Bulbectomy Effects

In inexperienced nonpregnant females, bilateral olfactory bulbectomy produces cannibalism in 60–80% (Fleming and Rosenblatt, 1974b; Schlein *et al.*, 1972). However, the same operation performed before mating or during pregnancy appears to have less of an effect on the onset of maternal behavior postpartum. Schlein *et al.* (1972) found that bulbectomy on the 22nd day of pregnancy less than 24 hr before parturition did not disrupt the normal onset of maternal behavior when pregnancy was terminated by Caesarean section at the same time, and the females were given foster pups 24 hr later. Apparently the hormones that stimulate maternal behavior can override the effect on the female of the irritability normally associated with this operation (which also may not develop fully within 24 hr).

On the other hand, Benuck and Rowe (1975) and Schwartz and Rowe (1976) compared bilaterally bulbectomized virgin females that were mated 2 weeks after the operation, unilaterally bulbectomized, and intact females, and found that the bilaterally bulbectomized animals were more apt to cannibalize their pups at parturition, and had more dead pups at the end of the first 24 hr, apparently as a result of deficiencies in maternal attention during this period.

Neonatal bulbectomy, which does not lead to heightened irritability in the adult animal, does not appear to affect maternal behavior (Pollack and Sachs, 1975).

3. Reinstatement of Maternal Behavior in Experienced Bulbectomized Females

In contrast to the deficits in maternal responsiveness shown by some bulbectomized females at their first parturition, females that have had one or more litters prior to removal of the bulbs do not appear to differ in any parameter of maternal behavior from sham-operated or unoperated multiparous females (Schlein et al., 1972; Schwartz and Rowe, 1976).

There is, in fact, a strong tendency for females on reexposure to pups to repeat the same pattern of behavior that they had manifested on first exposure, whether it be cannibalism or maternal behavior, and whether the females are nulliparous or have previously given birth. Fleming and Rosenblatt (1974b) mated three nonpregnant females that had cannibalized pups after bilateral bulbectomies, and three that had exhibited short-latency maternal behavior. Cannibalism reappeared at parturition among all three of the females that had cannibalized while nonpregnant, and none displayed maternal behavior. Females that had initially exhibited maternal behavior instead of cannibalism, in contrast, all were maternal toward their newborn pups.

Schlein et al. (1972) bulbectomized primiparous females 1 month after they had weaned their litters and then exposed them to young pups. Only 1 of 10 females cannibalized the pups, while 4 exhibited short-latency maternal behavior; sham-operated controls exhibited the same distribution of responses to pups.

Ongoing maternal behavior in sensitized nonpregnant females also is not disrupted by olfactory bulbectomy (Schlein et al., 1972). All but one female (of 14) continued to display maternal behavior after bulbectomy, and the one that cannibalized several pups then resumed maternal behavior. Thus, it appears that maternal behavior in the female rat, once established either by exposure to pups or hormonally at parturition in interaction with pups, is resistent to disruption and readily reinstated.

4. Olfactory Influences at the Onset of Maternal Behavior

Although it would *seem* that olfaction is not involved in the onset of maternal behavior at parturition, since olfactory bulbectomized females are not prevented from exhibiting maternal behavior, it may be that this is not an adequate procedure for studying this question. Our studies on virgin females show that bulbectomy, when it does not result in cannibalism, causes a rapid onset of maternal behavior; all other procedures we have used to reduce or eliminate olfactory sensitivity have the same effect. If the olfactory system (or systems) inhibits the initiation of maternal behavior in the nonpregnant animal, as these studies suggest, then it may be that the hormones and/or the predictable experiences of pregnancy act to ensure the rapid appearance of maternal behavior in part by reducing this inhibition.

We have recently completed a preliminary study which indicates that, even when nonpregnant females are hormonally stimulated to show short-latency maternal responsiveness, the olfactory system impedes the onset of maternal behavior (Mayer and Rosenblatt, unpublished observations). A group of 23 nonpregnant females were ovariectomized, hysterectomized, and given 100 μg/kg of estradiol benzoate to stimulate maternal behavior (Siegel and Rosenblatt, 1975b). Before the first test, 13 of the females were treated intranasally with zinc sulfate to reduce olfactory sensitivity, while the remaining females were injected intranasally with air. At 48 hr after surgery and estradiol benzoate injections (24 hr after intranasal treatments) the females were presented with 4 3-8-day-old pups; 5 of the females with reduced olfaction responded immediately, and 7 more responded within 7 hr. (The 13th female failed to become maternal within 5 days, after which pup exposure was terminated.) Females treated intranasally with a sham injection of air showed the usual distribution of latencies to become maternal, which ranged from 5 hr to 5 days (and averaged 2.5 days). Thus, reducing olfaction facilitated the onset of maternal behavior stimulated by estradiol benzoate. This, of course, does not rule out the possibility that estrogen and/or other hormones act directly to alter the olfactory inhibition of maternal behavior prior to parturition, although a single injection of estradiol benzoate, in combination with ovariectomy and hysterectomy, apparently leaves the olfactory inhibition largely intact. Another possibility, suggested first by Birch (1956) and later by Noirot (1972), is that prepartum experience with vaginal odors obtained by self-licking (Roth and Rosenblatt, 1967) adapts females to the odors that they then encounter in their initial contacts with pups, so that they are enabled to respond to other pup stimuli that promote maternal behavior. It is not unlikely that hormonal and experiential factors work together to ensure that postparturient females care for their young without delay.

B. OLFACTION AND THE MAINTENANCE OF MATERNAL BEHAVIOR

1. Ongoing Effects of Anosmia

Benuck and Rowe (1975) and Schwartz and Rowe (1976) observed the maternal behavior of primiparous females that had been bilaterally or unilaterally bulbectomized, and compared it with that of normal primiparous females. Bulbectomies were performed 2 weeks before mating, and the females were observed from parturition through Day 16 of lactation. In addition, Benuck and Rowe (1975) treated females intranasally with zinc sulfate on Day 20 of gestation, and observed them through Day 5 of lactation, a period during which there was clear evidence of olfactory deficit. As mentioned earlier, bilaterally bulbectomized females, compared with sham-operated and unilaterally bulbectomized females, had more dead pups 24 hr after parturition, were markedly deficient in retrieving scattered pups on a first retrieval test 24 hr after birth, and retrieved fewer pups during alternate-day tests throughout the period of observation. Schwartz and Rowe (1976) also found that the bilaterally bulbectomized females spent less time on their nests. In both studies, pup weights on Days 11–15 were significantly less if the mothers had been bilaterally bulbectomized. Females treated intranasally with zinc sulfate, on the other hand, were different in only one parameter from saline-treated control females; although their latencies to retrieve scattered pups were as short, they consistently retrieved fewer pups during (daily) 30-min tests. The greater deficits found among bulbectomized than among zinc sulfate-treated mothers suggest that the anosmia produced by both treatments is not the main cause of disordered maternal behavior after bulbectomy. The observation that even peripherally hyposmic females retrieve fewer pups than females with intact olfaction may reflect the motivating and directing roles of pup odors in normal retrieving behavior. Smotherman *et al.* (1978) have found that pup odors in the absence of pup vocalizations do not cause lactating females to select the arm of a Y maze that contains a pup, but that the combination of pup odors and pup ultrasonic vocalizations is more likely to lead to correct choices.

The studies of anosmic mothers by Benuck and Rowe (1975) and Schwartz and Rowe (1976) supplement two earlier studies in which the maternal behavior of bulbectomized females was observed on a short-term basis only. Beach and Jaynes (1956) did not find that multiparous bulbectomized females were significantly different from intact controls in retrieving three pups during a single test. Herrenkohl and Rosenberg (1972) bilaterally bulbectomized females two days before delivering their pups by Caesarean section, and compared their behavior with foster pups to the maternal behavior of intact females that also had been made maternal as a result of Caesarean delivery late in pregnancy. The bulbectomized females differed from the intact females only in spending less time retrieving. Apparently the bulbectomized female, if she has not had prior maternal experience, displays deficits in maternal behavior during or shortly after parturition, which later are overcome to a degree, although more subtle long-term effects of bulbectomy become evident in a somewhat slower rate of weight gain by her pups as compared with the pups of normal mothers.

The reduced amount of time spent in the nest by bulbectomized mothers (Schwartz and Rowe, 1976) may be a consequence of impairment of internal temperature control at the external temperature at which they were maintained (23°C); Söderberg and Larsson (1976) have reported that males bulbectomized at 30 days of age maintain a significantly higher core temperature at 22°C than sham-operated controls. Studies by Leon et al. (1978) on thermoregulatory control of nest occupation in maternal females lead to the prediction that periods of nest occupation will be shorter at higher maternal core temperatures. Söderberg and Larsson found that peripherally induced anosmia had the same effect as bulbectomy on male core temperature, whereas we noted earlier that zinc sulfate-treated mothers did not spend less time on their nests than saline-treated females (Benuck and Rowe, 1975). This difference in results, however, may arise from the more complete surgical deafferentation performed by Söderberg and Larsson (1976).

2.   Olfaction and the Elicitation of Maternal Behavior

Olfactory stimuli appear to play important roles in eliciting and directing maternal behavior during the normal course of maternal care. Smotherman et al. (1978) have reported that pup odors and ultrasonic calls by pups interact to produce localization and retrieval of pups that are out of the nest. Odors are not effective directional cues, whereas ultrasounds lead to correct localization, but odors increase the speed of searching by the female. In retrieving tests in which the pups could not be viewed directly by the mother, differential responding to the arms of a two-alley choice apparatus was increased when odor cues were added to ultrasounds.

Charten et al. (1971) have implicated olfaction in the mother rat's licking of the pups' perineal region during the first 17 days following parturition. Females that were bulbectomized after parturition showed a significant reduction, but not complete elimination, of perineal licking. The secretions from this area are attractive to rats of both sexes, as shown by the fact that they prefer a water solution including substances wiped from the perineal regions of pups to pure water (Charten et al., 1971). Fleming and Cummings (unpublished observations) have noted essentially the same phenomenon.

3.   Olfaction and Other Aspects of Maternal Care

Olfaction plays a role in other aspects of maternal care than the elicitation of behavior directed at the pups. One of these is milk production required for nursing, which is under the control of a complex of hormones, which includes prolactin and adrenocorticotrophic hormone. Suckling is the initial stimulus that causes the release of the hormones of the lactogenic complex, but Grosvenor (1965), Grosvenor et al. (1967), and Mena and Grosvenor (1971) have reported that pups placed under mothers separated from her by a wire mesh cage floor are capable of causing the release of prolactin from the anterior pituitary in amounts equal to that caused by suckling over a 30-min period. Pups therefore are able to overcome the inhibitory effects on prolactin release caused by prolactin inhibitory factor. Olfaction appears to be the principal sensory route through which pups exert their effect on prolactin release (Mena and Grosvenor, 1971). Blocking olfactory, visual, and auditory stimuli either singly or in combinations, either at their source or by surgery on the females, gave results which indicated that olfactory stimuli were most effective in stimulating release of prolactin, but visual stimuli could serve at times when olfaction was blocked.

The role of olfaction in releasing prolactin only emerges between the 7th and 14th day in primiparous mothers but it continues from then on until weaning and it is retained through a second litter, appearing early during the care of that litter (that is, 7th day).

Zarrow et al. (1972) have reported exteroceptive release of adrenocorticotrophic hormone measured by an increase in plasma corticosterone levels in lactating rats after exposure to pups in the absence of suckling. Olfactory bulbectomy eliminated this response to pups as did preventing pup odors from reaching the mothers by enclosing the pups in a sealed transparent container. Since corticosterone contributes to milk production, as does prolactin, olfaction plays an important role in maternal care through its regulation of lactation in conjunction with the stimulation provided by suckling.

VII. Developmental Aspects of Maternal Behavior

Until recently only postpartum females exhibited maternal behavior reliably in the laboratory, but with the discovery of pup-induced maternal behavior and its nonhormonal basis it has been possible to elicit maternal behavior from adult males as well as from females, and more recently it has been found that juveniles of both sexes can be sensitized to exhibit maternal behavior (Bridges et al., 1974). This has opened up the possibility of studying the ontogeny of maternal behavior from its early appearance through adulthood.

A. Origins and Early Onset of Maternal Behavior

In our first experiments (Mayer and Rosenblatt, 1979b) we studied the early appearance of maternal behavior by observing the responses of prepubertal 18–

30-day-old juveniles to pups less than a week old. We found that as early as 18 days of age juveniles of both sexes were attracted to 3–5-day-old pups when they were placed together in a testing arena, apart from their littermates and mothers. They exhibited sniffing, approach, licking, forepaw manipulation, and lying in contact with the pups, the latter resembling the licking and lying-in-contact behavior of maternal adult rats. The frequency of contacting pups and exhibiting one or another of these responses toward them increased between the 18th and 22nd day, then decreased sharply on the 24th day; only 20% or less of the 15-min test was spent with pups by 24-day old juveniles, compared with 75–95% of the test by 20–22-day-olds (Fig. 26).

The older juveniles tended to avoid the pups after initially exploring them. By exposing juveniles to pups initially before the 24th day, however, the development of avoidance could be delayed several days: the juveniles tended to maintain their initial level of responding to pups when retested at 4–6-day intervals, even when they reached the age of 24 days during an interval between tests (Fig. 27). This effect of prior experience was short lived, and by 30 days of age these juveniles also declined in their responsiveness to pups.

When daily exposure to pups was extended from 15 min to 22 hr (allowing the preweanlings 2 hr per day suckling with their own mothers) 22-day-old young became sensitized and exhibited retrieving, nestbuilding, and lying in contact

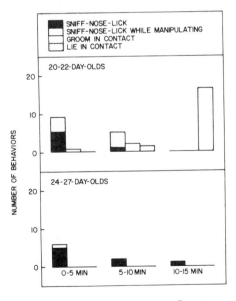

FIG. 26. Distributions of pup contact scores during first pup exposure in groups of differing age. (From Mayer and Rosenblatt, 1978b.)

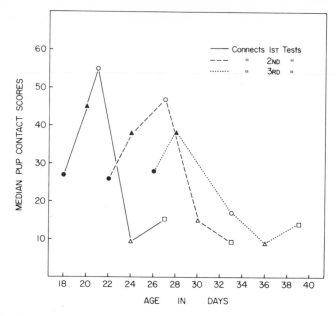

FIG. 27. Median pup contact scores at first, second, and third observations. Identical symbols identify repeated observations of the same group. (From Mayer and Rosenblatt, 1978b.)

with pups often draped over the pups as in nursing. About 50% of both males and females were sensitized, with an average latency of 1 day. Bridges *et al.* (1974) reported a similar short latency for sensitizing juveniles at 24 days, using a different strain of rats (Purdue-Wistar).

The positive responses of 20–22-day-old juveniles to young pups were highly discriminant ones, and when their responses changed at 23–24 days, they were also highly discriminant. At the earlier age juveniles did not respond simply to the warmth of pups but rather to their activity: they preferred live pups to either an empty warm nest bowl or dead pups, and they showed a preference for more active 8-day-old pups over 2-day-old pups. The 8-day-old pup was, in fact, more attractive to them than their own agemate but a 2-day-old pup was not. Juveniles that were 24 and 25 days of age continued to prefer live pups over an empty warm nest bowl, but they preferred agemates to the younger pups.

Our studies have led us to speculate that maternal behavior has its origin in juveniles as a specialized form of gregariousness—a gregariousness that also is evident in their huddling with agemates (Alberts, 1978a,b; Alberts and Brunjes, 1978; Cosnier, 1965) and with the mother. Indeed gregariousness continues to be a conspicuous characteristic of rats throughout life (Latané, 1969; Latané *et al.*, 1972; Latané and Steele, 1975). The pup stimuli to which juveniles respond with contact seeking seem to include movement, vocalizations, warmth, and the smells and tastes of their anogenital secretions, which Charten *et al.* (1971) and Fleming and Cummings (unpublished observations) have shown are attractive to adult rats of both sexes. As noted, the effectiveness of these stimuli in eliciting approach rapidly declines in 24-day-old juveniles, however, and the pups begin to elicit avoidance. Avoidance grows stronger with age and is present in most adults at their initial exposure to a litter of pups. One can, however, elicit approach, sniffing, and licking and even retrieving of a single pup by an adult that is presented for the first time with a pup and a small plastic toy (Rosenblatt, 1975b).

B. CHANGES IN MATERNAL RESPONSIVENESS AND THE EMERGENCE OF SEX DIFFERENCES

Between 22 and 30 days of age the increase in avoidance of pups upon initial exposure to them resulted in an increase in sensitization latencies (Bridges *et al.*, 1974; Mayer *et al.*, 1979). The change occurred equally in males and females and latencies in Charles River CD (Sprague-Dawley) rats rose to a median of 6 days from the earlier median latency of 1 day. A clear sex difference emerged between prepuberty at 30 days and postpuberty at 45 days; females showed a decrease in sensitization latency from their earlier long latency, and maternal behavior appeared after 3.5 days of pup exposure, while males showed an increase and their median latency rose to 10 days. From then until 90 days female latencies remained low (around 4 days) while male latencies remained high (10 days or longer), and a large percentage of 90-day-old males (up to 67%) failed to show maternal behavior during 2 weeks of pup exposure. Over 90% of females were sensitized during this same period, and their latencies continued at about 3.5-4 days of pup exposure at least through 146 days of age.

Adult males of the Long-Evans strain also have been found to have a lower level of maternal responsiveness than females (McCullough *et al.*, 1974; Quadagno *et al.*, 1973; Quadagno and Rockwell, 1972). Some strains, however, may show less dimorphism; within a Purdue-Wistar strain, for example, Bridges *et al.* (1974) found that the sensitization latencies of males and females were about the same at least through 54 days of age.

C. TIMIDITY, FEAR OF NOVELTY, AND OLFACTORY AVERSION INFLUENCING THE DEVELOPMENT OF MATERNAL RESPONSIVENESS

The appearance of avoidance reactions to young pups in both Charles River CD and Purdue-Wistar rats at around 23-24 days of age suggests that emotional responses to younger pups play an important role in the sensitization process. An increase in sensitization latencies during development may indicate that juveniles have become more timid and fearful of the novel stimuli encountered when young pups are placed in their cages (that is, the test arena), or that they more readily avoid particular stimuli, for example, pup odors; both of these changes may occur and interact with one another. Also, weaning normally occurs between 21 and 28 days, and during this period the olfactory characteristics of the mother, the nest, and the juveniles themselves are modified so that the contrast between nest–sibling–self odors and the young pup odors must become sharper. (Complete weaning, however, is not necessary for the first appearance of avoidance reactions to young pups, since in our experiments we saw pup avoidance by 24- and 25-day-old juveniles that were still suckling.) Moreover, olfactory stimuli become increasingly important in the behavior of rat juveniles from the 16th day onward (Alberts, 1978b), and their capacity to make fine olfactory discriminations increases at around 25 days of age (Leon and Behse, 1977).

We have used two procedures to investigate the roles of timidity and olfaction in the developmental changes in maternal responsiveness from 30 to 90 days of age: (1) brief daily handling (from the 21st day onward), which reduces timidity, and (2) intranasal infusion of zinc sulfate. At 30, 45, 60, and 90 days of age, males and females that had been handled were compared with nonhandled animals for differences in sensitization latencies. At several of these ages, animals were treated with zinc sulfate one day before pup exposure began, and some groups received both treatments before they were sensitized (Mayer et al., 1979).

Brief daily handling has been shown to increase activity and exploration, and to decrease defecation, in an open field, particularly when begun at about the time of weaning (Bronstein, 1972; Thompson and Lippman, 1972; Wild and Hughes, 1972). These changes are in the direction of lowered fearfulness or timidity. In our experiments, handling significantly reduced the sensitization latencies of both males and females tested at 30 days of age (Fig. 28), but only males continued to be affected by handling at 45 days of age; there was a minimal effect in males at 60 days and no effect at 90 days of age. Females were more affected than males by brief handling when they were tested at 30 days of age (latencies of 2 versus 4.5 days), but when males were given an additional amount of handling (5 min per day versus a few seconds), their latencies were further reduced and they became as responsive as the females, whereas intensive handling did not further reduce the latencies of the females.

Intranasal infusion of zinc sulfate has been used to investigate the contribution of olfactory sensitivity in this context (see the Appendix). At 30 days of age zinc sulfate infusion 24 hr prior to pup exposure did not significantly reduce sensitization latencies, although treated females had shorter latencies than those of sham-treated females (Fig. 28). By 45 and 60 days, however, blocking olfaction with zinc sulfate had a dramatic effect on female latencies, which were reduced from 3.5 or 4 days to 1 or 2 days of pup exposure. Males were not given zinc sulfate

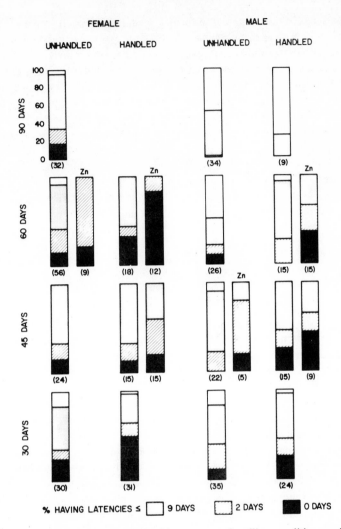

FIG. 28. Percentages of groups defined by sex, age, handling condition, and intact or impaired olfaction (Zn) that became maternal with latencies of 0, 2, or less, and 9 or less days. Numbers in parentheses represent sample sizes. (From Mayer *et al.*, 1978.)

infusion at 30 days of age. At 45 and 60 days, zinc sulfate reduced male latencies, but the effect was marked only when the intranasal treatment was combined with handling.

These studies have led us to believe that the initial increase in latencies for sensitization between 24 and 30 days of age may be based largely upon an increase in timidity or reactivity to fear stimuli, and that later increases may be based more specifically upon an olfactory-mediated aversion to pups. In developmental studies of open field behavior the first increase in defecation, presumably an automatic reaction accompanying fear, has been found at about 24 days of age, and defecation increases sharply between 24 and 30 days (Candland and Campbell, 1962). Similarly, reactivity to strong light and to noise in the open field have been reported to commence between 24 and 45 days of age (Livesey and Egger, 1970). Bronstein and Hirsch (1976) have found 30-day-old rats to be relatively unresponsive to stimuli that provoke defensive reactions in animals 30 days of age and older, including mild foot shock, a caged cat, and a suddenly moving object. Thus, the absence of an aversive reaction to pups by juveniles younger than 24 days of age, and its relatively sudden appearance around 24 days, parallels the development of reactivity to a wide variety of stimuli that generally arouse fearful or aversive responses in adults.

The olfactory inhibition of maternal behavior has been discussed in detail earlier (see Section VI,A).

D. HORMONAL INFLUENCES DURING THE DEVELOPMENT OF
    MATERNAL RESPONSIVENESS IN FEMALES

Although sensitization through pup exposure can occur without hormonal stimulation, nevertheless, hormones may play a role in determining the characteristic level of maternal responsiveness of the adult, as reflected in the sensitization latency. The postpubertal fall in sensitization latencies among females, and the rise among males, suggests that ovarian hormones, and perhaps also testicular hormones, begin to influence maternal responsiveness at puberty. In a series of studies we have begun to investigate this problem, chiefly with reference to females (Mayer and Rosenblatt, 1979a).

In our first experiments we attempted to dissociate chronological age and gonadal maturation, in order to determine whether the commencement of ovarian cycling at puberty has an effect on sensitization latencies that is independent of other aspects of maturation. First, we advanced puberty by injections of gonadotrophins (pregnant mares' serum); the treated females experienced their first ovulation and began cycling by 26–28 days of age, or about 10–12 days earlier than usual. They were exposed to pups when they were 30 days of age, and were compared to saline-injected females of the same age that had not yet ovulated. In the next experiment we prevented puberty by ovariectomizing females on their 21st day, and compared their sensitization latencies when they were 45 and 60 days old to those of sham-operated females that by these ages had attained reproductive maturity. Knowing that handling sharply reduces the sensitization latencies of prepubertal, 30-day-old females but does not affect those of postpubertal females, we observed both the prematurely pubertal and the nonpubertal ovariectomized females under handled and nonhandled conditions.

We found that both advancing puberty and preventing puberty had effects on sensitization latencies: the nonhandled 30-day-old females that were prematurely cycling had sensitization latencies that were between the longer 6-day latency of (nonhandled) prepubertal 30-day-olds and the shorter 3.5-day latency of (nonhandled) postpubertal 45-day-olds, and were not significantly different from either. Handling, however, did *not* reduce their sensitization latencies, and in this respect they were like 45- and 60-day-old females and unlike prepubertal females of their own age. Ovariectomy on the 21st day, on the other hand, prevented the decrease in latencies that normally occurs after puberty; nonhandled ovariectomized females at 45 and 60 days of age had sensitization latencies of 10 and 12 days of pup exposure, which were significantly longer than the 3.5–4-day latencies of intact, cycling females at those ages. Handling, however, *reduced* the latencies of the ovariectomized females tested at 45 and 60 days to about the level of intact females, much as handling reduces the sensitization latencies of prepubertal 30-day-olds. These findings suggest that ovarian secretions, which rise sharply at puberty, are an important factor in establishing the relatively short sensitization latency that characterizes the adult female.

In our next experiments we attempted to determine whether ovarian hormones are necessary to maintain as well as to establish the reduced sensitization latencies of cycling females. We ovariectomized females prepubertally at 21 days of age, postpubertally at 60 and 90 days of age, and at two ages within the pubertal transition period: 32 days of age when a few females are entering puberty but most are prepubertal, and 42 days when most have entered puberty but a few are still prepubertal.

Leon et al. (1973) have reported that adult females ovariectomized and exposed to pups 8 weeks later have sensitization latencies that are *lower* than the latencies of either intact females or females tested 4 weeks after ovariectomy. Moreover, these experimenters were able to reestablish relatively long latencies in 8-week-ovariectomized females by giving them daily injections of estradiol benzoate for 7 days before commencing pup exposure, leading them to conclude: (1) that maternal behavior in the intact nulliparous females is inhibited by estrogen and/or other steroids (Leon et al., 1975), and (2) that nearly 8 weeks are required after ovariectomy for the neuroendocrine status of the female to restabilize, and thus for the waning of estrogen inhibition of maternal behavior to become apparent. In order to be able to compare our findings to theirs, and to avoid testing during a period of possible neuroendocrine disequilibrium, we delayed sensitization tests in our experiments for 8 or more weeks after ovariectomy.

In these studies we broadened our sampling of sensitized maternal behavior. In addition to obtaining sensitization latencies (in days of pup exposure), we measured the compactness of the nests that the females constructed around themselves and the pups after becoming sensitized, and, for purpose of comparison, the compactness of the sleeping mats built before pup exposure was begun. Several groups ovariectomized both pre- and postpubertally were given an additional maternal behavior test after sensitization, in which a T maze containing test pups in each arm was added to the home cage of the sensitized female and she was allowed a series of daily 10-min trials to determine whether she would enter the maze and retrieve the pups back to the home cage (Bridges *et al.*, 1972). Again for purposes of comparison, we pretested the females for T-maze emergence behavior before beginning pup exposure by attaching an empty maze to each home cage for 10 min.

Results of these experiments established that ovarian hormones are necessary to maintain the relatively *short* sensitization latencies of adult females. Ovariectomies as early as 21 days and as late as 90 days of age increases sensitization latencies when pup exposure was begun 8 or more weeks after the operation. Moreover, the maternal behavior of ovariectomized females after sensitization was qualitatively somewhat different from that of cycling females; ovariectomized females built less compact nests around themselves and the foster pups, and a much smaller percentage of ovariectomized females retrieved pups from the T maze (32% versus 94%), although they entered the maze to the same extent as intact females, both before pup exposure and after sensitization.

In the next phase of these studies we asked whether injecting long-term-ovariectomized females with estradiol benzoate would reduce their sensitization latencies and affect their maternal nest building and T-maze retrievals. We injected three groups of females, ovariectomized at 32, 60, and 90 days of age, with 100 μg/kg of estradiol benzoate 24 hr before exposing them to pups, 8 or more weeks after ovariectomy. To duplicate the procedures of Leon *et al.* (1973) we injected additional females, ovariectomized at 90 days and tested 8 weeks later, with 12 μg of estradiol benzoate per day for 7 days prior to sensitization (which began 24 hr after the last injection).

The sensitization latencies of females ovariectomized at 32 and 60 days and tested at 120 days of age were reduced by a single injection of estradiol benzoate to the same level as the median latency of sham-operated, cycling females. Also, within these groups the percentages of animals retrieving from the T maze were increased following estradiol benzoate, although there still were fewer retrievers than among estrous cycling females. Among females ovariectomized at 90 days and tested 8 weeks later, a single estradiol benzoate injection had little effect on latencies; however, the ability of estradiol benzoate to alter the sensitization behavior of such females was demonstrated by a marked reduction in latencies among those that received seven daily injections. This result was in the opposite direction to that reported by Leon *et al.* (1973). Moreover, 75% of these females retrieved pups from the T maze, a higher percentage than in any other ovariectomized, estradiol benzoate-treated group. Maternal nest building also appeared to be altered by the estradiol benzoate injections; the scores (reflecting compactness) of sensitized estradiol benzoate-treated females were between those of oil-injected ovariectomized females and (oil-injected) cycling females. Thus, estrogen appears to play a major role in maintaining the relatively *short* sensitization latencies of adult cycling females; although ovariectomy lengthens sensitization latencies, responsiveness to pup stimulation can be reestablished to a significant degree by injecting estradiol benzoate, even 9½ weeks after the operation.

Many questions remain unresolved. For example, we do not know how soon after ovariectomy maternal responsiveness begins to wane. In previous experiments we did not find ovariectomized females to have longer latencies than intact females when sensitization was begun up to 2 weeks after the operation; however, T-maze tests might have revealed a difference that was not evident in sensitization latencies alone. Also, we would like to know what doses of estradiol benzoate, and in what temporal relation to pup exposure, will best restore the maternal responsiveness of ovariectomized females to the level of cycling females.

E. HORMONAL INFLUENCES ON THE DEVELOPMENT OF MATERNAL
    RESPONSIVENESS IN MALES AND THE PROBLEM OF SEXUAL
    DIFFERENTIATION IN THE DEVELOPMENT OF MATERNAL
    BEHAVIOR

Although prepubertally ovariectomized females have sensitization latencies at 45 and 60 days that are as long as those of similar aged males, and they respond to handling and reduced olfaction in the same way, at 90 days these females are more responsive than 90-day-old males. Males from 45 days onward are significantly less responsive to pups than females, and castration at 25 days or 45 days does not alter their adult latencies (Mayer and Rosenblatt, unpublished; McCullough *et al.*, 1974; Quadagno and Rockwell, 1972; Rosenberg and Herrenkohl, 1976).

There is some evidence that males also are less responsive to hormone treatments that stimulate maternal behavior in females. Lubin *et al.* (1972) found that an 11-day regimen of estradiol benzoate, progesterone, and prolactin injections, which reduced sensitization latencies from 6 to about 2 days in females, had no effect in castrated males, but doubling the dosages was somewhat effective: latencies were shorter than in untreated males but were longer and more variable than in females given the lower doses. On the other hand, Bridges *et al.* (1973), using a somewhat different schedule of the same hormones and a different strain of rat, were able to effect nearly the same reduction of sensitization latencies in castrated males as in females. Koranyi *et al.* (1976) injected maternal plasma into males and found them equally responsive as females: both exhibited maternal behavior in about 18 hr.

It has been suggested that maternal responsiveness in the rat differentiates in a masculine or feminine direction under the influence of perinatal hormones, particularly androgens, which, of course, have been shown to profoundly affect adult sexual behavior. In support of this hypothesis, neonatally castrated males of Long–Evans and Sprague–Dawley strains are more likely to show maternal behavior in response to pup stimulation than intact (sham-operated) males when both groups are tested as adults (Long–Evans: McCullough et al., 1974; Quadagno and Rockwell, 1972; Sprague–Dawley: Rosenberg and Herrenkohl, 1976). Bridges et al. (1973) did not find that neonatally castrated males of their strain sensitized more rapidly than males castrated as adults, but established that the two groups responded somewhat differently in one respect to hormone treatment; after hormonal induction of maternal behavior, neonatally castrated males were more likely to retrieve pups from a T-maze extension of the home cage than were males castrated as adults. It appears likely, therefore, that androgens secreted by the testes of the infant male act in some way or ways to reduce the likelihood that he will respond maternally to pups as an adult. The strength of the effect, however, seems to be small compared to the organizing effect of perinatal androgens on sex behavior. Also, the period during which androgens must be present to reduce maternal responsiveness probably is less circumscribed, since Rosenberg and Herrenkohl (1976) have found that males castrated as late as Day 10 are somewhat more responsive to pup stimulation as adults than are sham castrates.

Attempts to influence the maternal responsiveness of females in a masculine direction by administering testosterone during the perinatal period have yielded mixed results. Bridges et al. (1973) treated newborn females with testosterone propionate and tested them for maternal behavior as adults, following ovariectomy and a 21-day regimen of hormone injections. Compared with nonandrogenized females, testosterone propionate-treated females became maternal as rapidly in the home cage, but a significantly smaller percentage went on to retrieve pups from a T maze (6.5% of testosterone proprionate-treated females, compared with 42.9% of oil-injected controls). In this respect androgenized females were unlike males castrated at birth; the same proportion of the latter retrieved from the maze as nonandrogenized females (adult castrate males, 11.6%; neonatal castrate males, 42.9%; females, 48.0%).

Quadagno et al. (1973) and Quadagno et al. (1974) were unable to influence the maternal responsiveness of adult females by giving testosterone neonatally (100 μg or 1 mg, on Day 1 or 4), although the effects of the testosterone propionate treatment were evident in lower levels of sexual receptivity, after priming with estrogen and progesterone, and in the occurrence of polyfollicular ovaries. More recently, Quadagno and co-workers (1977) administered testosterone propionate to pregnant females (Days 16–21 or gestation), and found that the female offspring were virilized in external genitalia, but were not different from the female offspring of oil-treated controls in maternal or sex behavior. On the basis of these data, the authors have proposed that in the female rat a maternal mediating system differentiates under hormonal controls prior to Day 16 of gestation, but that the maternal mediation system in the male remains sensitive to hormonal influence for a longer period extending into early postnatal life.

We do not consider that the question of whether there is sexual dimorphism in the neurophysiological substrate of maternal behavior has been answered definitively. It would appear that the differences between females and males in sensitization latencies and other aspects of sensitized maternal behavior are based in part on differences between females and males in various kinds of nonmaternal behavior. A number of kinds of sexually dimorphic behavior have been identified in emergence tests, open fields, avoidance learning, and exploration tests. Females generally are more active and exploratory (Archer, 1975), they are faster to learn an active avoidance task (see Quadagno et al., 1977), possibly slower to learn passive avoidance (Denti and Epstein, 1972), and faster to extinguish a passive avoidance response (Beatty et al., 1971, 1973). Several of these types of behavior have been shown to differentiate, at least in part, under the influence of androgens during the pre- and early postnatal period (see Quadagno et al., 1977). The relation between these kinds of sexually dimorphic, nonmaternal behavior and tests of sensitized maternal behavior is particularly clear with regard to the T-maze test, which requires that the sensitized animal enter a novel apparatus and continue to behave maternally while there (that is, retrieve the test pups from the maze). We can predict from already established gender differences that males will be slower than females to enter and explore the maze, under identical conditions, and previous work also suggests that androgenized females will behave, in this respect, more like males than like nonandrogenized females. Even within the home cage, sensitization latencies probably are influenced to a degree by general traits that are sexually dimorphic. Such differences, for example, in activity level or timidity, of course, may be relevant to the special demands that care of the young places on females of the species. However, whether they arise in conjunction with or are entirely independent of differences between females and males in the substrate of maternal behavior remains to be determined.

VIII. Concluding Statement

Our studies have shown that the maternal behavior cycle is a developmental product of hormonal events during pregnancy, especially at its termination, and of behavioral stimulation received during interactions between the mother and her young. Investigation of females whose pregnancies were terminated prematurely by hysterectomy and/or ovariectomy have shown that the rise in estrogen, primarily, is responsible for the onset of maternal behavior under these conditions. Further studies revealed that to be effective this rise must occur free of the inhibiting influence of high levels of progesterone and that the decline in progesterone (progesterone withdrawal) in addition to its permissive action with respect to estrogen may itself facilitate a short-term increase in maternal responsiveness. We could find no role for either prolactin or adrenocortical hormones in the onset of maternal behavior in these studies.

All our findings in late-pregnancy-terminated females could be duplicated in nonpregnant females that are hysterectomized and ovariectomized, with the exception of the effect of progesterone withdrawal. It would be premature, however, to rule out this effect in these females. There were several differences in estrogen effects between nonpregnant and late-pregnant females but these were quantitative in nature.

There is considerable evidence that our findings in the late-pregnant female are relevant to what happens normally at the prepartum onset of maternal behavior. The decline in progesterone around 30 hr before parturition allows the earlier and subsequent rise in estrogen to stimulate maternal behavior with prolactin playing no role in this process.

However, following parturition the female rat enters a brief sexual phase during which she may become pregnant, but, whether or not she becomes pregnant, she undergoes the same hormonal changes in response to suckling and nonsuckling stimulation from the young. These consist of a prolonged diestrous phase characterized by high levels of progesterone secretion, inhibition of pituitary gonadotrophin hormone release, and the secretion of prolactin and other hormones of the lactogenic complex. The main point is that this hormonal complex is very different from that which gave rise to the onset of maternal behavior, and the suspicion that it does not support postpartum maternal behavior is confirmed by the failure to interfere substantially with this behavior by procedures that alter the postpartum secretion of circulating hormones. This does not rule out important effects of these hormones in postpartum maternal care, which have recently come to light with respect to the modulation of nursing behavior and the release of maternal pheromone and have long been known with respect to lactation and milk ejection.

It is apparent from our studies, therefore, that a new phase in the regulation of maternal behavior arises after its prepartum hormonal onset. A large body of research reported in this chapter, which need not be summarized, strongly indicates that during this phase maternal behavior is regulated by sensory stimulation that the female receives from her young. Suckling is not crucial but it may play a role under normal circumstances. Schneirla's concept of reciprocal stimulation (trophallaxis) is applicable here since undoubtedly it is the stimulative interactions between the mother and young that initially consolidate maternal behavior and later maintain it, and again it is through changes in the nature of these interactions that maternal behavior declines and the young are weaned.

The concept that the female needs to establish a relationship with her young by licking them during parturition, thereby transferring to the young the licking relationship developed with her own body, which was proposed by Birch (1956), we believe, contains the germ of the concept of a *transition period*, which we have proposed to deal with the shift from hormonal to nonhormonal regulation of maternal behavior around parturition. In his study using collared rats to prevent self-licking during pregnancy, Birch may have underestimated the strong impetus to initiate maternal behavior that derives from hormonal stimulation, and although he reported failure to take care of the newborn fetuses when the collar was removed at the start of parturition, other investigators have not been able to replicate these findings (Christophersen and Wagman, 1965; Eibl-Eibesfeldt and Kramer, 1958; Friedlich, 1962; Kirby and Horvath, 1968). Moreover, in several studies, although females were delivered by Caesarean section, nevertheless they exhibited maternal behavior when presented with newborn several hours later.

Our hormonal studies in both virgins and late-pregnancy-terminated females show that hormonal effects may persist for 72 hr after a single injection of estradiol benzoate at a dose that, though relatively high, may not be nonphysiological in view of the sharp rise in circulating levels of estradiol in the days before parturition. When stimulation is also introduced, the hormonal effect may still be detected at 96 hr. Thus, hormonal effects may continue beyond parturition and postpartum females deprived of pup stimulation during parturition may still be able to initiate maternal behavior.

The importance of pup contact during the early postpartum period was shown in studies in which females deprived of pups during parturition showed rapid waning of maternal behavior during the first four days. Conversely, females that were allowed contact with pups *only* during the first 2 days retained a readily arousable maternal responsiveness as long as 25 days after their last contact with pups. Also, Caesarean-section-delivered females given pups for 48 hr after they have recovered from surgery are rapidly sensitized 25 days later. We interpret these findings as evidence that maternal behavior had become established on a nonhormonal basis during these two days and therefore could easily be reinstated on the same basis many weeks later. This interpretation is supported by the findings that females that had already received normal hormonal stimulation sufficient to initiate maternal behavior but were prevented from receiving pup stimulation (and therefore of establishing a nonhormonal basis for maternal behavior) failed to be easily sensitized 25 days later and in fact showed no effects of the previous hormonal arousal.

Parturition may, nevertheless, be crucial for the transition to occur under normal circumstances. This concept is supported by the Bridges studies (1975,

1977) in which contact with as little as half the newly born litter during parturition is sufficient to enable females 25 days later to reinstate (or initiate) maternal behavior with latencies no greater than those of females allowed 48 hr of postpartum contact with pups.

The concept of a postpartum nonhormonal phase of maternal behavior cycle receives support from our studies of sensitization in which nonpregnant females, either intact, ovariectomized, or hypophysectomized, are induced to exhibit maternal behavior simply through prolonged exposure to young pups. Although not all aspects of maternal behavior can readily be elicited during sensitization (for example, aggression toward an intruder) most aspects can, and these appear quite similar to the maternal behavior of postpartum females.

Our studies on the neural basis of maternal behavior suggest a neural site at which both hormonal and sensory stimuli act to elicit maternal behavior. Implanting estrogen at this site stimulates maternal behavior, and lesioning this site prevents sensitization of nonpregnant females and interrupts ongoing maternal behavior in postpartum females. The successive interaction of hormonal and sensory stimuli at this site may account for the effect that sensory stimuli have in maintaining maternal behavior postpartum once it has been initiated under the influence of estrogen in the prepartum period.

Which stimuli mediate the transition between the two principal stages of maternal behavior? Birch (1956) suggested that olfactory–gustatory stimuli played an important role in this process, but evidence was not presented. It was not clear in Birch's presentation why females required previous habituation or had to *develop* positive responses to pup odors and taste. Our studies on the effects of anosmia on sensitization in nonpregnant females suggest one reason: pup odors are initially either aversive to inexperienced females or their novelty make evoke avoidance responses or perhaps both. We have shown this in several different ways that need not be reviewed here. In nonpregnant females overcoming of this aversion is a precondition for responding to other pup stimuli (for example, visual, sound, or tactile), which, we believe, accounts for the readiness with which anosmic females are sensitized.

What applies to the nonpregnant female, responding on a nonhormonal basis, may not apply to the parturient female whose behavior is under hormonal control, which may affect either olfactory responses or aversive responding. It has only recently been possible to show in preliminary studies that even hormonally stimulated females are accelerated in the onset of maternal behavior if they are made anosmic before exposure to pups. Such females nearly all respond immediately or within a few hours, whereas females that are able to smell the pups require, on the average, 24 hr. This suggests that even while under hormonal stimulation an adjustment to the odor of the pups occurs but that it may proceed more rapidly than without hormonal stimulation, or, perhaps, as Birch suggested, it is unnecessary because of the pregnant female's prior adaptation to her own vaginal secretions, which are also present on the pups when they are born.

Further study of the rates at which hormonal effects decline after parturition and nonhormonal regulation is established is required, but existing evidence indicates that by the third day postpartum the latter is well established.

Our studies of the ontogeny of maternal behavior, of interest in themselves, also have a bearing on the factors that are involved in the appearance of maternal behavior in adulthood. These studies reveal a tonic influence of estrogen upon maternal responsiveness that begins around puberty in the female and continues into adulthood. Ovariectomy removes this tonic influence, and sensitization latencies are elevated as a result. Latencies are also elevated between prepuberty and early adulthood by psychological factors that include an increase in timidity in response to novel stimuli and the emergence of olfactory aversion to pup odors.

The increase in timidity is not confined to pups and therefore to maternal behavior but is evident in many different kinds of behavioral tests of females (and of males) during this age period. Nor are the behavioral characteristics of females that enable them to have shorter sensitization latencies compared to males specific to maternal behavior. Thus, although maternal behavior in the female is a hormone-dependent behavior pattern, the psychological capacities underlying this behavior are not sharply different from those involved in other nonmaternal activities of the female.

There is no doubt, however, that maternal behavior is a specialized pattern of behavior having a function with respect to care of the young that is different from the functions of other behavior patterns. Retrieving and nest building may be less specialized than nursing and pup licking since the components of these patterns are seen in other contexts (for examples, food pellet retrieving, or mat building), whereas nursing (in contrast to lactation, which of course is not present except in maternal females) is not seen at other times and licking of objects was not found except when the objects were pups (Rosenblatt, 1975b). It is likely each of these components is stimulated by one or several sensory stimuli emanating either from within the female, for example, arising from thermoregulatory processes (Leon et al., 1978) or from the pups, but these have not yet been specified in the rat as Noirot (1972) has attempted to do in the mouse.

In view of the above it was particularly interesting to observe the emergence of these specialized responses to young pups in 18–22-day-old juveniles. The behavior of these young was a mixture of their customary social responses to littermates, as in huddling, and responses to the special properties of the pups, which led them to progress from sniffing, nosing, and licking them, to manipulating them with their forepaws and eventually to lying in contact with them, usually draped over them as in nursing. When exposure to these younger pups was extended over 22 hr, the components of maternal behavior gradually appeared in many of the juveniles. It would be worthwhile to study, as we are now engaged in doing, the developmental origin of maternal behavior in the gregariousness of young rats in relation to their littermates and mother during the suckling period.

APPENDIX: EVALUATION OF THE ZINC SULFATE TECHNIQUE FOR INTERRUPTING OLFACTION, WITH PARTICULAR REFERENCE TO ITS USE IN STUDIES OF RAT MATERNAL BEHAVIOR

Zinc sulfate applied in aqueous solution to the linings of the nose produces necrosis and sloughing of the olfactory epithelium while having little or no effect on nonsensory epithelium. The potential usefulness of this poorly understood phenomenon in behavioral studies of olfaction was first recognized by Alberts and Galef (1971), who described a technique of intranasal lavage, which rendered rats temporarily anosmic, as judged by inability to locate buried food by its odor. A hooked catheter was inserted through the mouth into the posterior opening of the nasal cavities, and zinc sulfate solution, injected through the catheter, flooded the nasal chambers and drained from the external nares. Since publication of their paper, zinc sulfate has been used widely in both anatomical and behavioral studies of the primary olfactory system in various rodents. There is as yet, however, no consensus regarding its appropriateness for behavioral research. Questions have been raised regarding the completeness and duration of the olfactory loss that results, and regarding the possibility that systemic poisoning with zinc is responsible for some or all of the effects on behavior. Our own experiences with the technique, however, have convinced us that it does have a place in behavioral research, particularly in our own area of maternal behavior in the rat.

It is somewhat difficult to compare the results of the many studies in the literature that have employed intranasal zinc sulfate, owing to wide variations in methods of application and in the procedures used to evaluate olfactory loss after treatment. Histological studies of rodent olfactory epithelium after single applications of zinc sulfate, nevertheless, have consistently reported extensive destruction in all or nearly all animals, despite differences in composition of the zinc sulfate solution (from 1 to 5% zinc sulfate, with or without sodium chloride) and in methods of intranasal lavage (Mulvaney and Heist, 1971; Schultz, 1960; Singh et al., 1976; Winans and Powers, 1977). Patches of normal-appearing sensory epithelium also have been found consistently. Winans and Powers (1977), for example, estimated that from 3 to 22% of the olfactory epithelium of hamsters appeared normal when viewed through a light microscope 48 hr after intranasal infusion, the remaining sensory epithelium necrotic and undergoing sloughing. Consistent with a 70% or greater loss of olfactory receptors, a number of studies have reported that the behavioral reactions of rodents to such odors as freshly smoked cigar butt (Vandenbergh, 1973), female urine (Edwards and Burge, 1973), and vaginal secretions (Devor and Murphy, 1973; Powers and Winans, 1973) are altered for three or more days after treatment. On the other hand, at least some animals in some studies have been found to retain considerable ability to respond to olfactory cues after single applications of zinc sulfate, and olfactory function appears to return to normal, at least in rats, in a few days to two weeks (Alberts and Galef, 1971; Fleming and Rosenblatt, 1974b; Slotnick and Gutman, 1977). Our own study (Mayer and Rosenblatt, 1975) showed that 24 hr after treatment with zinc sulfate only 40% of trained, food-deprived females were unable to locate a strongly scented, flavored food (chocolate bits buried under wood shavings in one of two identical feeders). A less odorous (and less favored) food, guinea pig pellets, was not found by 70% of the same females. The ability to choose the feeder containing food returned rapidly among those females first failing to do so, and seemed complete 5 days after treatment. This finding has been confirmed by Slotnik and Gutman (1977), using more refined methods of testing for olfactory function, indicating that single applications of zinc sulfate (2.9–5% solutions) cannot be relied upon to produce complete or long-lasting anosmia in rats. (Since there are species differences in the anatomy of the nasal chambers, it perhaps is not justified to extend this conclusion to other rodents.) Recently, however, Thor et al. (1976) and Thor and Flannelly (1977) have described a technique of multiple applications of zinc sulfate that does impair olfaction in rats for an extended time, perhaps permanently; multiple applications, then, may extend the usefulness of intranasal zinc sulfate to research situations in which complete anosmia, and/or long-term loss of function, are critical.

As Slotnick and Gutman (1977) have observed, the usefulness of a technique that produced *partial* loss of olfaction must be decided empirically for each type of investigation. Since pretreating nonpregnant rats with intranasal zinc sulfate reliably facilitates the induction of maternal behavior through pup exposure, the partial anosmia produced by the procedure must be sufficient, either because it selectively eliminates pup odors, or because it lowers olfactory input generally. If an additional nonolfactory mechanism is involved, what it can be is not yet apparent. Fleming et al. (1979) recently have confirmed that partial disruption of olfaction resulting from surgical cuts into but not through the olfactory bulbs facilitates the sensitization of nonpregnant rats, even though the ability to locate hidden food pellets by their odor is not lost.

Perhaps the more serious question regarding the zinc sulfate technique is whether systemic poisoning can account for the behavioral effects of the procedure. Concern that systemic poisoning occurs is based upon the observation that treated animals lose weight in the days immediately following treatment; zinc toxicity in rats is characterized by depressed growth and anemia (Underwood, 1971; however, Underwood attributes the depressed growth rats accompanying toxicity in part to the unpalatability of high zinc diets.) Also, similarities have been found between the open-field behavior of animals treated intranasally with zinc sulfate and animals given an intraperitoneal injection of the same solution. After both routes of administration, animals show diminished activity levels, which in the open field results in a decrease in locomotion and rearing, and an increase in freezing (Mayer and Rosenblatt, 1977; Sieck and Baumbach, 1974).

Zinc generally is considered to be a relatively nontoxic element; it is nutritionally essential in trace quantities, and can be ingested by rats in excessive quantities over generations without noticeable effect (Underwood, 1971). In order to attribute the acute weight loss and lethargy sometimes seen after intranasal zinc sulfate to systemic poisoning, it is necessary to assume that most of the zinc flushed through the nose somehow is absorbed through the nasal mucosa. Weight loss and lethargy, of course, are nonspecific symptoms. Since intranasally applied zinc sulfate causes necrosis of the olfactory epithelium, the procedure inevitably is followed by some degree of nasal congestion and discomfort, which may depress activity and eating in the absence of systemic toxicity. With regard to more serious posttreatment illness, an alternative explanation to systemic poisoning is contained in the histological observation that aspirated solution causes a spotty or confluent hemorrhagic consolidation of lung tissue (Singh *et al.*, 1976). In our own experience, changes in technique designed to diminish the possibility that small amounts of solution are aspirated have reduced greatly both weight loss and posttreatment lethargy, without diminishing the effectiveness of the procedure in facilitating the onset of sensitized maternal behavior. In one of our early studies, for example, we found that intranasally treated females lost an average of 12.4 g during the first 48 hr after lavage, representing a 5.2% loss of initial body weight, whereas air-injected females gained about 1% of their initial weight. At that time we reported that about 20% of zinc sulfate-injected animals died within 24 hr (Mayer and Rosenblatt, 1975). In our most recent study using improved methods of intranasal injection, the average weight loss 48 hr after treatment was 2% of initial weight, and there were no deaths among 28 injected females. The facilitation of maternal behavior among these females, however, was as great or greater than we had found in earlier studies. Therefore, we feel that most of the undesirable, debilitating aftereffects of the intranasal zinc sulfate technique can be controlled by improved injection procedures.

It also is important to note that, although both intraperitoneal and intranasal injections of zinc sulfate may depress activity and produce similar changes in rearing, locomotion, and freezing as observed in an open field, only intranasal applications have been shown to alter behaviors that are known to depend on olfaction. For example, in every study in which we have treated adult rats with intranasal zinc sulfate before exposing them to pups, sensitization latencies of the intranasally treated animals were significantly reduced relative to those of sham-injected animals. When we gave intraperitoneal injections of zinc sulfate to nonpregnant females, however, their sensitization latencies were found to be as long as those of females not receiving the chemical (and significantly longer than those of the intranasally treated females in the same experiment; Mayer and Rosenblatt, 1977). The zinc sulfate technique, insofar as it interrupts the olfactory system at the level of its receptors, may prove to have an important advantage for some types of investigation over the surgical techniques that attempt to deafferent the system at the level of the posterior or cribriform plate (the anterior poles of the olfactory bulbs). Not only do the olfactory bulbs remain intact and functional after intranasal zinc sulfate, but it appears that the vomeronasal or accessory olfactory system is entirely spared. The vomeronasal nerve fibers pass caudally from the organs (which lie within the nasal septum) beneath the epithelium of the septum, through the cribriform plate, and between the main olfactory bulbs before synapsing in the accessory olfactory bulbs (Negus, 1958). Therefore, techniques of olfactory deafferentation that involve ablating the olfactory mucosa, scraping the cribriform plate, or transecting or ablating the olfactory bulbs inevitably damage the vomeronasal nerves, even if the accessory bulbs are left intact. Lateral olfactory tract lesions also usually destroy projections from both main and accessory bulbs (Winans and Powers, 1977). Intranasal zinc sulfate, on the other hand, has been found to leave the sensory epithelium of the vomeronasal organs normal in appearance (as studied through a light microscope) in rat pups (Singh et al., 1976), and in adult hamsters (Winans and Powers, 1977). In our laboratory we have found that the vomeronasal organs of adult rats also appear quite normal after intranasal zinc sulfate (Mayer and Winans, unpublished observations). We do not yet have direct evidence in rats that the vomeronasal system functions in a normal manner after intranasal zinc sulfate; however, it has been established in hamsters that intranasal zinc sulfate does not reproduce the effects of vomeronasal nerve transections on male sex behavior (Powers and Winans, 1973; Winans and Powers, 1977). It seems probably that zinc sulfate, were it to reach the interior of the vomeronasal organs, would damage the sensory epithelium. Therefore, it is important that the organs be examined histologically before it is concluded that a particular method of intranasal lavage spares the vomeronasal system. Also, we would like to see in rats as well as in hamsters that a behavior which is known to depend on the vomeronasal system is not disrupted by zinc sulfate. With these reservations in mind, however, it does seem that the zinc sulfate technique holds more promise of selectively deafferenting only the primary olfactory system than others currently available.

Acknowledgments

The authors wish to thank Drs. Alison S. Fleming and Michael Numan for their critical reading of the manuscript, Ms. Cynthia Banas for preparation of the illustrations, and Ms. Winona Cunningham and Ms. Nancy Jachim for secretarial assistance. The research was supported by USPHS Grant MH-08604 to J. S. R. and a Biomedical Research Support Grant. Publication No. 330 of the Institute of Animal Behavior, Rutgers—The State University.

References

Alberts, J. R. 1978a. Huddling by rat pups: Group behavioral mechanisms of temperature regulation and energy conservation. *J. Comp. Physiol. Psychol.* **92**, 231–245.

Alberts, J. R. 1978b. Huddling by rat pups: Multisensory control of contact behavior. *J. Comp. Physiol. Psychol.* **92**, 220–230.

Alberts, J. R., and Brunjes, P. C. 1978. Ontogeny of thermal and olfactory determinants of huddling in the rat. *J. Comp. Physiol. Psychol.* **92**, 897–906.

Alberts, J. R., and Galef, B. G. 1971. Acute anosmia in the rat: A behavioral test of peripherally induced olfactory deficit. *Physiol. Behav.* **6**, 619–621.

Archer, J. 1975. Rodent sex differences in emotional and related behavior. *Behav. Biol.* **14**, 451–479.

Bast, J. D., and Melampy, R. M. 1972. Luteinizing hormone, prolactin and ovarian 20α-hydroxysteroid dehydrogenase levels during pregnancy and pseudopregnancy in the rat. *Endocrinology* **91**, 1499–1505.

Baum, M. J. 1978. Failure of pituitary transplants to facilitate the onset of maternal behavior in ovariectomized virgin rats. *Physiol. Behav.* **20**, 87–89.

Beach, F. A. 1937. The neural basis of innate behavior. I. Effects of cortical lesions upon the maternal behavior pattern in the rat. *J. Comp. Psychol.* **24**, 393–436.

Beach, F. A., and Jaynes, J. 1956. Studies of maternal retrieving in rats. III. Sensory cues involved in lactating females response to her young. *Behavior* **10**, 104–125.

Beach, F. A., and Wilson, J. 1963. Effects of prolactin, progesterone, and estrogen on reactions of nonpregnant rats to foster young. *Psychol. Rep.* **13**, 231–239.

Beatty, W. W., and Beatty, P. A. 1970. Hormonal determinants of sex differences in avoidance behavior and reactivity to electric shock in the rat. *J. Comp. Physiol. Psychol.* **3**, 446–455.

Beatty, W. W., Beatty, P. A., and Bowman, R. G. 1971. A sex difference in the extinction of avoidance behavior in rats. *Psychonom. Sci.* **23**, 213–214.

Beatty, W. W., Gregoire, K. L., and Parmiter, L. L. 1973. Sex differences in retention of passive avoidance behavior in rats. *Bull. Psychonom. Soc.* **2**, 99–100.

Ben-David, M., Danon, A., and Sulman, F. G. 1971. Evidence of antagonism between prolactin and gonadotrophin secretion: Effect of methallibure on perphenazine-induced prolactin secretion in ovariectomized rats. *J. Endocrinol.* **51**, 719–725.

Benuck, I., and Rowe, F. A. 1975. Centrally and peripherally induced anosmia: Influences on maternal behavior in lactating female rats. *Physiol. Behav.* **14**, 439–447.

Bernstein, H., and Moyer, K. E. 1970. Aggressive behavior in the rat: Effects of isolation and olfactory bulb lesions. *Brain Res.* **20**, 75–84.

Bintarningsih, Lyons, W. R., Johnson, R. E., and Li, C. H. 1958. Hormonally-induced lactation in hypophysectomized rats. *Endocrinology* **63**, 540–547.

Birch, H. G. 1956. Sources of order in the maternal behavior of animals. *Am. J. Orthopsychiat.* **26**, 279–284.

Blake, C. A., Norman, R. L., and Sawyer, C. H. 1972. Effects of estrogen and/or progesterone on serum and pituitary gonadotropin levels in ovariectomized rats. *Proc. Soc. Exp. Biol. Med.* **141**, 1100–1103.

Blaustein, J. D., and Wade, G. N. 1977. Concurrent inhibition of sexual behavior, but not brain [3H] estradiol uptake, by progesterone in female rats. *J. Comp. Physiol. Psychol.* **91**, 742–751.

Bridges, R. S. 1975. Long-term effects of pregnancy and parturition upon maternal responsiveness in the rat. *Physiol. Behav.* **14**, 245–249.

Bridges, R. S. 1977. Parturition: Its role in the long term retention of maternal behavior in the rat. *Physiol. Behav.* **18**, 487–490.

Bridges, R. S. 1978. Retention of rapid onset of maternal behavior during pregnancy in primiparous rats. *Behav. Biol.* **24**, 113–117.

Bridges, R., Zarrow, M. X., Gandelman, R., and Denenberg, V. H. 1972. Differences in maternal responsiveness between lactating and sensitized rats. *Dev. Psychobiol.* **5**, 127–137.

Bridges, R. S., Zarrow, M. X., and Denenberg, V. H. 1973. The role of neonatal androgen in the expression of hormonally induced maternal responsiveness in the adult rat. *Horm. Behav.* **4**, 315–322.

Bridges, R. S., Zarrow, M. X., Goldman, B. D., and Denenberg, V. H. 1974. A developmental study of maternal responsiveness in the rat. *Physiol. Behav.* **12**, 149–151.

Bridges, R. S., Feder, H. H., and Rosenblatt, J. S. 1977. Induction of maternal behaviors in primigravid rats by ovariectomy, hysterectomy, or ovariectomy plus hysterectomy: Effect of length of gestation. *Horm. Behav.* **9**, 156–169.

Bridges, R. S., Rosenblatt, J. S., and Feder, H. H. 1978a. Stimulation of maternal responsiveness after pregnancy termination in rats: Effect of time of onset of behavioral testing. *Horm. Behav.* **10**, 235–245.

Bridges, R. S., Rosenblatt, J. S., and Feder, H. H. 1978b. Serum progesterone levels and maternal behavior after pregnancy termination in rats: Behavioral effects of progesterone maintenance and withdrawal. *Endocrinology* **102**, 258–267.

Bronstein, P. M. 1972. Open-field behavior in the rat as a function of age: Cross-sectional and longitudinal studies. *J. Comp. Physiol. Psychol.* **80**, 335–341.

Bronstein, P. M., and Hirsch, S. M. 1976. Ontogeny of defensive reactions in Norway rats. *J. Comp. Physiol. Psychol.* **90**, 620–629.

Buckle, J. W., and Nathanielsz, P. W. 1975. A comparison of the characteristics of parturition induced by prostaglandin $F_{2\alpha}$, infused intra-aortically, with those following ovariectomy in the rat. *J. Endocrinol.* **64**, 257–266.

Cain, D. P. 1974. Olfactory bulbectomy: Neural structures involved in irritability and aggression in the male rat. *J. Comp. Physiol. Psychol.* **86**, 213–220.

Caligaris, L., Astrada, J. J., and Taleisnik, S. 1974. Oestrogen and progesterone influence on the release of prolactin in ovariectomized rats. *J. Endocrinol.* **60**, 205–215.

Candland, D. K., and Campbell, B. A. 1962. Development of fear in the rat as measured by behavior in the open field. *J. Comp. Physiol. Psychol.* **55**, 593–596.

Catala, S., and Deis, R. P. 1973. Effect of oestrogen upon parturition, maternal behaviour and lactation in ovariectomized pregnant rats. *J. Endocrinol.* **56**, 219–225.

Charten, D., Adrien, J., and Cosnier, J. 1971. Declencheurs chimiques du comportement du léchage des petits par la ratte parturiente. *Rev. Comp. Anim.* **5**, 89–94.

Chester, R., Dukes, M., Slater, S. R., and Walpole, A. L. 1972. Delay of parturition in the rat by anti-inflammatory agents which inhibit the biosynthesis of prostaglandins. *Nature (London)* **240**, 37–38.

Christophersen, E. R., and Wagman, W. 1965. Maternal behavior in the albino rat as a function of self-licking deprivation. *J. Comp. Physiol. Psychol.* **60**, 142–144.

Cosnier, J. 1963. Quelques problèmes posés par le "comportement maternel provoqué" chez la ratte. *C. R. Soc. Biol.* **157**, 1611–1613.

Cosnier, J. 1965. Le comportement grégaire de rat de'élevage. Doctoral dissertation, University of Lyon.

Cosnier, J., and Couturier, C. 1966. Comportement maternal provoqué chez les rattes adultes castrées. *C. R. Soc. Biol.* **160**, 789–791.

Denenberg, V. H., Rosenberg, K. M., and Zarrow, M. X. 1969. Mice reared with rat aunts: Effects in adulthood upon plasma corticosterone and open-field activity. *Physiol. Behav.* **4**, 705–707.

Denti, A., and Epstein, A. 1972. Sex differences in the acquisition of two kinds of avoidance behavior in rats. *Physiol. Behav.* **8**, 611–615.

Devor, M., and Murphy, M. R. 1973. The effect of peripheral olfactory blockade on the social behavior of the male golden hamster. *Behav. Biol.* **9**, 31–42.

Didiergeorges, F., Vergnes, M., and Karli, P. 1966. Privation des afférences olfactives et aggresivité interspécifique du rat. *C. R. Soc. Biol.* **160**, 866–868.

Douglas, R., Isaacson, R., and Moss, R. 1969. Olfactory lesions, emotionality and activity. *Physiol. Behav.* **4**, 379–381.

Edwards, D. A., and Burge, K. G. 1973. Olfactory control of the sexual behavior of male and female mice. *Physiol. Behav.* **11**, 867–872.

Eibl-Eibesfeldt, I., and Kramer, S. 1958. Ethology: The comparative study of animal behavior. *Q. Rev. Biol.* **33**, 181–211.

Erskine, M. S. 1978. Hormonal and experiential factors associated with the expression of aggression during lactation in the rat. Ph.D. dissertation, University of Connecticut, Storrs.

Fleming, A., and Rosenblatt, J. S. 1974a. Maternal behavior in the virgin and lactating rat. *J. Comp. Physiol. Psychol.* **86**, 957–972.

Fleming, A., and Rosenblatt, J. S. 1974b. Olfactory regulation of maternal behavior in rats: I. Effects of olfactory bulb removal in experienced and inexperienced lactating and cycling females. *J. Comp. Physiol. Psychol.* **86**, 221–232.

Fleming, A., and Rosenblatt, J. S. 1974c. Olfactory regulation of maternal behavior in rats: II. Effects of peripherally induced anosmia and lesions of the lateral olfactory tract in pup-induced virgins. *J. Comp. Physiol. Psychol.* **86**, 233–246.

Fleming, A. S., Vaccarino, F., Tambosso, L., and Chee, P. 1979. Vomeronasal and olfactory system modulation of maternal behavior in the rat. *Science* **203**, 372–374.

Friedlich, O. B. 1962. A study of maternal behavior in the albino rat as a function of self-licking deprivation. Master's Degree Thesis, Southern Illinois University, Carbondale.

Gala, R. R. 1970. Studies on maintaining the lactational diestrum after early litter weaning. *Proc. Soc. Exp. Biol. Med.* **133**, 164–167.

Grosvenor, C. E. 1965. Evidence that exteroceptive stimuli can release prolactin from the pituitary gland of the lactating rat. *Endocrinology* **76**, 340–342.

Grosvenor, C. E., Mena, F., Dhariwal, A. P. S., and McCann, S. M. 1967. Reduction of milk secretion by prolactin-inhibiting factor: Further evidence that exteroceptive stimuli can release pituitary prolactin in rats. *Endocrinology* **81**, 1021–1028.

Herrenkohl, L. R. 1971. Effects on lactation of progesterone injections administered during late pregnancy in the rat. *Proc. Soc. Exp. Biol. Med.* **138**, 39–42.

Herrenkohl, L. R., and Lisk, R. D. 1973. Effects on lactation of progesterone injections administered before and after parturition in the rat. *Proc. Soc. Exp. Biol. Med.* **142**, 506–510.

Herrenkohl, L. R., and Rosenberg, P. A. 1972. Exteroceptive stimulation of maternal behavior in the naive rat. *Physiol. Behav.* **8**, 595–598.

Hinde, R. A. 1965. Interaction of internal and external factors in integration of canary reproduction. *In* Sex and Behavior'' (F. A. Beach, ed.), pp. 381–415. Wiley, New York.

Hull, E. M., and Homan, H. D. 1975. Olfactory bulbectomy, peripheral anosmia, and mouse killing and eating in rats. *Behav. Biol.* **14**, 481–488.

Johnson, N. P. 1972. Postpartum ovulation in the rat. Ph.D. dissertation, Purdue University, Lafayette, Indiana.

Jost, A. 1959. Développement des foetus, accouchement et allaitement chez des Rattes costrées en fin de gestation. *Arch. Anat. Microsc.* **48**, 133–140.

Kalra, P. S., Fawcett, C. P., Krulich, L., and McCann, S. M. 1973. The effects of gonadal steroids on plasma gonadotropins and prolactin in the rat. *Endocrinology* **92**, 1256–1268.

Karli, P. 1956. The Norway rat's killing responses to the white mouse: An experimental analysis. *Behavior* **10**, 81–103.

Karli, P. 1961. Róle des afferences sensorielles dans le déclenchement du comportement de'aggression interspécifique rat-souris. *C. R. Soc. Biol.* **155**, 644–646.

Karli, P., Vergnes, M., and Didiergeorges, R. 1969. Rat-mouse interspecific aggressive behavior and its manipulation by brain ablation and by brain stimulation. *In* "Aggressive Behavior" (S. Garattini and E. B. Sand, eds.), pp. 47–55. Wiley, New York.

Kinder, E. F. 1927. A study of the nestbuilding activity of the albino rat. *J. Exp. Zool.* **47**, 117–161.

Kirby, H. W., and Horvath, T. 1968. Self-licking deprivation and maternal behaviour in the primiparous rat. *Can. J. Psychol.* **22**, 369–375.

Koller, G. 1956. Hormonale und psychische Steuerung beim Nestbau weisser Mäuse. *Zool. Anz. Suppl.* **19** *(Verh. Dtsch. Zool. Ges. 1955)*, 123–132.

Koranyi, L., Lissak, K., Tamasy, V., and Kamaras, L. 1976. Behavioral and electrophysiological attempts to elucidate central nervous system mechanisms responsible for maternal behavior. *Arch. Sex Behav.* **5**, 503–510.

Koranyi, L., Phelps, C. P., and Sawyer, C. H. 1977. Changes in serum prolactin and corticosterone in induced maternal behavior in rats. *Physiol. Behav.* **18**, 287–292.

Krehbiel, D., and LeRoy, L. M. 1978. The quality of maternal behavior induced in the rat by various treatments. Paper presented at the Eastern Conference on Reproductive Behavior, Madison, Wisconsin.

Kuo, E. Y. H., Cobb, W. R., Esber, H. H., and Bogden, A. E. 1974. Effects of hysterectomy on milk secretion and serum levels of prolactin, growth hormone, estrogen, and progesterone in rhesus monkeys with hormone-induced uterine hypertrophy. *Am. J. Obstet. Gynecol.* **120**, 368–375.

Labriola, J. 1953. Effects of caesarian delivery upon maternal behavior in rats. *Proc. Soc. Exp. Biol. Med.* **83**, 556–567.

Latané, B. 1969. Gregariousness and fear in laboratory rats. *J. Exp. Soc. Psychol.* **5**, 61–69.

Latané, B., and Steele, C. 1975. The persistence of social attraction in socially deprived and satiated rats. *Anim. Learn. Behav.* **3**, 131–134.

Latané, B., Joy, V., Meltzer, J., Lubell, B., and Cappell, H. 1972. Stimulus determinants of social attraction in rats. *J. Comp. Physiol. Psychol.* **79**, 13–21.

Leblond, C. P. 1940. Nervous and hormonal factors in the maternal behavior of the mouse. *J. Genet. Psychol.* **57**, 327–344.

Leblond, C. P., and Nelson, W. O. 1937. Maternal behavior in hypophysectomized male and female mice. *Am. J. Physiol.* **120**, 167–172.

Lehrman, D. S. 1956. On the organization of maternal behavior and the problem of instinct. *In* "L'Instinct dans le Comportement des Animaux et de l'Homme" (P.-P. Grasse, ed.), pp. 475–514. Masson, Paris.

Lehrman, D. S. 1961. Gonadal hormones and parental behavior in birds and infrahuman mammals. *In* "Sex and Internal Secretions" (W. C. Young, ed.), pp. 1268–1382. Williams & Wilkins, Baltimore, Maryland.

Lehrman, D. S. 1965. Interaction between internal and external environments in the regulation of the reproductive cycle of the ring dove. *In* "Sex and Behavior" (F. A. Beach, ed.), pp. 355–380. Wiley, New York.

Leon, M., and Behse, J. 1977. Dissolution of the pheromonal bond: Waning of approach responses by weanling rats. *Physiol. Behav.* **18**, 393–397.

Leon, M., Numan, M., and Moltz, H. 1973. Maternal behavior in the rat: Facilitation through gonadectomy. *Science* **179**, 1018-1019.

Leon, M., Numan, M., and Chan, A. 1975. Adrenal inhibition of maternal behavior in virgin female rats. *Horm. Behav.* **6**, 165-171.

Leon, M. L., Croskerry, P. G., and Smith, G. K. 1978. Thermal control of mother-young contact in rats. *Physiol. Behav.* **21**, 793-811.

Linkie, D. M., and Niswender, G. D. 1972. Serum levels of prolactin, luteinizing hormone, and follicle stimulating hormone during pregnancy in the rat. *Endocrinology* **90**, 632-637.

Livesey, P. J., and Egger, G. J. 1970. Age as a factor in open-field responsiveness in the white rat. *J. Comp. Physiol. Psychol.* **73**, 93-99.

Lott, D., and Fuchs, S. 1962. Failure to induce retrieving by sensitization or the injection of prolactin. *J. Comp. Physiol. Psychol.* **55**, 1111-1113.

Lubin, M., Leon, M., Moltz, H., and Numan, M. 1972. Hormones and maternal behavior in the male rat. *Morm. Behav.* **3**, 369-374.

McCullough, J. Quadagno, D. M., and Goldman, B. D. 1974. Neonatal gonadal hormones: Effect on maternal and sexual behavior in the male rats. *Physiol. Behav.* **12**, 183-188.

MacKinnon, D. A., and Stern, J. M. 1977. Pregnncy duration and fetal number: Effects on maternal behavior in rats. *Physiol. Behav.* **18**, 793-797.

Marinari, K. T., and Moltz, H. 1977. Disruption of vaginal cyclicity and maternal behavior. Paper presented at the Eastern Conference on Reproductive Behavior, University of Connecticut, Storrs.

Mayer, A. D., and Rosenblatt, J. S. 1975. Olfactory basis for the delayed onset of maternal behavior in virgin female rats: Experiential effect. *J. Comp. Physiol. Psychol.* **89**, 701-710.

Mayer, A. D., and Rosenblatt, J. S. 1977. Effects of intranasal zinc sulfate on open field and maternal behavior in female rats. *Physiol. Behav.* **18**, 101-109.

Mayer, A. D., and Rosenblatt, J. S. 1979a. Hormonal influences during the ontogeny of maternal behavior in female rats. *J. Comp. Physiol. Psychol.* (in press.)

Mayer, A. D., and Rosenblatt, J. S. 1979b. Ontogeny of maternal behavior in the laboratory rat: Early origins in 18 to 27 day old young. *Dev. Psychobiol.* (in press).

Mayer, A. D., Freeman, N. G., and Rosenblatt, J. S. 1979. Ontogeny of maternal behavior in the laboratory rat: Factors underlying changes in responsiveness from 30 to 90 days. *Dev. Psychobiol.* (in press).

Mena, F., and Grosvenor, C. E. 1971. Release of prolactin in rats by exteroceptive stimulation; sensory stimuli involved. *Horm. Behav.* **2**, 107-116.

Moltz, H. 1972. The ontogeny of maternal behavior in some selected mammalian species. *In* "Ontogeny of Vertebrate Behavior" (H. Moltz, ed.), pp. 263-313. Academic Press, New York.

Moltz, H. and Wiener, E. 1966. Effects of ovariectomy on maternal behavior of the primiparous and multiparous rat. *J. Comp. Physiol. Psychol.* **62**, 382-387.

Moltz, H., Robbins, D., and Parks, M. 1966. Caesarean delivery and maternal behavior of primiparous and multiparous rats. *J. Comp. Physiol. Psychol.* **61**, 455-460.

Moltz, H., Levin, R., and Leon, M. 1969. Differential effects of progesterone on the maternal behavior of primiparous and multiparous rats. *J. Comp. Physiol. Psychol.* **67**, 36-40.

Moltz, H., Lubin, M., and Numan, M. 1970. Hormonal induction of maternal behavior in the ovariectomized nulliparous rat. *Physiol. Behav.* **5**, 1373-1377.

Moltz, H., Leon, M. Numan, M., and Lubin, M. 1971. Replacement of progesterone with a phenothiazine in the induction of maternal behavior in the ovariectomized nulliparous rat. *Physiol. Behav.* **6**, 735-737.

Mori, J., Nagasawa, H., Yanai, R., and Masaki, J. 1974. Changes in serum levels of follicle stimulating hormone and luteinizing hormone shortly before and after parturition in rats. *Acta Endocrinol.* **75**, 491-496.

Morishige, W. K., and Rothchild, I. 1974. Temporal aspects of the regulation of corpus luteum function by luteinizing hormone, prolactin, and placental luteotrophin during the first half of pregnancy in the rat. *Endocrinology* **95**, 260–274.

Morishige, W. K., Pepe, G. J., and Rothchild, I. 1973. Serum luteinizing hormone (LH), prolactin and progesterone levels during pregnancy in the rat. *Endocrinology* **92**, 1527–1530.

Moyer, K. E. 1968. Kinds of aggression and their physiological basis. *Commun. Behav. Biol.* **2**, 65–87.

Mulvaney, B. D., and Heist, H. E. 1971. Regeneration of rabbit olfactory epithelium. *Am. J. Anat.* **131**, 211–252.

Negus, V. 1958. "Comparative Anatomy and Physiology of the Nose and Paranasal Sinuses." Livingstone, London.

Nissen, H. W. 1930. A study of maternal behavior in the white rat by means of the obstruction method. *J. Genet. Psychol.* **37**, 377–393.

Noirot, E. 1972. The onset and development of maternal behavior in rats, hamsters and mice. *In* "Advances in the Study of Behavior" (D. S. Lehrman, R. A. Hinde, and E. Shaw eds.), Vol. 4. Academic Press, New York.

Numan, M. 1974. Medial preoptic area and maternal behavior in the female rat. *J. Comp. Physiol. Psychol.* **87**, 746–759.

Numan, M. 1978. Progesterone inhibition of maternal behavior in the rat. *Horm. Behav.* **11**, 209–231.

Numan, M., Leon, M., and Moltz, H. 1972. Interference with prolactin release and the maternal behavior of female rats. *Horm. Behav.* **3**, 29–38.

Numan, M., Rosenblatt, J. S., and Komisaruk, B. R. 1977. The medial preoptic area and the onset of maternal behavior in the rat. *J. Comp. Physiol. Psychol.* **91**, 146–164.

Obias, M. D. 1957. Maternal behavior of hypophysectomized gravid albino rats and the development and performance of their progeny. *J. Comp. Physiol. Psychol.* **50**, 120–124.

Paul, L., and Kupferschmidt, J. 1975. Killing of conspecific and mouse young by male rats. *J. Comp. Physiol. Psychol.* **88**, 755–763.

Pepe, G., and Rothchild, I. 1972. The effect of hypophysectomy on day 12 of pregnancy on the serum progesterone level and the time of parturition in the rat. *Endocrinology* **91**, 1380–1385.

Pepe, G. J., and Rothchild, I. 1974. A comparative study of serum progesterone levels in pregnancy and in various types of pseudopregnancy in the rat. *Endocrinology* **95**, 275–279.

Plume, S., Fogarty, C., Grota, L. J., and Ader, R. 1968. Is retrieving a measure of maternal behavior in the rat? *Psychol. Rep.* **23**, 627–630.

Pollak, E. I., and Sachs, B. D. 1975. Male copulatory behavior and female maternal behavior in neonatally bulbectomized rats. *Physiol. Behav.* **14**, 337–343.

Powers, J. B., and Winans, S. S. 1973. Sexual behavior in peripherally anosmic male hamsters. *Physiol. Behav.* **10**, 361–368.

Quadagno, D. M., and Rockwell, J. 1972. The effect of gonadal hormones in infancy on maternal behavior in the adult rat. *Horm. Behav.* **3**, 55–62.

Quadagno, D. M., McCullough, J., Ho, G. K., and Spevak, A. M. 1973. Neonatal gonadal hormones: Effect on maternal and sexual behavior in the female rat. *Physiol. Behav.* **11**, 251–254.

Quadagno, D. M., Debold, J. F., Gorzalka, B. B., and Whalen, R. E. 1974. Maternal behavior in the rat: Aspects of concaveation and neonatal androgen treatment. *Physiol. Behav.* **12**, 1071–1074.

Quadagno, D. M., Briscoe, R., and Quadagno, J. S. 1977. Effect of perinatal gonadal hormones on selected nonsexual behavior patterns: A critical assessment of the nonhuman and human literature. *Psychol. Bull.* **84**, 63–80.

Reisbick, S., Rosenblatt, J. S., and Mayer, A. D. 1975. Decline of maternal behavior in the virgin and lactating rat. *J. Comp. Physiol. Psychol.* **89**, 722-732.

Richards, M. 1967. Maternal behaviour in rodents and lagomorphs. *In* "Advances in Reproductive Physiology" (A. McClaren, ed.), Vol. 2. Academic Press, New York.

Richter, C. P. 1927. Animal behavior and internal drives. *Q. Rev. Biol.* **2**, 307-343.

Riddle, O., Lahr, E. L., and Bates, R. W. 1942. The role of hormones in the initiation of maternal behavior. *Am. J. Physiol.* **137**, 299-317.

Rodriguez-Sierra, J. F. 1977. Neural and hormonal factors mediating sexual behavior in female rats. Doctoral dissertation, Rutgers University, Newark, New Jersey.

Rodriguez-Sierra, J., and Rosenblatt, J. S. 1977. Does prolactin play a role in estrogen-induced maternal behavior in rats: Apomorphine reduction of prolactin release. *Horm. Behav.* **9**, 1-7.

Rosenberg, P. A., and Herrenkohl, L. R. 1976. Maternal behavior in male rats: Cirtical times for the suppressive action of androgens. *Physiol. Behav.* **16**, 293-297.

Rosenblatt, J. S. 1965. The basis of synchrony in the behavioural interaction between the mother and her offspring in the laboratory rat. *In* "Determinants of Infant Behaviour - III" (B. M. Foss, ed.). Methuen, London.

Rosenblatt, J. S. 1967. Nonhormonal basis of maternal behavior in the rat. *Science* **156**, 1512-1514.

Rosenblatt, J. S. 1970. Views on the onset and maintenance of maternal behavior in the rat. *In* "Development and Evolution of Behavior: Essays in Memory of T. C. Schneirla" (L. R. Aronson, E. Tobach, D. S. Lehrman, and J. S. Rosenblatt, eds.), pp. 489-515. Freeman, San Francisco.

Rosenblatt, J. S. 1975a. Prepartum and postpartum regulation of maternal behaviour in the rat. (Ciba Foundation Symposium, No. 33) Assoc. Scientific Publ., Amsterdam.

Rosenblatt, J. S. 1975b. Selective retrieving by maternal and nonmaternal female rats. *J. Comp. Physiol. Psychol.* **88**, 678-686.

Rosenblatt, J. S., and Lehrman, D. S. 1963. Maternal behavior in the laboratory rat. *In* "Maternal Behavior in Mammals" (H. L. Rheingold, ed.), pp. 8-57. Wiley, New York.

Rosenblatt, J. S., and Siegel, H. I. 1975. Hysterectomy-induced maternal behavior during pregnancy in the rat. *J. Comp. Physiol. Psychol.* **89**, 685-700.

Roth, L., and Rosenblatt, J. S. 1967. Changes in self-licking during pregnancy in the rat. *J. Comp. Physiol. Psychol.* **63**, 397-400.

Rothchild, I. 1960. The corpus luteum-pituitary relationship: The association between the cause of luteotrophin secretion and the cause of follicular quiescence during lactation; the basis for a tentative theory of the corpus luteum-pituitary relationship in the rat. *Endocrinology* **67**, 9-41.

Rothchild, I. 1962. Corpus luteum-pituitary relationship: The effect of progesterone on the folliculotropic potency of the pituitary in the rat. *Endocrinology* **70**, 303-313.

Rothchild, I., Billiar, R. B., Kline, I. T., and Pepe, G. 1973. The persistence of progesterone secretion in pregnant rats after hypophysectomy and hysterectomy: A comparison with pseudopregnant, deciduomata-bearing pseudopregnant, and lactating rats. *J. Endocrinol.* **57**, 63-74.

Sachs, B. D., Warden, A. F., and Pollak, E. 1971. Studies of postpartum estrus in rats. Paper presented at the Eastern Conference on Reproductive Behavior, Haverford, Pennsylvania.

Scalia, F., and Winans, S. S. 1975. The differential projections of the olfactory bulb and accessory olfactory bulb in mammals. *J. Comp. Neurol.* **161**, 31-56.

Schlein, P. A., Zarrow, M. X., Cohen, H. A., Denenberg, V. H., and Johnson, N. P. 1972. The differential effect of anosmia on maternal behavior in the virgin and primiparous rat. *J. Reprod. Fert.* **30**, 39-42.

Schneirla, T. C. 1952. A consideration of some conceptual trends in comparative psychology. *Psychol. Bull.* **49**, 559-597.

Schultz, E. W. 1960. Repair of the olfactory mucosa. *Am. J. Pathology* **37**, 1-19.

Schwartz, E., and Rowe, F. A. 1976. Olfactory bulbectomy: Influences on maternal behavior in primiparous and multiparous rats. *Physiol. Behav.* **17**, 879–883.

Shaikh, A. A. 1971. Estrone and estradiol levels in the ovarian venous blood from rats during the estrous cycle and pregnancy. *Biol. Reprod.* **5**, 297–307.

Sieck, M. H., and Baumbach, H. D. 1974. Differential effects of peripheral and central anosmia producing techniques on spontaneous behavior patterns. *Physiol. Behav.* **16**, 407–425.

Siegel, H. I., and Rosenblatt, J. S. 1975a. Hormonal basis of hysterectomy-induced maternal behavior during pregnancy in the rat. *Horm. Behav.* **6**, 211–222.

Siegei, H. I., and Rosenblatt, J. S. 1975b. Estrogen induced maternal behavior in hysterectomized-ovariectomized virgin rats. *Physiol. Behav.* **14**, 465–471.

Siegel, H. I., and Rosenblatt, J. S. 1975c. Progesterone inhibition of estogen-induced maternal behavior in hysterectomized-ovariectomized virgin rats. *Horm. Behav.* **6**, 223–230.

Siegel, H. I., and Rosenblatt, J. S. 1975d. Latency and duration of estrogen induction of maternal behavior in hysterectomized-ovariectomized virgin rats: Effects of pup stimulation. *Physiol. Behav.* **14**, 473–476.

Siegel, H. L., and Rosenblatt, J. S. 1977. Effects of pregnancy termination on maternal behavior, lordosis, ovulation, and progesterone levels in the rat. *East. Conf. Reprod. Behav., Univ. Conn., Storrs.* (Abstr.)

Siegel, H. I., and Rosenblatt, J. S. 1978a. Duration of estrogen stimulation and progesterone inhibition of maternal behavior in pregnancy-terminated rats. *Horm. Behav.* **11**, 12–19.

Siegel, H. I., and Rosenblatt, J. S. 1978b. Effects of adrenalectomy on maternal behavior in pregnancy-terminated rats. *Physiol. Behav.* **21**, 831–833.

Siegel, H. I., Doerr, H., and Rosenblatt, J. S. 1978. Further studies on estrogen-induced maternal behavior in hysterectomized-ovariectomized nulliparous rats. *Physiol. Behav.* **21**, 99–103.

Siegel, H. I., Ahdieh, H. B., and Rosenblatt, J. S. 1978. Hysterectomy-induced facilitation of lordosis behavior in the rat. *Horm. Behav.* **11**, 273–278.

Singh, P. J., Tucker, A. M., and Hofer, M. A. 1976. Effects of nasal AnSo₄ irrigation and olfactory bulbectomy on rat pups. *Physiol. Behav.* **17**, 373–382.

Slob, A. K., Snow, C. E., and de Natris-Mathot, E. 1973. Absence of behavioral deficits following neonatal undernutrition in the rat. *Dev. Psychobiol.* **6**, 177–186.

Slotnick, B. M., and Gutman, L. A. 1977. Evaluation of intranasal zinc sulfate treatment on olfactory discrimination in rats. *J. Comp. Physiol. Psychol.* **91**, 942–950.

Slotnick, B. M., Carpenter, M. L., and Fusco, R. 1973. Initiation of maternal behavior in pregnant nulliparous rats. *Horm. Behav.* **4**, 53–59.

Small, W. S. 1899. Notes on the psychic development of the young white rat. *Am. J. Psychol.* **11**, 80–100.

Smotherman, W. P., Bell, R. W., Hershberger, W. A., and Coover, G. D. 1978. Orientation to rat pup cues: Effects of maternal experiential history. *Anim. Behav.* **26**, 265–273.

Söderberg, U., and Larsson, K. 1976. Impaired temperature regulation in rats after anosmia induced peripherally or centrally. *Physiol. Behav.* **17**, 993–995.

Spector, S. A., and Hull, E. M. 1972. Anosmia and mouse killing by rats: A nonolfactory role for the olfactory bulb. *J. Comp. Physiol. Psychol.* **80**, 354–356.

Stern, J. M. 1977. Effects of ergocryptine on postpartum maternal behavior, ovarian cyclicity and food intake in rats. *Behav. Biol.* **21**, 134–140.

Stern, J. M., and MacKinnon, D. S. 1976. Postpartum, hormonal, and nonhormonal induction of maternal behavior in rats: Effects on t-maze retrieval of pups. *Horm. Behav.* **7**, 305–316.

Stern, J. M., and MacKinnon, D. A. 1978. Sensory regulation of maternal behavior in rats: Effects of pup age. *Dev. Psychobiol.* **11**, 579–586.

Stern, J. M., and Siegel, H. I. 1978. Prolactin release in lactating, primiparous and multiparous thelectomized, and maternal virgin rats exposed to pup stimuli. *Biol. Reprod.* **19**, 177–182.

Stone, C. P. 1925. Preliminary note on maternal behavior of rats living in parabiosis. *Endocrinology* **9**, 505-512.

Strauss III, J. F., Sikoloski, J., Caploe, P., Duffy, P., Mintz, G., and Stambaugh, R. L. 1975. On the role of prostaglandins in parturition in the rat. *Endocrinology* **96**, 1040-1043.

Sturman-Hulbe, M. and Stone, C. P. 1929. Maternal behavior in the albino rat. *J. Comp. Psychol.* **9**, 203-237.

Szechtman, H., Siegel, H. I., Rosenblatt, J. S., and Komisaruk, B. R. 1977. Tail-pinch facilitates onset of maternal behavior. *Physiol. Behav.* **19**, 807-809.

Takayama, M., and Greenwald, G. S. 1973. Direct luteotropic action of estrogen in the hypophysectomized-hysterectomized rat. *Endocrinology* **92**, 1405-1413.

Terkel, J. 1970. Aspects of maternal behavior in the rat with special reference to humoral factors underlying maternal behavior at parturition. Doctoral dissertation, Rutgers University.

Terkel, J. 1972. A chronic cross-transfusion technique in freely behaving rats by use of a single heart catheter. *J. Appl. Physiol.* **33**, 519-522.

Terkel, J., and Rosenblatt, J. S. 1968. Maternal behavior induced by maternal blood plasma injected into virgin rats. *J. Comp. Physiol. Psychol.* **65**, 479-482.

Terkel, J., and Rosenblatt, J. S. 1971. Aspects of nonhormonal maternal behavior in the rat. *Horm. Behav.* **2**, 161-171.

Terkel, J., and Rosenblatt, J. S. 1972. Humoral factors underlying maternal behavior at parturition: Cross transfusion between freely moving rats. *J. Comp. Physiol. Psychol.* **80**, 365-371.

Terkel, J., Bridges, R. S., and Sawyer, C. H. 1979. Effects of transecting lateral neural connections of the medial preoptic area on maternal behavior in the rat: Nest building, pup retrieval, and prolactin secretion. *Brain Res.* (in press).

Thoman, E. B., and Levine, S. 1970. Effects of adrenalectomy on maternal behavior in rats. *Dev. Psychobiol.* **3**, 237-244.

Thompson, R. W., and Lippman, L. G. 1972. Exploration and activity in the gerbil and rat. *J. Comp. Physiol. Psychol.* **80**, 439-448.

Thor, D. H., and Flannelly, K. J. 1977. Anosmia and toxicity of topical intranasal zinc. *Physiol. Psychol.* **5**, 261-269.

Thor, D. H., Carty, R. W., and Flannelly, K. J. 1976. Prolonged peripheral anosmia in the rat by multiple intranasal applications of zinc sulfate solution. *Bull. Psychon. Soc.* **7**, 41-43.

Underwood, E. J. 1971. "Trace Elements in Human and Animal Nutrition." Academic Press, New York.

Vandenbergh, J. G. 1973. Effects of central and peripheral anosmia on reproduction of female mice. *Physiol. Behav.* **10**, 257-261.

Van Der Schoot, P., Lankhorst, R. R., De Roo, J. A., and De Greef, W. J. 1978. Suckling stimulus, lactation, and suppression of ovulation in the rat. *Endocrinology* **103**, 949-956.

Vergnes, M., and Karli, P. 1963. Declenchement du comportement d'agression interspecifique rat-souris par ablation bilaterale des bulbes olfactifs. Action de l'hydroxyzine sur cette agressivife provoqué. *C. R. Soc. Biol.* **157**, 1061-1063.

Vergnes, M., and Karli, P. 1965. Etude des voies nerveuses d'une influence inhibitrice s'exerçant sur l'agressivité interspecifique du rat. *C. R. Soc. Biol.* **157**, 176-180.

Vermouth, N. T., and Deis, R. P. 1974. Prolactin release and lactogenesis after ovariectomy in pregnant rats: Effect of ovarian hormones. *J. Endocrinol.* **63**, 13-20.

Wiesner, B. P., and Sheard, N. M. 1933. "Maternal Behavior in the Rat." Oliver & Boyd, London.

Wild, J. M., and Hughes, R. N. 1972. Effects of postweaning handling on locomotor and exploratory behavior in young rats. *Dev. Psychol.* **7**, 76-79.

Winans, S. S., and Powers, J. B. 1977. Olfactory and vomeronasal deafferentiation of male hamsters: Histological and behavioral analyses. *Brain Res.* **126**, 325-344.

Ying, S.-Y., and Greep, R. O. 1973. Effect of caesarean section on postpartum ovulation and changes in serum LH in postpartum rats. *Proc. Soc. Exp. Biol. Med.* **142,** 61–63.

Ying, S. -Y., Gove, S., Fang, V. S., and Greep, R. O. 1973. Ovulation in postpartum rats. *Endocrinology* **92,** 108–116.

Zarrow, M. X., Sawin, P. B., Ross, S., and Denenberg, V. H. 1962. Maternal behavior and its endocrine basis in the rabbit. *In* "Roots of Behavior" (E. L. Bliss, ed.), pp. 187–197. Harper, New York.

Zarrow, M. X., Gandelman, R., and Denenberg, V. H. 1971. Prolactin: Is it an essential hormone for maternal behavior in mammals? *Horm. Behav.* **2,** 343–354.

Zarrow, M. X., Schlein, P. A., Denenberg, V. H., and Cohen, H. A. 1972. Sustained corticosterone release in lactating rats following olfactory stimulation from the pups. *Endocrinology* **91,** 191–196.

Subject Index